Deregulated Electricity Structures and Smart Grids

Deregulated Electricity Structures and Smart Grids

Edited by
Baseem Khan, Om Prakash Mahela,
Sanjeevikumar Padmanaban, and
Hassan Haes Alhelou

CRC Press
Taylor & Francis Group
Boca Raton London New York

CRC Press is an imprint of the
Taylor & Francis Group, an **informa** business

MATLAB® and Simulink® are trademarks of The MathWorks, Inc. and are used with permission. The MathWorks does not warrant the accuracy of the text or exercises in this book. This book's use or discussion of MATLAB® and Simulink® software or related products does not constitute endorsement or sponsorship by The MathWorks of a particular pedagogical approach or particular use of the MATLAB® and Simulink® software.

First edition published 2022
by CRC Press
6000 Broken Sound Parkway NW, Suite 300, Boca Raton, FL 33487-2742

and by CRC Press
4 Park Square, Milton Park, Abingdon, Oxon, OX14 4RN

CRC Press is an imprint of Taylor & Francis Group, LLC

© 2022 selection and editorial matter, Baseem Khan, Om Prakash Mahela, Sanjeevikumar Padmanaban and Hassan Haes Alhelou; individual chapters, the contributors

ISBN: 978-0-367-75433-4 (hbk)
ISBN: 978-1-032-23511-0 (pbk)
ISBN: 978-1-003-27803-0 (ebk)

DOI: 10.1201/9781003278030

Typeset in Times
by codeMantra

Contents

Preface

In recent years, the electrical industry has shifted from a regulated to a deregulated structure. In the energy industry, this revolution introduced efficacy and market competition. Different segments of the electricity industry, such as GENCO, TRANSCO, and DISCOMS, have been deregulated and developed into separate corporations. Smart grid technology will also be adopted in other parts of the world. The rapidly expanding energy demand will be met by smart grid technology in the generation, transmission, and distribution sectors. These technologies are speedily altering modern energy systems, which are becoming more complex. Energy utilities' major functions are no longer power generation, transmission, and distribution; instead, system stability, dependability, efficacy, and security have emerged as significant concerns. These issues aid energy utilities in the long-term management, control, and operation of generation sources, lowering their environmental and social impacts. The energy sector's planning and operation necessitate efficiency in order to maximize the benefits for both utilities and consumers at the same time. Furthermore, energy conservation can be referred to as power generation, which helps to reduce carbon emissions in the atmosphere. Various techniques are always evolving to overcome the challenges mentioned above. The implementation of these smart grid technologies encourages the discovery and optimal exploitation of renewable energy sources. However, there is still a significant disparity between current electricity generation and load demand. Utilities are integrating old, stable energy sources with intermittent renewable energy sources such as solar, wind, and fuel cells to reduce the energy gap. Traditional sources of energy production are thus being explored in various parts of the world for the reliable running of the power sector. The environment suffers as a result of our reliance on traditional energy sources. Hence, meeting demand with renewable energy sources and techniques is a critical issue for enhancing our energy security. Green energy sources have piqued the interest of researchers all around the world. These sources will aid in the integration of renewable energy sources into the electric grid at a faster rate. The fundamental problem with renewable energy sources is that their generation is highly erratic and dependent on environmental circumstances. As a result, energy generation from solar and wind sources fluctuates in nature, affecting generating frequency, voltage, and waveform, as well as the quality and quantity of energy supplied to the integrated grid system. Energy demand will increase by 70% by 2030 compared to current levels. Traditional energy supplies, on the other hand, are rapidly depleting, posing severe worries about per capita energy use. Energy consumption per capita is a measure of a country's standard of living. Thus, a variety of renewable energy sources must be researched to see which are locally available and which may be optimized to increase energy generating efficiency. The challenges with the deregulated smart grid system clearly show that the current generating, planning, and operation procedures require more efficient solutions.

The numerous issues of the deregulated electricity market are discussed in this book. Furthermore, including renewable energy sources into the system necessitates a systemic engineering approach to measuring the many alternatives at multiple levels of generation, planning, control, operation, and use.

DESCRIPTION OF THE BOOK

By analysing various research difficulties in these fields, the major goal of this book is to give key resources in the current literature for the core areas of smart power grid and renewable energy generation, planning, controlling, and operation. This book serves as a valuable resource for scientists, researchers, students, and academicians all over the world who are interested in adopting and applying cutting-edge research in the field of smart power grids and renewable energy systems. As a result, the Deregulated Electricity Structures and Smart Grids conference will serve as a platform for discussing cutting-edge research in the fields of smart power grids and renewable energy systems. The following are the book's key objectives:

1. With planning and operational concerns, this book identifies and studies the deregulated electricity market issues.
2. The breadth of various renewable energy technologies, as well as their operation and control difficulties, are identified and investigated.
3. The different power quality challenges related with renewable energy generation's grid interface, as well as their remedies, are identified.
4. The numerous integration challenges of the energy storage system and electric vehicles into the utility grid are identified and investigated.

Deregulated electric power systems, their structures, operation, and control, renewable energy resources, power quality difficulties with renewable energy sources, and the integration of energy storage systems and electric cars are all hot topics in current research.

This book will serve as a source of inspiration for scholars, and it will allow them to share their expertise of energy efficiency, power quality, and renewable energy integration for better resource utilization. Researchers are working on the integration of renewable energy sources, electric vehicles, and synchronized operation of solar power, batteries, and the grid supply for the development of smart power systems in order to satisfy load demand and enhance energy efficiency.

MATLAB® is a registered trademark of The MathWorks, Inc. For product information, please contact:
The MathWorks, Inc.
3 Apple Hill Drive
Natick, MA 01760-2098 USA
Tel: 508-647-7000
Fax: 508-647-7001
E-mail: info@mathworks.com
Web: www.mathworks.com

Acknowledgements

This book on Deregulated Electricity Structures and Smart Grids is an outcome of the inspiration and encouragement given by many individuals for whom these words of thanks are only a token of our gratitude and appreciation.

Our sincere gratitude goes to the people who contributed their time and expertise to this book. We highly appreciate their efforts in achieving this project. We would like to acknowledge the help of all the people involved in this project, and, more specifically, we would like to thank each one of the authors for their contributions and the editorial board/reviewers regarding the improvement of quality, coherence, and the content presentation of this book.

Second, we would like to express our sincere thanks to CRC Press/Taylor & Francis Group for their continuous support and for giving us an opportunity to edit this book.

We are thankful to our family members for their prayers, encouragement, and care shown towards us during the completion of this book on Deregulated Electricity Structures and Smart Grids. Thank you all!!!

We also express our gratitude to the GOD for all the blessings!!!

Baseem Khan
Hawassa University, Hawassa, Ethiopia

Om Prakash Mahela
Power System Planning Division, RRVPNL, Jaipur, India

Sanjeevikumar Padmanaban
KPR Institute of Engineering and Technology, Tamilnadu, India

Hassan Haes Alhelou
Tishreen University, Lattakia, Syria

Editors

Baseem Khan (M'16) earned his Bachelor of Engineering degree in Electrical Engineering from Rajiv Gandhi Technological University, Bhopal, India, in 2008. He earned his Master of Technology and Doctor of Philosophy degrees in Electrical Engineering from the Maulana Azad National Institute of Technology, Bhopal, India, in 2010 and 2014, respectively. Currently, he is an assistant professor at Hawassa University, Ethiopia. His research interests include power system restructuring, power system planning, smart grid technologies, meta-heuristic optimization techniques, reliability analysis of renewable energy system, power quality analysis, and renewable energy integration.

Om Prakash Mahela (M'13) was born in Sabalpura, Kuchaman City, Rajasthan, India, in 1977. He earned his B.E. degree from the College of Technology and Engineering, Udaipur, India, in 2002; M.Tech. degree from Jagannath University, Jaipur, India, in 2013; and Ph.D. from Indian Institute of Technology Jodhpur, India, in 2018, all in Electrical Engineering. From 2002 to 2004, he was an assistant professor with the Rajasthan Institute of Engineering and Technology, Jaipur, India. From 2004 to 2014, he was a junior engineer with the Rajasthan Rajya Vidyut Prasaran Nigam Ltd., India, and assistant engineer since July 2014. He has authored more than 100 research articles and book chapters. His research interests include power quality, power system planning, grid integration of renewable energy sources, FACTS devices, transmission line protection, and condition monitoring. He was a recipient of the University rank certificate in 2002, Gold Medal in 2013, best research paper award from IEEE ICSEDPS 2018, C.V. Raman Gold Medal in 2019, and outstanding Reviewer Awards from Elsevier journals.

Sanjeevikumar Padmanaban is SMIEEE'15, FIETE'18, FIE'18, FIET'19 and Chartered Engineer (CEng., India) who earned his Bachelor's (First Class), Master's (Distinction), and Ph.D. degrees in Electrical Engineering from the University of Madras, India (2002), Pondicherry University, India (2006), and University of Bologna, Italy (2012), respectively. His research work is focused in the field of power electronics (multi-phase/multilevel converters). He is on the editorial board and is Associate Editor of *IEEE Systems Journal* and *IET PEL* and subject editor of *IET RPG, IET GTD, IEEE Access, Turkish Journal of Electrical Engineering & Computer Science, Journal of Power Electronics* (JPE-Korea), and *FACETS* (Canada).

Hassan Haes Alhelou (S'15) is a faculty member at Tishreen University, Lattakia, Syria. He is also a Ph.D. researcher at the Isfahan University of Technology (IUT), Isfahan, Iran. He is included in the 2018 Publons list of the top 1% best reviewers and researchers in the field of engineering. He was the recipient of the Outstanding Reviewer Award from *Energy Conversion and Management Journal* in 2016, *ISA Transactions* journal in 2018, *Applied Energy* journal in 2019, and many other awards. He was the recipient of the best young researcher in the Arab Student Forum Creative among 61 researchers from 16 countries at Alexandria University, Egypt, 2011. He has published more than 30 research papers in high-quality, peer-reviewed journals and international conferences. He has also performed more than 160 reviews for highly prestigious journals including *IEEE Transactions on Industrial Informatics, IEEE Transactions on Industrial Electronics, Energy Conversion and Management, Applied Energy*, and *International Journal of Electrical Power & Energy Systems*. He has participated in more than 15 industrial projects. His major research interests are power systems, power system dynamics, power system operation and control, dynamic state estimation, frequency control, smart grids, microgrids, demand response, load shedding, and power system protection.

Contributors

Behrooz Adineh
Electrical Engineering Department
Semnan University
Semnan, Iran

Sunil Agarwal
Department of Electrical Engineering
Apex Institute of Engineering and
    Technology
Jaipur, India

Karm Veer Arya
MIS Research Group
Atal Bihari Vajpayee-Indian Institute
    of Information Technology and
    Management (ABV-IIITM)
Gwalior, India

Frede Blaabjerg
Department of Energy Technology
Aalborg University
Aalborg, Denmark

Aashish Kumar Bohre
Department of Electrical Engineering
National Institute of Technology (NIT)
Durgapur, India

Pooya Davari
Department of Energy Technology
Aalborg University
Aalborg, Denmark

Hooman Firoozi
Flexible Energy Resource, School of
    Technology and Innovations
University of Vaasa
Vaasa, Finland

Vijayakumar Gali
Department of Electrical & Electronics
    Engineering
Poornima University
Jaipur, India

Akhil Ranjan Garg
Department of Electrical Engineering,
    Faculty of Engineering and
    Architecture
J.N.V. University
Jodhpur, India

Rahul Garg
Department of Electrical Engineering
Government Mahila Engineering
    College
Ajmer, India

Manish Gaur
Department of Computer Science and
    Engineering, Institute of Engineering
    and Technology Lucknow
Dr. A.P.J. Abdul Kalam Technical
    University
Lucknow, India

T. Ghose
Department of Electrical & Electronics
    Engineering
Birla Institute of Technology
Ranchi, India

Manoj Gupta
Department of Electrical & Electronics
    Engineering
Poornima University
Jaipur, India

Praveen Kumar Gupta
Electrical Engineering Department
Sardar Vallabhbhai National Institute of
    Technology
Surat, India

Sunil Kumar Gupta
Department of Electrical & Electronics
    Engineering
Poornima University
Jaipur, India

Hossein Hafezi
Flexible Energy Resource, School of
    Technology and Innovations
University of Vaasa
Vaasa, Finland

Arvind Kumar Jain
Department of Electrical Engineering
National Institute of Technology
    Agartala
Tripura, India

Ritu Jain
Electrical Engineering Department
Sardar Vallabhbhai National Institute of
    Technology
Surat, Gujarat

Vineet Kansal
Department of Computer Science and
    Engineering, Institute of Engineering
    and Technology Lucknow
Dr. A.P.J. Abdul Kalam Technical
    University
Lucknow, India

Reza Keypour
Electrical Engineering Department
Semnan University
Semnan, Iran

Hosna Khajeh
Flexible Energy Resource, School of
    Technology and Innovations
University of Vaasa
Vaasa, Finland

Baseem Khan
Department of Electrical and Computer
    Engineering
Hawassa University
Hawassa, Ethiopia

Hannu Laaksonen
Flexible Energy Resource, School of
    Technology and Innovations
University of Vaasa
Vaasa, Finland

Sushma Lohia
Department of Electrical Engineering
Jagan Nath University
Jaipur, India

Vasundhara Mahajan
Electrical Engineering Department
Sardar Vallabhbhai National Institute of
    Technology
Surat, India

Om Prakash Mahela
Power System Planning Division
Rajasthan Rajya Vidyut Prasaran Nigam
    Ltd.
Jaipur, India

Ashok G. Matani
Government College of Engineering
Jalgaon [M.S.] India

Pankaj Mishra
Department of Electrical & Electronics
    Engineering
Birla Institute of Technology
Deoghar, India

Soumya Mudagal
Electrical Engineering Department
Sardar Vallabhbhai National Institute of
    Technology
Surat, India

Chethan Parthasarathy
Flexible Energy Resource, School of
    Technology and Innovations
University of Vaasa
Vaasa, Finland

Vijaykumar K. Prajapati
Electrical Engineering Department
Government Engineering College
Patan, India

Miadreza Shafie-khah
Flexible Energy Resource, School of
    Technology and Innovations
University of Vaasa
Vaasa, Finland

Anand Sharma
Department of Electrical & Electronics
    Engineering
Poornima University
Jaipur, India

Surender Kumar Sharma
Department of Electrical and
    Electronics Engineering
Poornima University
Jaipur, India

Abhishek Singh
Department of Computer Science and
    Engineering
Institute of Engineering and Technology
Lucknow, India

Ashish Singh
Department of Electrical Engineering
National Institute of Technology (NIT)
Durgapur, India

Neeraj Kumar Singh
Electrical Engineering Department
Sardar Vallabhbhai National Institute of
    Technology
Surat, India

Madisa V. G. Varaprasad
Department of Electrical & Electronics
    Engineering
Poornima University
Jaipur, India

Sanju Verma
Department of Electrical Engineering
Apex Institute of Engineering and
    Technology
Jaipur, India

Atul Kumar Yadav
Electrical Engineering Department
Sardar Vallabhbhai National Institute of
    Technology
Surat, India

Kiran Yadav
Department of Electrical Engineering
Jagan Nath University
Jaipur, India

1 Optimal Decision Making under Uncertainty Using Heuristic Approach in Restructured Power System

Arvind Kumar Jain

CONTENTS

DOI: 10.1201/9781003278030-1

1

1.1 INTRODUCTION

Over the past two decades, several countries around the world have restructured their power sectors. The basic objective behind restructuring of the power industry has been to introduce competition for improving efficiency. However, due to several barriers caused by huge capital cost, transmission congestion and losses, the competitive power markets are imperfect due to oligopoly nature. This flawed nature of the electricity markets has forced the participants to adopt a new way of understanding their participation in the market. In the competitive environment, the suppliers' revenue depends on their ability to sell the energy, and buyers' saving depends on their active participation in the Electricity Market (EM). To facilitate competitive trading of energy, most of the day-ahead electricity markets require their participants to submit their bids. The System Operator (SO) or Power Exchange (PX) clears the market based on their bids [1–5].

In the last two decades, some research has been carried out to obtain profitable strategies of a power supplier. While developing the bidding strategies, a supplier estimates the market clearing price (MCP) and rivals' bidding strategies and utilizes the game theory approach. Due to technical and regulatory constraints, various bidding models, forecasted demand and rivals' bid strategies make bidding strategy problem a stochastic optimization problem. Application of heuristic methods to the strategic bidding problem has been reported in the literature. Because heuristic techniques are less affected by the size and non-linearity of the problem and can converge to the optimal solution, where most of the analytical methods fail to converge. However, application of Artificial Bee Colony (ABC) algorithm, which has several advantages over similar population-based heuristic methods, has not been reported for developing the optimal bidding strategy in electricity markets. Separate energy and ancillary services markets have been developed in various parts of the world. In most of the countries, the procurement of ancillary services is contracted to the SO. However, in some countries, the SO purchases some of the ancillary services through market mechanism. Generally, energy market is a forward market, and ancillary services market is close to real-time market. During the actual time of delivery, imbalance between supply and demand is experienced due to frequent change in load. This imbalance may disturb the system frequency. Therefore, matching of supply and demand is continuously required. Energy-balancing services are traded in both the day-ahead and the real-time balancing market.

This chapter includes background of restructured power system along with technical and policy issues, congestion management methods, strategic bidding, generators financial risk and brief introduction about the Pennsylvania-New Jersey-Maryland (PJM), Nord Pool and Indian electricity markets. ABC algorithm-based approach for optimal decision making under uncertainty has been discussed to solve stochastic optimization problem. A case study to show the impact of ABC algorithm on decision making is presented in ensuing sections along with future directions and conclusion.

1.2 BACKGROUND: RESTRUCTURED POWER SYSTEM

Deregulation of power sector has forced the suppliers to adopt a new way of understanding their business. In vertically integrated structure, the electric utilities were guaranteed regulated rates of the electric supply with some profit above the production cost.

However, in competitive power markets, suppliers' revenue depends on their ability to sell electricity with maximum possible profit. Further, with the introduction of the competitive EMs, area of optimal bidding strategy has gained attention of the researchers to analyse the techno-economic issues related to the EMs and estimate the risk associated with the strategic behaviour of the participants. The complexity in electricity markets arises due to large number of entities, contractual obligations, separation of energy and ancillary services markets, and different market models [6–10].

1.2.1 COMPETITIVE ELECTRICITY MARKETS: AN OVERVIEW

The electricity markets have been established globally to facilitate economical operation through competition. Power system security is the significant feature of the system operation for reliable wheeling of electrical energy. In a competitive environment, ancillary services are utilized for managing secure operation of power system. A general introduction to the EM models and entities is given below.

- *EM models*
  Market models can be classified, based on the nature of transactions, as following.
- *Pool market model*
  In this model, the SO or market administrator electronically receives price quantity bids from power suppliers and buyers in the energy exchange. The SO clears the market using merit order dispatch and sends the information to the qualified participants.
- *Bilateral market model*
  In bilateral market model, suppliers and buyers enter into power purchase agreement directly. Settlement of energy price, quantity and duration depends on the discretion of supplier and buyer. The SO doesn't interfere in between. Quantities traded and trade prices are at the discretion of these parties and not a matter of the SO.
- *Pool + bilateral market model*
  In this model, suppliers and buyers submit their bids into the day-ahead pool market as well as sign bilateral contract with each other for a certain period. Hence, this model provides more flexible options for transmission access.
- *Key market participants*
  - Gencos
    Gencos, the owners of generation facilities, are responsible for construction, operation and maintenance of their generating plants. They can bid in a competitive power market to provide energy and ancillary services to customers. They may also enter into bilateral contracts to provide these services. During the transmission congestion, the SO may ask the Gencos to reschedule their generators for managing power balance. Each Genco tries to maximize profit in power market by strategic bidding, at their own risk.

- Transcos

   Transcos are the owners of the transmission facilities. They are wire companies and are answerable for planning, construction, operation and maintenance of transmission network to allow the flow of power from the generating companies to the network of wholesalers through their transmission network. In almost all the restructured markets, the wire companies have the regulated natural monopoly to provide nondiscriminatory services.

- Discos

   Discos are responsible for construction and maintenance of distribution network for connecting transmission grid to end-users. They are responsible for maintaining system reliability and taking care of power quality.

- SO

   Primary responsibility of SO is to ensure wheeling of electricity from source to sink reliably. SO participates neither in wire business nor in generation. It posts non-confidential information like MCP, buy and sell quantity, available transfer capability, transmission congestion, etc. to all the participants. Further, SO coordinates day-ahead scheduling and purchases ancillary services for managing power balance.

- Retailcos

   Retailco is an entity in the competitive market having legal approval to sell retail electricity. A retailer purchases electricity from pool market and resells it to the customers.

- Power traders

   Power traders assist the purchasers/sellers in finalizing the short- and long-term contracts on charge basis. They also provide some sort of guarantee to the suppliers as a security for their payments.

- Customers

   Customers are connected to the distribution network or transmission network depending on its size. It is an end user of electric energy. In regulated structure, customers were bound to buy power from a utility. However, in deregulated structure, customers are free to choose their service providers.

- *Types of electricity markets*

   The EMs, based on trading, can be divided into three categories, viz., energy market, ancillary services market and transmission market. The markets can also be classified as day-ahead market (DAM) or hour-ahead market and spot market. In addition, options can be traded through PX or on a bilateral basis.

   - Energy market

      The competitive trading of electric energy between the buyers and sellers occurs in energy market. The SO accepts supply and demand bids from market participants, aggregates these bids into supply-demand curve and calculates MCP for the energy. In case of transmission congestion, adjustments are made using congestion charges decided in the ancillary market.

- Ancillary services market

  These services are required to ensure the power quality, security and reliability of the power system. In the monopolistic structure, ancillary services were bundled with the power supply. In the restructured situation, these are opened to market competition. Types of ancillary services are voltage control, real power balancing, frequency control, spinning reserve, etc.

- Transmission market

  Transmission rights are traded in transmission market through centralized auction conducted by the SO. The objective of transmission market is to finalize optimum bids to maximize revenue keeping in view the transmission congestion management. Competing suppliers, customers and traders use the transmission market to support bilateral contract and commercial market transactions.

- Forward market

  Trading of power in forward market may begin years in advance and continue until actual delivery of power. Day-ahead and hour-ahead forward markets are used in most of the electricity markets to schedule resources at each hour of the following day and to fulfil the deviations from the day-ahead schedule, respectively.

- Real-time market

  To ensure the reliable operation of electric grid, power balance should be maintained at every instance. The real-time power production, load and transmission capacity differs from forward market schedule, and this necessitates the establishment of a real-time market. Sudden generator outages, line outages, transmission congestion, extreme weather conditions and other events can trigger unexpected imbalance between supply and demand. Thus, a spot market is necessary to ensure nondiscriminatory access to the grid and to support competitive energy market.

1.3 TECHNICAL AND POLICY ISSUES IN THE ELECTRICITY MARKETS

Some of the major technical issues are related with the system operation, whereas the policy issues are associated with the EM operation. Some of the issues are given below:

- Electricity prices are highly volatile due to several factors such as low demand price elasticity, transmission congestion, market model, imperfect competition, market power, intermittent nature of renewable energy resources, etc.
- Transmission congestion occurs when all parties want to purchase electricity from cheaper sources without caring for the distance between the source and delivery point, which may lead to the overloading of certain transmission lines and the equipment. Transmission congestion may jeopardize the system security and reliability and can give market power to some participants.

- To ensure satisfactory operation of electric grid, ancillary services like reserve capacity margin, and balancing/regulating services, are required.
- Demand-side participation into market design needs to be incorporated to achieve the most efficient and effective market performance.
- The characteristics of EM make the auction mechanism vulnerable for gaming, resulting in artificial increase in the MCP. Suppliers may also create the transmission congestion. Demand responsiveness may reduce the gaming opportunities.

The above issues have to be addressed carefully in the EM. In a restructured market, the aim is to provide quality power to the customers at the minimum cost, while ensuring secure operation of the power system. While meeting these objectives, each of the market participants tries to maximize its profit. Under competitive scenario, in certain markets, each supplier and also buyers try to submit bids strategically to maximize their profit under varying system operating conditions.

1.4   STRATEGIC BIDDING

In a competitive electricity market, single-sided bidding market or double-sided bidding market model, as shown in Figures 1.1 and 1.2 respectively, is used for power trading. In a single-sided bidding market, suppliers submit supply price-quantity bids and buyers submit demand in MW for each interval (an hour or half an hour or 15 min DAM). However, in a double-sided bidding market, buyers also submit the price-quantity bids. The bid/offer prices are shown ₡/MWh, where ₡ is a currency unit.

The system/market operator clears the market using merit order dispatch procedure. If congestion does not exist, MCP is same at each node/location, otherwise it is different at each node/location which is known as Locational Marginal Price (LMP). The optimal bidding problem, in electricity markets, has been attempted mainly for the suppliers. However, the active demand response is also gaining importance under smart grid scenario. Demand response empowered the customers to shift their demand to enhance their saving.

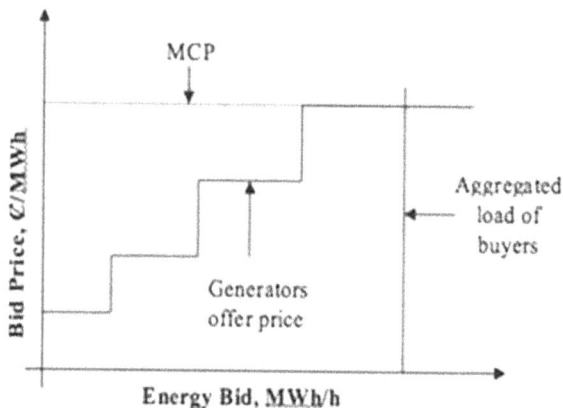

FIGURE 1.1   Single auction market.

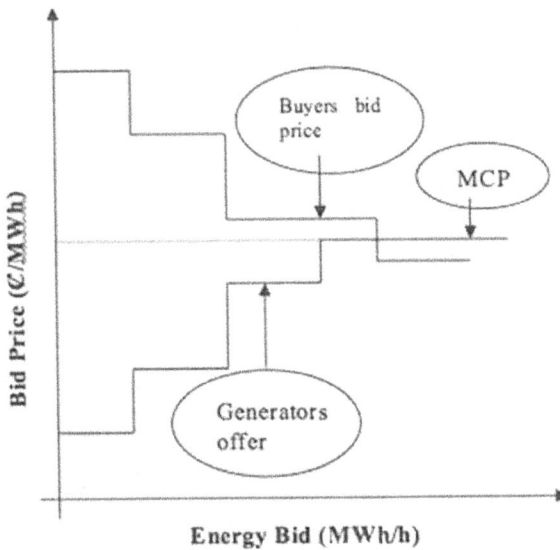

FIGURE 1.2 Double auction market.

Less number of power producers, huge investment, losses, congestion, etc. make EM imperfect. Therefore, action of one player can change the energy price and profit of the suppliers [11]. In perfect competitive scenario, to obtain high profit, all the power producers should submit bid close to marginal cost. However, in oligopolistic markets, power suppliers increase their profit by submitting price more than marginal cost. When a power producer submits bid more than the marginal cost to maximize profit, it is called strategic bidding. Demand variation, generator cost characteristics, rivals' bidding strategies, and operating and regulatory constraints can affect the bidding strategy of an individual supplier. Among these, the most uncertain factor [12] is the rivals' bidding behaviour. Therefore, the aim of the optimal bidding by a market participant is to drive an explicit model of bidding strategy considering its own cost and constraints, rivals' bidding strategy and the market guidelines.

1.5 RISK OF GENERATORS

The electricity markets have been, or are in the process of being, deregulated worldwide. The purpose of deregulation is to lower the electricity price and introduce state-of-the-art technology to foster innovation and development. However, with deregulation and competition in electricity markets, the suppliers face various risks, such as electricity-price volatility, fuel price risk, generator and line outages, transmission constraints and regulatory risk. Main concern of a generation bidder is to formulate optimal bidding strategy to qualify in EM that will minimize the risk that the bidder will not be supplying energy to the market. Other problems are operational risks such as transmission constraints, generator and line outages, which lead to financial loss to generators. Risk management is necessary in competitive electricity markets to reduce the participant's losses as well as to avoid the inefficiency in markets. Risk management consists of identifying losses, measuring losses and minimizing losses.

1.6 CONGESTION MANAGEMENT

In the EM, overloading of transmission corridors can affect the power trading. Therefore, transmission system congestion management and pricing are important tasks of the SO for efficient functioning of the competitive electricity market. Since the transmission corridor has finite capacity, it may become obligatory for the SO to select the expensive bids to avoid transmission overloading. In a vertically integrated utility, optimal power flow approach has been used for managing the transmission congestion and, at the same time, to find the economic load dispatch. With the restructuring of the electric utility and increase in the number of market players, it is essential to develop a congestion management mechanism and provide the open access to all market players. Methods to mitigate congestion under restructured environment are given below.

1.6.1 LMP

The values of the Lagrange multipliers obtained after solving the market-clearing problem are considered as the LMP of real power at the respective node. LMPs at different nodes will be different on including the power flow constraints in the optimization problem in case it reaches its limit; otherwise they will be same at all the locations/nodes. Thus, the LMP method is an effective approach for pricing as well as managing congestion in various markets like PJM Interconnection, California, New England and New York [13,14].

Although LMP is an efficient and fair method to manage congestion, in order to obtain it, many iterations of optimal power flow need to be performed. Further, the LMP calculation is more complex for market players to understand and develop their bids. Also, few participants can intentionally create congestion to increase the price at some nodes. This fact has been evinced by PJM market [13]. If the network is interconnected such that the effect of each generator's output on each transmission line is minimum, then this method is found to be more effective.

1.6.2 ZONAL PRICE

This approach [15] is like the LMP method, which considers the price at each node. However, in zonal price approach, the network is divided into few zones. Zones are selected in a fashion so that the intra-zonal congestion is minimum. Zonal price is calculated considering inter-area congestion. Intra-zonal congestion is managed by using generators re-dispatch procedure. When a transmission path within a zone is regularly congested, a new zone is created. This transforms a zonal interface into an inter-zonal interface, and congestion over this interface is managed by using adjustment bids. However, creation of new zone(s) may enhance the market power.

Therefore, to improve the market efficiency, the new zones are created only if whole sale generation markets on both sides of the interface are competitive.

The main objective of this approach is to give information about the congested areas to the market participants so that they can make a decision to invest in the new generation wisely. But it faces a difficulty in managing the intra-area congestion with counter trading method. However, it also attracts generation companies to exercise market power wherein they can bid a very low price and intentionally cause congestion. In this way, they can get money without production of electricity.

1.6.3   MARKET SPLIT

The transmission congestion and its management is relevant when different control areas of power system are interconnected. The interconnected networks were developed to maintain frequency. However, in electricity markets, interconnections are also helping electricity trading. Therefore, in competitive environment, the reason for transmission congestion is due to conflict of economic interest among the market participants. To relieve the congestion, this method splits the control areas in electricity sub-markets [16,17].

In this method, PX calculates system price without considering transmission constraints. After obtaining the solution of the market clearing problem, the power flow constraints are verified. In case of observing any constraint to be binding, the market is splitted into sub-markets and the problem is solved again to obtain the MCP and quantity. This gives rise to different prices in each sub-market. This method ensures the maximum utilization of capacity on the inter-connectors, when congestion occurs.

1.6.4   COUNTER FLOW RE-DISPATCH

A counter flow re-dispatch method has been used by various countries in order to avoid any congestion likely to occur due to zonal price or market splitting. In this method, few generation companies may be asked to modify their output such that a counter flow may be produced and overloading is relieved in few transmission lines. The generation companies will be paid/refunded the costs incurred in modifying their output at the last moment.

1.7   AN OVERVIEW OF THE FEW EM MODELS

The models of the Nord Pool, PJM Interconnection and Indian electricity markets have been considered as the case studies in this book. Different EMs have different business rules. Therefore, to gain a clear understanding of the EM, a brief overview of these markets is presented below.

1.7.1   NORD POOL ELECTRICITY MARKET

Nord Pool is a joint electricity market, comprising Sweden, Norway, Finland and Denmark. It is considered as a model PX in many evolving PX designs [17]. Nord Pool is jointly operated by two SOs – Statnett in Norway and Svenska Kraftnat in Sweden.

It has two market segments: Elspot and Elbas. Elspot offers day-ahead auction for the next day delivery. Market splitting method is used for determining area price and congestion management. In the Elbas market, energy is traded continuously 24×7 up to 1 h prior to supply. The Elbas provides an opportunity to all the participants for fine-tuning their portfolio [17] (Figure 1.3).

1.7.2 PJM ELECTRICITY MARKET

As a regional transmission organization, PJM Interconnection manages the world's largest competitive wholesale EM [18]. The PJM energy market consists of DAM, which serves as energy spot market, and real-time balancing market. In the DAM, hourly LMPs are calculated for the next operating day based on generation offers, demand bids and scheduled bilateral transactions.

In the real-time balancing market, LMPs are calculated at 5-min intervals based on actual grid operating conditions. If the lowest-priced electricity can reach all locations, prices are the same across the entire grid. When there is transmission congestion, power cannot flow freely to certain locations. In that case, more-expensive generators are asked to meet that demand. As a result, the LMP is higher at those locations. All qualified Gencos and Load Serving Entities are paid the LMP at their locations. Congestion charges for bilateral transactions are equal to the difference between the source and the sink LMPs. In PJM, zonal LMPs are load-weighted nodal LMP within predefined load zones. Figure 1.4 provides the market time line in PJM.

The PJM receives buying and selling bids for day-ahead energy market until 12:00 and clears the market just prior to 16:00 of the trading day using security constrained unit commitment and economic dispatch. As shown in Figure 1.4, by 16:00, PJM announces day-ahead LMPs and hourly schedules for the market participants. Available capacity resources, which are not selected in the day-ahead scheduling, may alter their bids for use in the real-time market.

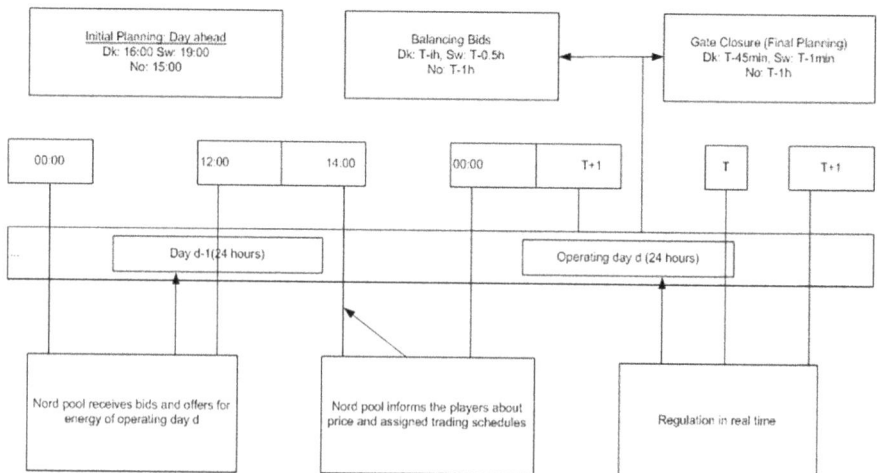

FIGURE 1.3 Market time line in the Nord Pool.

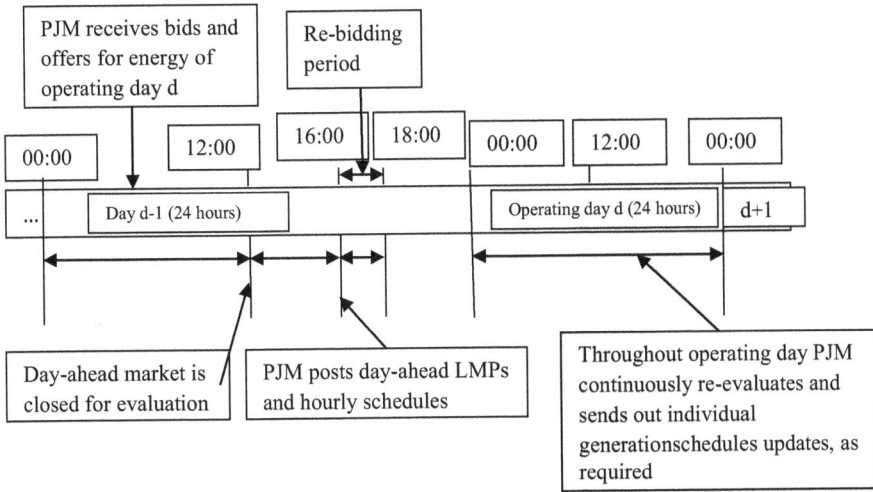

FIGURE 1.4 Market time line I in the PJM.

1.7.3 INDIAN ELECTRICITY MARKET

Indian electricity market, at present has two PXs, Indian Energy Exchange (IEX) at New Delhi and Power Exchange India Limited (PXIL) at Mumbai, where all the eligible participants submit their bids for each trading period. It has uniform price auction and double- sided bidding system. Indian EM has two sub-markets: DAM and Term-Ahead Market (TAM). In DAM, trading is performed 1 day in advance for every hour of the day when the electrical energy is to be supplied. Bids are invited from both the buyers and sellers to trade electricity. The equilibrium point of the demand and supply curve decides the MCP. The sellers as well as the buyers qualifying the market are asked to sell and purchase the electricity at this uniform MCP. Congestion management has been carried out using market-splitting methodology [16]. Term-Ahead Market (TAM) offers Intra-day, Day-ahead contingency, and Daily and Weekly contracts.

In India, real-time balancing EM does not exist. However, frequency linked Unscheduled Interchange (UI) mechanism is indirectly used as a real-time balancing market. The UI mechanism [19] was introduced in India in 2002–2003 as an integral part of Availability Based Tariff. It consists of three parts: (i) a fixed charge varying with the share of generation capacity allocated to the beneficiary and the availability level achieved by the generator, (ii) an energy charge based on daily scheduled supply and (iii) a charge for Unscheduled Interchanges (UI charges). In the case of UI, beneficiaries have to pay high rate during peak load hours, which discourages them from overdrawing and, thus, pulling down the frequency. The UI rate curve is shown in Figure 1.5 based on the values given in [19]. The net energy interchange of each party is metered for every 15 mins time block and compared with the scheduled energy for the same time block to determine the UI charges. DAM timeline of the Indian EM for actual delivery of power on next day are as following [16]. The National Load Dispatch Centre (NLDC) and Regional Load Dispatch Centres (RLDCs) act as SOs.

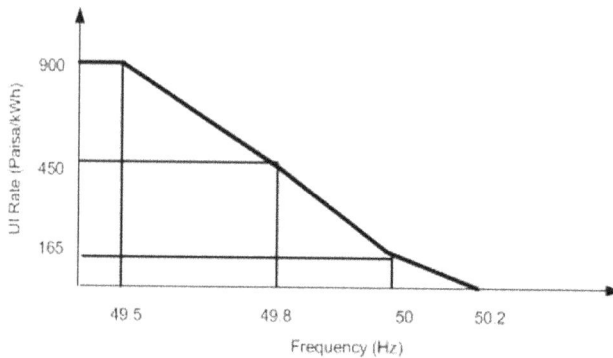

FIGURE 1.5   UI price curve.

10 AM–12 PM:	Sellers and buyers submit bid to the PX, either to IEX or PXIL.
11:00 AM	NLDC communicates to Indian PX the list of interfaces on which unconstrained flows are required.
1:00 PM	PX communicates to NLDC the interchange on various interfaces.
2:00 PM	In case of congestion, NLDC communicates to the PX the period of congestion and available limit for scheduling on respective interfaces.
3:00 PM	PX submits application for scheduling of Collective Transaction (CT).
4:00 PM	NLDC sends the details (scheduling request of collective transaction (CT)) to all the concerned RLDCs.
5:30 PM	NLDC/RLDCs confirm the accepted schedule to PX.
6:00 PM	RLDC issues the schedule.

1.8   ABC ALGORITHM

Optimization methods are primarily used to maximize or minimize certain objective by satisfying number of operating constraints. Various optimization techniques are available in the literature for solving non-linear optimization problem, such as classical optimization techniques and evolutionary techniques. Recently, evolutionary techniques are widely used to solve the non-linear engineering optimization problems, which do not require constraints and objective functions to be continuously differentiable.

The ABC algorithm [21] is a nature-inspired optimization algorithm, which works in a way similar to the one used by honey bees to search for their food. This technique divides artificial bees' colony into two parts, namely, employed bees and onlookers. Employed bees are considered in the first part and onlookers are considered as the second part. Each employed bee is assigned only one food source, which represents a possible solution. If the food source is abandoned, then the employed bee corresponding to it is known as scout. If a solution is not found to get better in a certain predefined number of trials, then this solution or the food source is abandoned. This predefined number of trials is called limit, which plays an important role in controlling the algorithm. Figure 1.6 depicts the flow chart of the ABC algorithm. One cycle of this

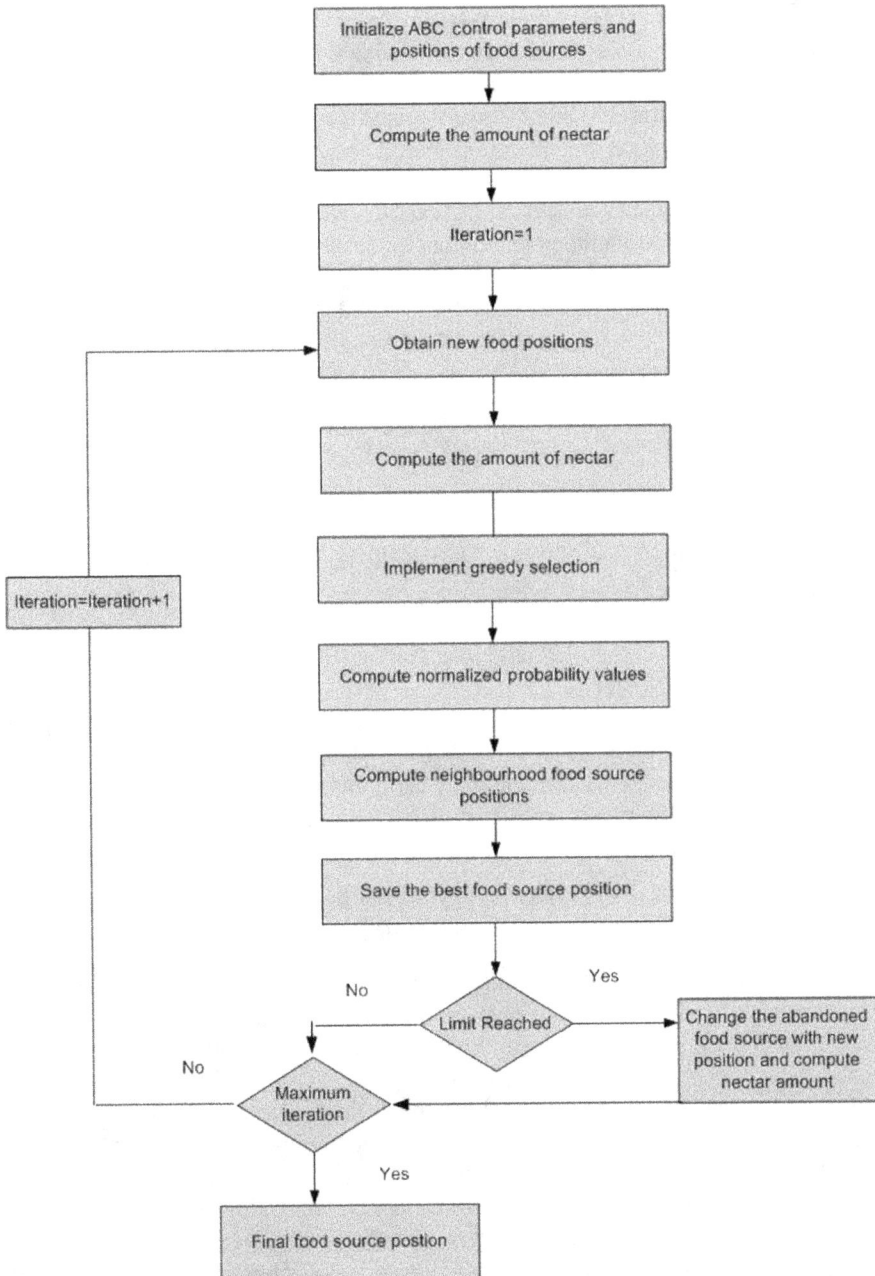

```
┌─────────────────────────────────┐
│ Initialize ABC control parameters │
│      and positions of food sources │
└─────────────────────────────────┘
                 │
                 ▼
┌─────────────────────────────────┐
│    Compute the amount of nectar   │
└─────────────────────────────────┘
                 │
                 ▼
┌─────────────────────────────────┐
│            Iteration=1            │
└─────────────────────────────────┘
                 │
                 ▼
┌─────────────────────────────────┐
│       Obtain new food positions   │
└─────────────────────────────────┘
                 │
                 ▼
┌─────────────────────────────────┐
│    Compute the amount of nectar   │
└─────────────────────────────────┘
                 │
                 ▼
┌─────────────────────────────────┐
│     Implement greedy selection    │
└─────────────────────────────────┘
                 │
                 ▼
┌─────────────────────────────────┐
│ Compute normalized probability values │
└─────────────────────────────────┘
                 │
                 ▼
┌─────────────────────────────────┐
│ Compute neighbourhood food source │
│            positions              │
└─────────────────────────────────┘
                 │
                 ▼
┌─────────────────────────────────┐
│   Save the best food source position │
└─────────────────────────────────┘
```

Iteration=Iteration+1

Limit Reached — No / Yes

Change the abandoned food source with new position and compute nectar amount

Maximum iteration — No / Yes

Final food source postion

FIGURE 1.6   ABC technique.

algorithm consists of the following three steps: (i) assignment of food sources to the employed bees and calculation of the corresponding nectar amount, (ii) assignment of onlooker to the food sources and calculation of the corresponding nectar amount and

(iii) determination of the scout bees and then random assignment of the possible food sources to these scouts. The quality of each solution is judged by calculating the nectar amount of the food source. This method is inherently suitable to solve a maximization problem, where the aim is to find the maximum value of the objective function $F(\theta)$, $\theta \varepsilon R^P$. Let θ_i be the position of the ith food source, and $F(\theta_i)$ is the nectar amount of the corresponding food source and is proportional to the energy $E(\theta_i)$.

Assume that $P(c) = \{\theta_i (c)$ $i = 1, 2, \dots S\}$ (c: cycle, S: number of food sources around the hive) represent the population of food sources being visited by bees.

An onlooker bee prefers a food source with the higher value of the nectar amount. With the increase in the nectar amount, the probability of preferring that food source by an onlooker bee also increases in same proportion. Hence, the probability of choosing a food source placed at θ_i by a bee is given as

$$Pr_i = fit_i \bigg/ \sum_{i=1}^{Sn} fit_i \qquad (1.1)$$

This probability helps an onlooker bee to decide about its visit to the food source placed at θ_i and to find a neighbour food source to take its nectar. The location of the chosen neighbour food source is determined by

$$\theta_i(c+1) = \theta_i(c) \pm \phi_i(c) \qquad (1.2)$$

$\phi_i(c)$ is a step generated randomly to locate a neighbour food source having higher nectar around θ_i. If the nectar amount $F(\theta_i(c+1))$ at $\theta_i(c+1)$ is higher than that at $\theta_i(c)$, then the bee goes to the hive and share her information with others, and the position $\theta_i(c)$ of the food source is replaced with $\theta_i(c+1)$, otherwise $\theta_i(c)$ is kept as it is.

1.9 STOCHASTIC PROGRAMMING PROBLEM FOR OPTIMAL DECISION MAKING

Due to certain special features of the competitive/restructured power system, like less number of power suppliers, huge capital investment, and uncertain nature of renewable energy, losses and transmission congestion, a generating company has to take optimal decisions throughout a decision horizon with inadequate information. The most uncertain factor that compounds the difficulties in decision process [12] is the rivals' bidding behaviour. Further, with restructured power system, the generating companies have different types of risks, such as electricity-price volatility, fuel price risk, generator and line outages, transmission constraints and regulatory risk which lead to financial loss to generators. Therefore, the aim of a generating company is to develop an explicit model of bidding strategy considering its own cost and constraints, rivals' bidding behaviour and the market rules.

A stochastic bi-level optimization problem (BLOP) is developed for optimal bid strategy. In this problem, price and quantity are decision variable vectors. The price and dispatch output obtained from the first stage are utilized by the generator-i in the second stage for the profit maximization. Therefore, lower level problem is modelled as constraints for the upper level problem. Further, the decision-making process of a generating company is as follows: Consider an EM consisting of N independent power generating companies, group of customers (loads), SO for controlling interconnected power system network and a PX to manage the forward electricity market. The objective of the generating company-i is to maximize its expected profit. For calculating the profit, the generating company-i needs to forecast the MCP either by forecasting it or by simulating the DAM and the balancing market clearing process. Further, Marginal cost $MC_i = 2a_iP_i + b_i$ of the generating company-i is derived from the generator cost characteristic. The strategic bidding price of the ith generator is assumed to be

$$\rho_i = s_i\left(2a_iP_i + b_i\right) \tag{1.3}$$

The multiplier s_i is a decision variable which is real number and used to develop the optimal bidding strategy of generating company.

Note that for decision to be optimal, a single optimization problem is required to be solved simultaneously. The general expression of a bi-level stochastic non-linear problem is as below.

1.9.1 BLOP

The bidding strategy optimization problem of a generating company-i is modelled as BLOP, in which the upper level optimization problem represents the profit maximization for the generator-i, while the lower level optimization problem is used by the SO for day ahead and balancing energy market clearing. Thus, the proposed optimization problem is described as

$$\text{Max} \sum_{t=1}^{24} \left[\begin{array}{l} \left\{ \text{MCP}^t \times P_{is}^t - C\left(P_{is}^t\right) \right\} + U_1^t \times \left\{ \text{MCP}^{t+} \times \Delta P_{is}^{t+} - C\left(\Delta P_{is}^{t+}\right) \right\} \\ + U_2^t \times \left\{ \text{MCP}^{t-} \times \Delta P_{is}^{t-} - C\left(\Delta P_{is}^{t-}\right) \right\} \end{array} \right] \tag{1.4}$$

subject to,

$$S_i^{\min} \le S_i \le S_i^{\max} \tag{1.5}$$

and the sub-problem defined as below,

$$\text{Max} \sum_{t=1}^{T} \left(\sum_{j=1}^{N_l} \rho_{jd}^t \times P_{jd}^t - \sum_{i=1}^{N_g} \rho_{is}^t \times P_{is}^t \right) \tag{1.6}$$

s.t.

$$\sum_{i=1}^{N_g} P_{is}^t - \sum_{j=1}^{N_l} P_{jd}^t = 0 \quad \forall t \tag{1.7}$$

$$P_{is\,\min}^t \le P_{is}^t \le P_{is\,\max}^t \quad \forall i, \forall t \tag{1.8}$$

$$P_{jd}^t \leq P_{jd\,max}^t \quad \forall j, \forall t \tag{1.9}$$

where

$$\text{Min} \sum_{i=1}^{N_g} \left(U_1^t \times \rho_{is}^{t+} \times \Delta P_{is}^{t+} + U_2^t \times \rho_{is}^{t-} \times \Delta P_{is}^{t-} \right) \tag{1.10}$$

subject to,

$$\sum_{i=1}^{N_g} \left(U_1^t \times \Delta P_{is}^+ + U_2^t \times \Delta P_{is}^- \right) = \sum_{j=1}^{N_l} \Delta P_{jd}^t \quad \forall t \tag{1.11}$$

$$0 \leq \Delta P_{is}^{t+} \leq P_{is\,max}^t - P_{is}^t \quad \forall i, \forall t \tag{1.12}$$

$$P_{is\,min}^t - P_{is}^t \leq \Delta P_{is}^{t-} \leq 0 \quad \forall i, \forall t \tag{1.13}$$

$$U_1^t = \begin{cases} 1 & \text{if } \Delta P_d^t > 0 \\ 0 & \text{otherwise} \end{cases} \tag{1.14}$$

$$U_2^t = \begin{cases} 1 & \text{if } \Delta P_d^t < 0 \\ 0 & \text{otherwise} \end{cases} \tag{1.15}$$

where

$$\Delta P_d^t = \sum_{j=1}^{N_l} \Delta P_{jd}^t \tag{1.16}$$

where ρ_{jd}^t and ρ_{is}^t are the bid prices of buyer-j and generating company-i at time t in \$/MW, respectively, P_{jd}^t is the demand of buyer-j to be fulfilled at time t, P_{is}^t is the dispatch output of the supplier-i at time t in the DAM, $P_{is\,min}^t$ and $P_{is\,max}^t$ are the minimum and the maximum generating capacity of the generating company-i at time t, $P_{jd\,max}^t$ is the maximum demand requirement of buyer-j at time t. ρ_{is}^{t+} and ρ_{is}^{t-} are the incremental and decremental bid prices of supplier-i at time t, ΔP_{is}^{t+} and ΔP_{is}^{t-} are the incremental and decremental dispatch output of supplier-i at time t, ΔP_{jd}^t is the change in the demand of buyer-j at time t and ΔP_d^t is the total change in demand at time t. N_g and N_l are the number of suppliers and buyers, respectively, and T is the scheduling horizon. The financial gain of generating company-i obtained by selling electricity in both the day-ahead and the balancing energy markets is given by Eq. (1.4).

The lower level optimization problem determines the MCPs and active power outputs of both the DAM and balancing energy market. This information is then used in maximizing the profit of the generator-i in the upper level optimization problem. Thus, the lower level problem acts as a constraint to the upper level problem. The bid prices and the output of the generator-i for the day-ahead and balancing energy market are obtained by solving the optimization problem given in Eq. (1.4). The bid prices of the generators are utilized while determining the MCPs of both the day-ahead as well as the balancing energy market, which are then used to calculate the profit of the company. A classical optimization technique would suffice to solve the lower level problem because of its linear nature. But the upper level problem being non-linear requires an efficient algorithm to determine the global optima [20].

The selection of either the up regulation or the down regulation market can be made in the proposed BLOP by using binary variables, and the model can be solved using a Mixed Integer Non-Linear Programming (MINLP) solver. But the use of binary variables and non-convexity of the problem may cause the solution method to stuck into sub-optimal or local solutions or may be in infeasible region. Hence, rather than using binary variables, an IF-THEN approach is utilized to select any one market, i.e. either the up regulation or the down regulation market, to fulfil the requirements of the optimization problem.

The lower level problem has been solved in sequence in the proposed methodology using the optimization function of MATLAB. The MCPs as well as the active power output of all the generators in both the day-ahead and balancing energy market form the output of the lower level optimization problem. These outputs are then provided as inputs to the upper level problem, whose solution gives the optimally coordinated bidding strategy. In this chapter, the upper level problem has been solved using the ABC algorithm [22].

1.9.2 UNCERTAINTY MODELLING THROUGH MONTE CARLO METHOD

One of the major problems in developing a profit maximization problem by generation companies in an EM is to model their rivals' bidding behaviour, which is quite uncertain. However, no participant would like to have any loss, so it is sure that his or her bid-price will be higher than his or her generation cost. This assumption works fairly in an EM because of its oligopolistic nature. Further, information about at least the bid quantity (MW) of rivals can be accessed utilizing the database available in the public domain. Assumption of normal distribution of their bidding prices is sufficient to model their bidding strategy. The rivals' bidding prices distribution can be given as [23,24].

$$\text{pdf}\left(\rho_n\right) = \frac{1}{\sqrt{2\pi}\sigma_n}\exp\left(-\frac{\left(\rho_n - \mu_n\right)^2}{2\left(\sigma_n\right)^2}\right)$$

where ρ_n, μ_n and σ_n are the bid price, mean value of the bid prices and the standard deviation of bid price of nth rival, respectively.

Mean and variance of the rivals' bid price have been estimated using the procedure assumed in [24]. The changes in the rivals' bidding strategies have been obtained by generating the random sample values of ρ_n ($n = 1, 2, ..., N$) based on their pdf, as given in Eq. (1.17).

The rivals' bidding price has been estimated using Monte Carlo method [25,26], which approximates a mathematical problem probabilistically using a statistical sampling technique. Its steps are given below.

- A large number of samples of bid prices of all the rivals are generated randomly in accordance with their pdfs.
- Large trial outcomes are obtained by solving the optimization problem with sample values of bid prices of all the rivals.
- The expectation values are calculated by averaging all the trial outcomes.

1.10 SOLUTION ALGORITHM

The block diagram, depicting the working of the coordinated bidding strategy, is shown in Figure 1.7. A combination of nature-inspired algorithm and a traditional optimization technique has been used to obtain the solution of the proposed BLOP. The upper level problem is solved using the ABC algorithm whereas the optimization function of MATLAB is used to solve the lower level problem.

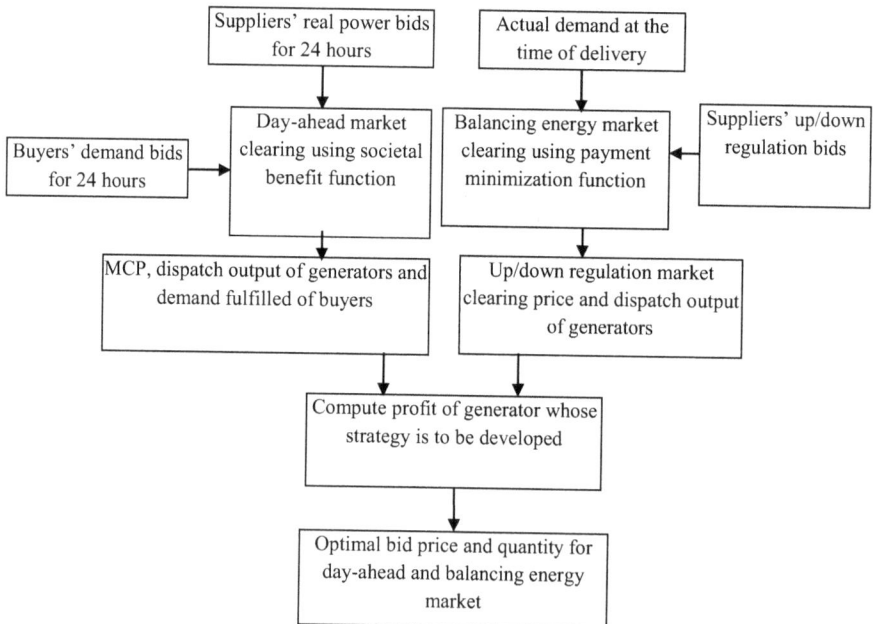

FIGURE 1.7   Proposed bidding technique.

1.11 CASE STUDY

A case study considering three different scenarios are discussed and analysed. In first scenario, it is assumed that rivals bid at their marginal cost in the Day Ahead Energy Market, 3 times of the marginal cost in the up regulation and 0.8 times of the marginal cost in the balancing energy market. Second scenario is same as first except that all the rivals are withholding 20% capacity. In last scenario, it is assumed that rivals bid strategically in both the markets, and all the rivals are withholding 20% capacity. The objective of this case study is to obtain the optimal bidding strategy of a generating company-1 in a given EM considering 24-h time horizon. Generating company-1, bid price lower and the upper bounds in day-ahead and balancing energy markets is considered as:

- DAM:
    Lower bound = marginal cost Upper bound = 1.5 × marginal cost
- Balancing energy market:
    - Incremental bid
        Lower bound = 1.5 × marginal cost Upper bound = 3 × marginal cost
    - Decremental bid
        Lower bound = 0.7 × marginal cost Upper bound = 0.8 × marginal cost

These assumptions are valid for several electricity markets, including the Nordic market.

The data corresponding to generating companies' generators cost coefficients, maximum generation capacity and load entities price and demand are given in Tables 1.1 and 1.2. Figure 1.8 depicts the 24-h forecasted demand and change in demand in a particular hour.

Using the algorithm described in Section 1.3 for coordinated bidding strategy, in the first scenario, generating companies 2 and 3 dispatch their full capacity in each

TABLE 1.1
Cost Coefficients and Generation Capacity of Generators

Generating Company	Generator	a_{ic} ($/MW²h)	b_{ic} ($/MWh)	P_{imax} (MW)
1	1	0.0156	7.92	200
2	2	0.0194	7.85	150
3	3	0.0482	7.97	150

TABLE 1.2
Price-Quantity Bid Details of Load Entities

Load Entities	Bid Price (Rs/MWh)	Bid Quantity (MW)
1	100.17	200
2	110.15	80

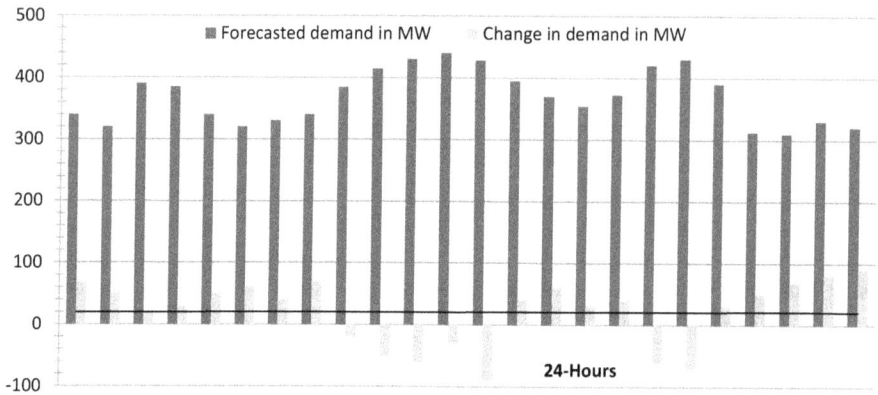

FIGURE 1.8   Forecasted demand in 5 bus system.

hour in the DAM because their bid price is less than the generating company-1, hence the generating companies 2 and 3 get preference over the generating company-1. Balancing energy market (BEM) results demonstrate that, despite the higher bid price of the generating company-1 than its rivals, it has dispatched in the up regulation Balancing energy market (BEM) during the hours 1–8, 14–17 and 20–24. It happens because the generating companies 2 and 3 have exhausted their full capacity in the DAM. Generating company-3 is dispatched in the down regulation balancing energy market due to bidding higher than its rivals.

In second scenario, due to capacity withholding strategy of rivals, dispatched output of the generating company-1 is more in the DAM as compared to case I. Further, despite the lower bid price in comparison to case I in the up regulation BEM, dispatch of generating company-1 has reduced due to capacity withholding strategy of the rivals. During the hours 2 and 22, the generating company-1 has dispatched 50 and 70 MW, respectively, in the up regulation market despite the higher bid price as compared during the other hours. The reason behind this is that, in these hours, the bid price of the generating company-1 is lesser than the bid prices of the generating companies 2 and 3. In this case, the generating company-2 is also participating in the up regulation market during hours 1, 3, 4, 8, 14–15, 17, 20 and 23–24 and in the down regulation market during hour 13 due to change in the bid strategy as compared to the case I.

In hour 13, the desired reduction in the forecasted load was 90 MW, as shown in Figure 1.8. Due to the highest bid price of the generating company-3 in the down regulation market, its 80 MW output is reduced first, and then second highest bidder, i.e. generating company-2 is called to reduce the remaining 10 MW. Generating company-3 participates in the up regulation market during hour 14. During hour 14, the demand of the up regulation was 40 MW. During this hour, the generating companies 1 and 2 are cheaper than the generating company-3, but their combined capacity for up regulation was 35 MW only. Therefore, the generating company-3 dispatched 5 MW in the up regulation and became the marginal generating company.

In the last scenario it is considered that the rivals are participating in both the markets strategically and also withholding 20% capacity. From the results, it has been observed

TABLE 1.3

Comparison of Profit for Coordinated and Uncoordinated Bidding Strategy of Generating Company 1

	Profit in $	
Cases	Coordinated Bidding Strategy	Uncoordinated Bidding Strategy
Case I	17,121.3	4,655.80
Case II	16,094.9	4,655.80
Case III	13,475.7	4,726.80

that the dispatched output of the generating companies 1–3 are same as in case II due to capacity withholding strategy of the rival. However, the DAM prices have increased and the up regulation market prices have decreased due to strategic bidding of the rivals.

To demonstrate the impact of proposed strategy on generating company-1 profit, simulating results obtained using uncoordinated bidding strategy, in which the generating company-1 is assumed to bid only in the DAM for all the cases, has been compared with those obtained using coordinated bidding strategy.

From Table 1.3, it is observed that the profit obtained using the coordinated bidding strategy is higher than the profit obtained using the uncoordinated bidding strategy in all the cases. The difference in profit is due to high up regulation prices.

1.12   FUTURE RESEARCH DIRECTIONS

Keeping in view the work mentioned in this chapter, the following topics are identified for future research in this area.

- Coordinated bidding strategy of a generating company has been developed for day-ahead and balancing energy markets considering conventional fossil fuel-based generators. The renewable generation like wind power may affect the bidding strategy of the conventional generators due to its intermittent nature. Thus, the impact of the wind power on bidding strategy of a supplier owning conventional generator may be examined.
- A combined methodology for building optimal bidding strategies for the buyers and the suppliers can be developed. Each of the individuals from these participant groups may choose appropriate bidding parameters to get more benefits, subject to power system and market constraints. The problem could be formulated in a way which can be solved using stochastic optimization technique to reflect the uncertainties involved in predicting the participants' bidding strategy.
- In this chapter, case study has considered bidding strategy for the balancing market along with the energy market. Bidding strategy can also be developed for reactive power market.

1.13 CONCLUSION

The prime aim of the work presented in this chapter has been to develop optimally coordinated bid strategy of a generating company considering Nord Pool type day-ahead and balancing energy markets. Bidding strategy problem has been formulated as a BLOP, and the ABC algorithm has been applied to obtain hourly the optimal coordinated bidding strategy for 24-h time horizon. Simulation results revealed that the proposed algorithm is suitable for developing bidding strategy of a generating company mainly because it combines the exploration and exploitation processes successfully. Scout bees performed the exploration process which is good for obtaining global solution, whereas the exploitation process performed by the employed and the onlooker bees is good for the local solution. Therefore, the ABC algorithm is capable to find the best optimum bidding strategy of a generating company. Further, it is seen that the profit obtained using the coordinated bidding strategy is more compared to the profit obtained using the uncoordinated bidding strategy.

REFERENCES

1. B. Khan, G. Agnihotri, S. E. Mubeen, and G. Naidu, "A TCSC Incorporated Power Flow Model for Embedded Transmission Usage and Loss Allocation," *AASRI Procedia*, vol. 7, pp. 45–50, 2014.
2. B. Khan, G. Agnihotri, G. Gupta, and P. Rathore, "A Power Flow Tracing based Method for Transmission Usage, Loss & Reliability Margin Allocation," *AASRI Procedia*, vol. 7, pp. 94–100, 2014.
3. B. Khan, G. Agnihotri, P. Rathore, A. Mishra, and G. Naidu, "A Cooperative Game Theory Approach for Usage and Reliability Margin Cost Allocation under Contingent Restructured Market," *International Review of Electrical Engineering*, vol. 9, no. 4, pp. 854–862, 2014.
4. B. Khan and G. Agnihotri, "A Comprehensive Review of Embedded Transmission Pricing Methods Based on Power Flow Tracing Techniques," *Chinese Journal of Engineering*, vol. 2013, Article ID 501587, 13 pages, 2013.
5. B. Khan, G. Agnihotri, and A. S. Mishra, "An Approach for Transmission Loss and Cost Allocation by Loss Allocation Index and Co-operative Game Theory," *Journal of The Institution of Engineers (India): Series B*, vol. 97, pp. 41–46, 2016.
6. B. Khan and G. Agnihotri, "A novel transmission loss allocation method based on transmission usage," *2012 IEEE Fifth Power India Conference*, 2012, pp. 1–3.
7. P. Rathore, G. Agnihotri, B. Khan, and G. Naidu, "Transmission Usage and Cost Allocation Using Shapley Value and Tracing Method: A Comparison," *Electrical and Electronics Engineering: An International Journal (ELELIJ)*, vol. 3, pp. 11–29, 2014.
8. B. Khan and G. Agnihotri, "An Approach for Transmission Usage & Loss Allocation by Graph Theory," *WSEAS Transactions on Power Systems*, vol. 9, pp. 44–53, 2014.
9. S. Khare, B. Khan and G. Agnihotri, "A Shapley Value Approach for Transmission Usage Cost Allocation under Contingent Restructured Market," *2015 International Conference on Futuristic Trends on Computational Analysis and Knowledge Management (ABLAZE)*, Noida, 2015, pp. 170–173.
10. B. Khan, G. Agnihotri, and G. Gupta, "A Multipurpose Matrices Methodology for Transmission Usage, Loss and Reliability Margin Allocation in Restructured Environment," *Electrical & Computer Engineering: An International Journal*, vol. 2, no. 3, p. 11, September 2013.

11. A. K. David and F. Wen, "Market Power in Electricity Supply," *IEEE Transactions on Energy Conversion*, vol. 16, pp. 352–360, 2001.

12. A. K. David and F. Wen, "Optimal Bidding Strategies and Modeling of Imperfect Information Among Competitive Generators," *IEEE Transactions on Power Systems*, vol. 16, pp. 15–21, 2001.

13. F. Li, "Continuous Locational Marginal Pricing (CLMP)," *IEEE Transactions on Power Systems*, vol. 22, pp. 1638–1646, 2007.

14. "A Brief Description of the Six Regional Transmission Organizations (RTOs)," American Public Power Association. Retrieved from http://www.APPAnet.org.

15. Retrieved from http://www.caiso.com.

16. Retrieved from http://www.iex.com.

17. Retrieved from http://www.nordpoolspot.com/The_Elspot_market.(105).

18. Scheduling operation, PJM manual II: PJM, 2007. Retrieved from http://www.pjm.com.

19. Retrieved from http://www.cercind.gov.in.

20. R. K. Mediratta, V. Pandya, and S. A. Khaparde. "Power Markets Across the Globe and Indian Power Market," *Paper presented at Fifteenth National Power Systems Conference (NPSC)*, IIT Bombay, India, December 2008.

21. D. Karaboga and B. Basturk, "On the Performance of Artificial Bee Colony (ABC) Algorithm," *International Journal of Applied Soft Computing*, vol. 8, pp. 687–697, 2008.

22. T. Peng and K. Tomsovic, "Optimal Bidding Strategies: An Empirical Conjectural Approach," *Proceeding of IFAC Symposium on Power Plants and Power Systems*, Seoul Korea, 2003.

23. A. K. Jain, S. C. Srivastava, S. N. Singh, and L. Srivastava, "Bacteria Foraging Optimization Based Bidding Strategy under Transmission Congestion," *IEEE Systems Journal*, vol. 9, no. 1, pp. 141–151, 2015.

24. H. L. Song, C. C. Liu, and J. Lawarree, "Decision Making of an Electricity Supplier Bid in a Spot Market," *Proceedings of IEEE Power Engineering Society Summer Meeting* (Vol. 1, pp. 692–696), 1999.

25. F. Author and G. S. Monte Carlo, *Concepts, Algorithm and Applications*. Berlin, Heidelberg: Springer-Verlag, 1995.

26. L. E. Ruff, "Stop Wheeling and Start Dealing: Resolving the Transmission Dilemma," *The Electricity Journal*, vol. 7, pp. 24–43, 1994.

2 Harmonic Mitigation Methods in Microgrids

Behrooz Adineh, Reza Keypour,
Pooya Davari, and Frede Blaabjerg

CONTENTS

2.1 INTRODUCTION

For decades, the most popular energy resources to produce electricity were fossil fuel-based sources. However, due to the enormous concerns arising from using the fossil fuel sources such as environmental problems, gradual decrement of these sources, the increasing cost of fuel, and energy demand, renewable energy sources (RES) have been widely used as an alternative solution to provide reliable, safe, inexpensive, and renewable energy for the modern power system and its customers [1–5]. RES like solar cells, wind energy, and fuel cells, along with power electronic devices and local controllers, comprise a distributed generation (DG). The main components of a smart microgrid are various types of DGs, local loads, and hierarchical controllers, as shown in Figure 2.1. A microgrid can provide electricity to the loads in two modes: grid-connected mode and islanding mode. Generally, the microgrid can be connected to the main network, and loads at the point of common coupling (PCC). The hierarchical control of the smart microgrid consists of three parts: primary, secondary, and tertiary levels. The main objectives of the primary level are sharing power between DGs,

DOI: 10.1201/9781003278030-2

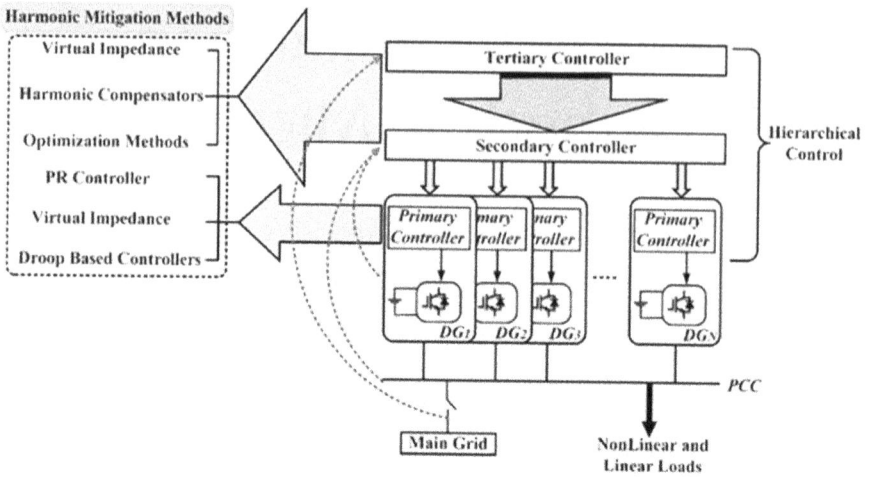

FIGURE 2.1   Main parts of the microgrid and harmonic mitigation methods at each level.

avoiding circulating current in the microgrid and maintaining voltage and frequency at a certain rate. Droop controller, which is responsible for active power sharing, voltage controller, and current controller are the basic controllers at the primary level. The secondary controller has to reduce voltage and frequency deviations induced by the primary level. At the tertiary level, the main task is to control the optimal and economical performances of the smart microgrid in the grid-tied mode.

Although the microgrids have many advantages such as reducing environmental issues and cost of electricity for local users, providing power for the local customers in case of the power outage, etc., they have brought some challenges to the power system like reliability issues, protection concerns, and power quality challenges [6–10]. A challenging problem that arises in smart microgrid's domain is harmonic distortion (HD) issue due to the presence of the nonlinear loads and power electronic-based equipment. There are methods available in the literature for solving the voltage and current HD problems, which are reviewed in [11–13].

As can be seen in Figure 2.1, the proportional resonant (PR) controller, virtual impedance (VI), and droop-based controller are mostly utilized on the primary controller of the microgrid's hierarchical control to compensate HD. Moreover, several authors have suggested methods such as VI, harmonic compensators, and optimization approaches at the secondary and tertiary smart levels of the microgrid hierarchical control to resolve the harmonic problems in microgrids. The basic concept of these methods is reviewed and explained in Sections 2.4 and 2.5.

This chapter begins with an explanation for why we have HD in microgrids. Moreover, the harmonic standards commonly used for microgrid harmonic concerns are reviewed in Section 2.3. Then, the main methods used in the hierarchical control of the microgrid for harmonic compensation purposes are summarized in Sections 2.4 and 2.5. In Section 2.6, a flowchart is proposed to show diagrammatically how the harmonic compensation methods can apply in microgrids. The simulation results of two harmonic mitigation methods are shown and discussed in Section 2.7. Finally, this chapter is concluded in Section 2.8.

2.2 HARMONICS IN MICROGRIDS

As human being has tried to transfer energy from one place to another through the transmission lines, harmonic components are generated in power system and would have deleterious effects on the different parts of the power system. Nowadays, due to the presence and penetration of power electronic devices, and nonlinear loads in the power system, the harmonics and their impacts on the power system and microgrids should be addressed. Hence, the main reasons of the harmonic generation and the main parts involved in producing harmonic components to the system are explained in this section.

The fundamental parts of an islanded microgrid including RESs, power electronic devices, filters, line and grid impedances, and loads are illustrated in Figure 2.2. One of the effective parts to be addressed and investigated in the harmonic genera-tion concept is power electronic-based devices. These devices are mainly responsible for generating and transferring power from the RES side to the load side. However, the switching operation of these power converters will provoke the HD problems and should be undertaken with great effort in microgrids. Moreover, the interactions between controls of different power electronic devices and other components might increase HD issues [14].

The presence of the nonlinear loads in microgrids is widely considered in literature as another crucial factor producing voltage and current HD in microgrids. These loads are power electronic-based loads, which inject harmonic currents at PCC. Therefore, the voltage waveform at PCC will be distorted by connecting the nonlinear loads to the microgrid.

Another factor in harmonic pollution of microgrids is the resonance phenomena. Due to the use of inductor-capacitor (LC) or inductor-capacitorinductor (LCL) filters, which have capacitance characteristics, and the inductance existing in the line or grid imped-ances, the resonance would appear in the microgrids. In addition to the three factors for the HD in microgrids, which are mentioned above, there would be other factors like changing the modes of operation from grid-tied mode to islanding mode or vice versa.

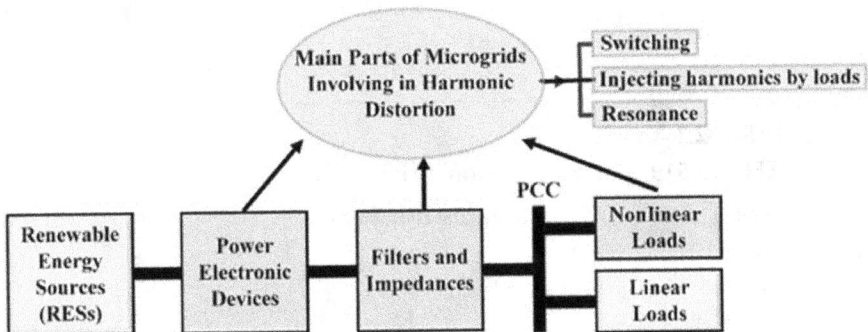

FIGURE 2.2 Harmonics in microgrids.

As a brief conclusion of this section, the main factors of harmonic generation, which are power electronic devices, filters and impedances, and nonlinear loads in microgrids, are introduced and explained. To overcome and reduce voltage and current HD in microgrids, the filter of each DG should be determined properly, and the controllers used in the microgrid hierarchical control should be improved efficiently. Hence, the main approaches suggested in literature to compensate HD in smart microgrids are reviewed in Sections 2.4 and 2.5. The standards and limitations for harmonic concerns in microgrids are discussed in Section 2.3.

2.3   HARMONIC STANDARDS

It is necessary to know what harmonic standards and limitations have been made by the electrical organizations, which should be met in harmonic mitigation purposes to keep the microgrid harmonics under certain values. Hence, these international standards are explained in this section. International Electrotechnical Commission (IEC) and Institute of Electrical and Electronics Engineers, (IEEE) are two well-known organizations that establish and provide guidelines for power system users and companies.

One of the useful recommendations for harmonic control in renewable energy-based systems is IEEE 519, which was released by IEEE Power and Energy Society in March 2014. The voltage harmonic limits of IEEE standards are shown in Table 2.1, which are different for various bus voltages at PCC (v_{PCC}). The specific values shown in the table must be met by the proposed harmonic mitigation methods for each individual harmonics (h) and total harmonic distortion (THD) of the microgrid.

Two other suitable standards are IEC 61000 Section 2.2 and EN 50160, which have designated limitations for both individual harmonics and THD of low voltage networks, and the last versions of these standards were produced in 2018 and 2004, respectively [15]. The specific values of voltage harmonics for these two standards are shown in Table 2.2. As can be seen, the limited criteria for all harmonic orders up to 25th and THD are the same in the two standards. Moreover, the IEC 61000 standard has recommended using the following equation for harmonic orders between 25 and 49:

$$0.2 + \frac{25}{n} \times 0.5 \qquad (2.1)$$

TABLE 2.1
IEEE Std 519: Voltage Harmonic Limitations

Rated v_{pcc} (kV)	Individual Harmonic (%)	THD (%)
$V \leqslant 1$	5	8
$1 < V \leqslant 69$	3	5
$69 < V \leqslant 161$	1.5	2.5
$161 < V$	1	1.5

TABLE 2.2
IEC 61000- 2- 2 and EN 50160: Voltage Harmonic Limitations

	Harmonic Order								
	5	7	11	13	17	19	23	25	THD (%)
IEC	6%	5%	3.5%	3%	2%	1.5%	1.5%	1.5%	8
EN	6%	5%	3.5%	3%	2%	1.5%	1.5%	1.5%	8

It is observed from Tables 2.1 and 2.2 that the harmonic limitations are different for each harmonic order in both IEC and ENC standards. However, the IEEE standards suggest to keep all individual harmonic less than 5%. Moreover, the IEC and EN standards are used for low voltage networks, and so far, they have not released any standards for harmonic control in microgrid or renewable energy-based networks. The discussed harmonic standards do not include high order harmonics, which are close to the switching frequency (e.g., 10 kHz) and may cause resonant issues in microgrids.

2.4 HARMONIC MITIGATION APPROACHES AT PRIMARY LEVEL

The parallel connection of N DGs connected to the PCC and also the local controllers and harmonic controllers including harmonic droop controller and VI in each DG of the microgrid are shown in Figure 2.3. Voltage of the filter capacitance v_c, and output current i_o of each DG are measured and used in the local controller to provide the gate signals of the voltage source inverter (VSI). The main controllers of the local control are droop controller, and voltage and current controllers implemented in the inner controller. The harmonic droop controller, VI, and PR controllers used in

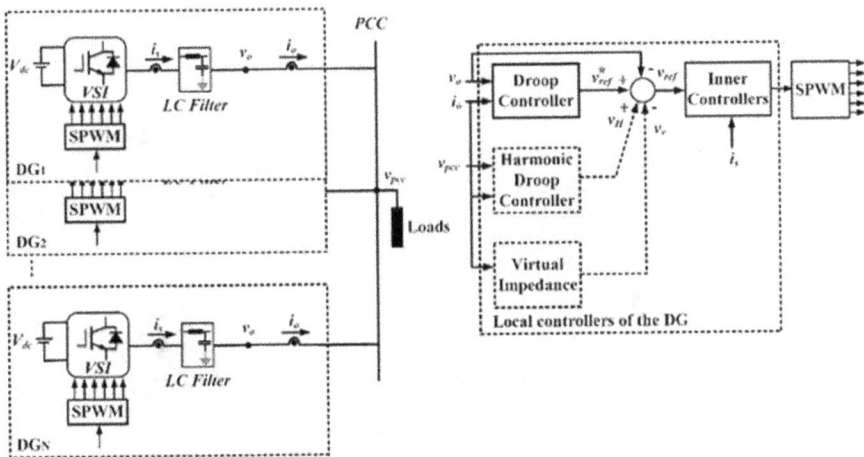

FIGURE 2.3 Local controllers along with the harmonic mitigation controllers in each DG of the microgrid.

the inner controller for harmonic mitigation purposes are described in the following subsections.

As can be seen from Figure 2.3, the voltage reference, which is the input of the inner controllers, can be obtained as follows:

$$v_{\text{ref}} = v_{\text{ref}}^* + v_v + v_H \tag{2.2}$$

where v_{ref}^*, v_v, and v_H are the voltage reference generated by the droop controller, voltage output of the harmonic droop controller, and voltage output of the VI, respectively. The main aim of the VI-based harmonic mitigation methods and harmonic droop controllers on the primary controller is to adjust the voltage reference.

2.4.1 PROPORTIONAL RESONANT (PR) CONTROLLER

The block diagram of the inner controllers utilized in the local controller of each DG is shown in Figure 2.4. Voltage and current controllers are the main controllers used in this controller. The PR controller can be utilized in both voltage and current controllers. As can be seen in Figure 2.4, the PR controller consists of three transfer functions, which are proportional (G_P), resonant (G_R), and harmonic (G_H). The main advantages of the PR controllers are reducing synchronization issues and steady-state errors in three-phase and single-phase microgrids, respectively. These controllers are implemented in stationary frame ($\alpha\beta$).

The transfer functions of an ideal PR controller in general term can be written as follows:

$$G_{Pi} = k_p, \quad G_{Ri} = \frac{k_{ri}s}{s^2 + \omega_n^2}, \quad G_{Hi} = \sum_h \frac{k_{hi}s}{s^2 + (\omega_n h)^2} \tag{2.3}$$

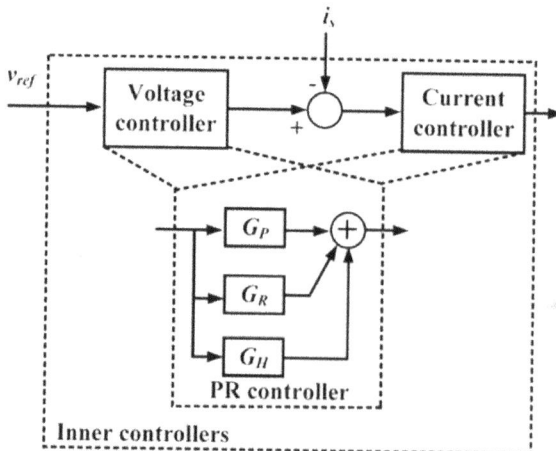

FIGURE 2.4 Inner controllers of each DG.

where k_p, k_{ri}, and k_{hi} are the proportional, resonant, and harmonic gains, respectively. ω_n and h are the system frequency and harmonic order, respectively. k_p is determined like the one in PI controller and controls the dynamics of the DG. The use of ideal PR would result in an infinite gain at ω_n. Therefore, a non-ideal PR transfer function is suggested as follows to overcome this issue:

$$G_{Pn} = G_{Pi} = k_p, \quad G_{Rn} = \frac{k_{rn}\omega_c s}{s^2 + 2\,\omega_c s + \omega_n^2}, \quad G_{Hn} = \sum_h \frac{k_{hn}\omega_c s}{s^2 + 2\,\omega_c s + (\omega_n h)^2} \quad (2.4)$$

where k_p, k_{rn}, and k_{hn} are the proportional, resonant, and harmonic gains for the non-ideal PR controller, respectively. ω_c is the cut-off frequency of the controller.

The bode diagrams of the ideal and non-ideal PR controllers with $k_{ri} = k_{hi} = k_{rn} = k_{hn} = 1$ (these values are just selected for the simplicity) are plotted in Figure 2.5 for $\omega_c = 1$, $\omega_n = 100\,\pi$ and $h = 5, 7, 11$, which are the most distinguished harmonic components in the microgrid. As can be seen, the ideal PR has sharper peaks at the system and harmonic frequencies than the non-ideal one. Moreover, the magnitudes of the ideal PR are much bigger than the non-ideal one.

The bode diagram of the non-ideal PR is sketched in Figure 2.6 for different values of $\omega_c = 1, 10, 20$. It is obvious that harmonic compensation at selected harmonic components can be achieved by using both ideal and non-ideal PR controllers, as shown in Figures 2.5 and 2.6. In addition to the selective harmonic compensation, it is observed that the resonant peaks become narrower with the decrease in ω_c. Although it would make the extraction process in inner controllers more accurate, the PR controller would be sensitive to the frequency changes, especially in grid-tied applications. Moreover, the transient response of the PR controller will be slower. Therefore, it is suggested to use the non-ideal PR controller with ω_c between 5 and 15 in order to have better filter performance [16].

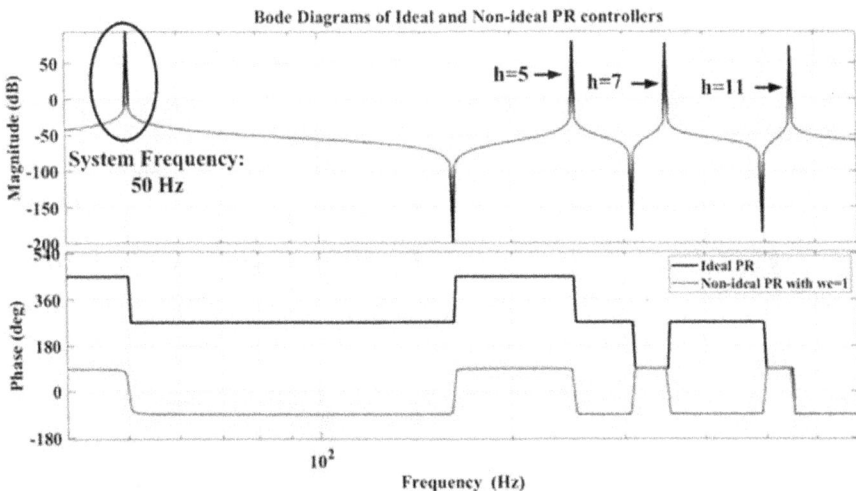

FIGURE 2.5 Bode diagram of the ideal and non-ideal PR controllers with $\omega_c = 1$.

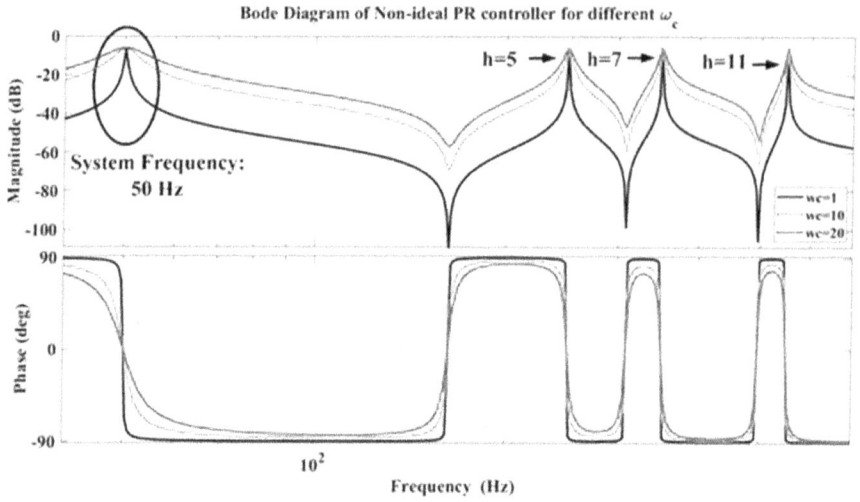

FIGURE 2.6   Bode diagram of the non-ideal PR controllers with different $\omega_c = 1, 10, 20$.

In conclusion of this subsection, the non-ideal PR controller has better performance and advantages such as adjustable resonant peaks and non-infinite gain compared to the ideal one in practical applications. In the next subsection, the VI loop as a tool for harmonic compensation purposes will be explained.

2.4.2   VIRTUAL IMPEDANCE

The VI concept is first used at the primary controller of DGs to make the inverters' impedance as possible as inductive. Thereby, the reactive and harmonic power sharing can be achieved by using VI Z_v as shown in Figure 2.7 in the local control loop. The VI can be inductive, resistive, or capacitive for different situations and purposes.

The block diagram of the DG local controllers (based on Figures 2.3 and 2.4) along with the VI loop is shown in Figure 2.8. The voltage of the VI loop can be written as follows:

$$V_v = Z_v i_o \tag{2.5}$$

To compensate selected harmonic components in the microgrid, Z_v can be used in the form of virtual resistive as follows [17]:

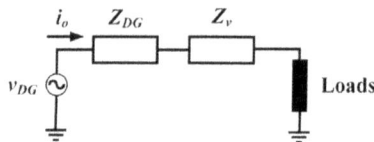

FIGURE 2.7   Simplified equivalent circuit of the DG.

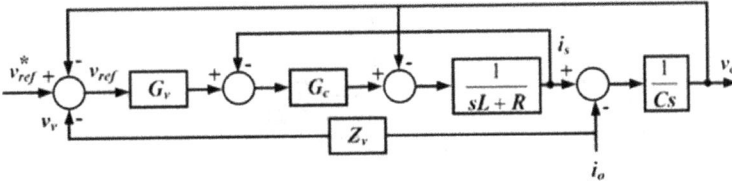

FIGURE 2.8   DG local controllers along with the VI loop.

$$Z_v = \sum_{h=5,7,9} \frac{R_h \omega_{ch} s}{s^2 + \omega_{ch} s + (\omega_{nh})^2} \tag{2.6}$$

The capacitive loop also can be used in selected harmonic compensation methods, which is given as follows [17]:

$$Z_v = \sum_{h=5,7,9} \frac{\omega_{ch} (k_{ph} s + k_{ih})}{s^2 + \omega_{ch} s + (\omega_{nh})^2} \tag{2.7}$$

where R_h, ω_{ch}, k_{ph}, and k_{ih} are the virtual resistive, bandwidth at the selected harmonic, proportional, and integral gains of each harmonic.

Although VI-based methods are used in harmonic mitigation applications, it is revealed that they have the following drawbacks:

1. They are complex and have computational burdens.
2. There are many parameters in such approaches, which should be designed and determined accurately to reduce HD.
3. These methods are sensitive to the DG impedances and frequency variations.
4. To compensate harmonic components and especially negative harmonic components, these methods strongly need accurate harmonic extraction methods, which make their structure more complex and their response slower.

2.4.3   DROOP-BASED CONTROLLERS

One of the main controllers on the primary controller of each DG in the smart microgrid is droop controller, which is responsible to share power properly between DGs. The block diagram of the conventional droop controller named as $P-\omega$ and $Q-V$ is shown in Figure 2.9. Based on the measured output current and voltage of each DG, the active and reactive power is determined. Then, through the low pass filters (LPFs), the voltage reference of droop controller can be calculated as follows:

$$\omega_i = \omega_n - G_{Pi} P_i, \quad E_i = E_i^* - G_{Qi} Q_i \tag{2.8}$$

where E_i^*, P_i, and Q_i are the magnitude of the voltage, active and reactive power for the inverter i in the multi-inverter microgrid, respectively. G_{Pi} and G_{Qi} are the droop controller

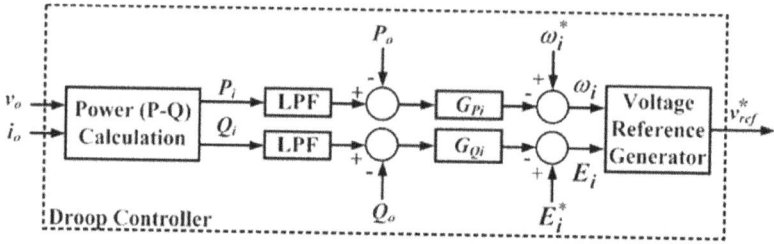

FIGURE 2.9   Block diagram of the conventional droop controller.

coefficients, which should be tuned accurately based on the rated power, voltage, and frequency of the each DG. To improve power sharing by the droop controller, the derivative and integrative terms can be added to G_{Pi} and G_{Qi} [18]. However, adaptive or modified droop controller can result in microgrid instability. Although the droop controller with some modification can be used for active power sharing, it is suggested to use VI-based approaches along with the droop controller for sharing reactive power between DGs.

Recent studies have indicated that the harmonic droop controller along with the conventional one can be used to suppress harmonics at the primary level of each DG [19–23]. A simplified block diagram of the harmonic droop controller is shown in Figure 2.10. The voltage harmonic generation process, in which the output voltage harmonic will be added to the droop controller voltage reference, is similar to the conventional droop controller. The measured voltage and current are used to calculate active and reactive harmonic power. After multiplying these calculated harmonic powers to the harmonic droop controller coefficients (n_h and m_h), the voltage harmonic is then generated at each desired harmonic frequency. Therefore, the harmonic droop controller equations can be written as follows:

$$E_h = -n_h P_h, \quad \omega_h = h\omega_n - m_h Q_h \tag{2.9}$$

where n_h and m_h are the harmonic droop controller coefficients. P_h and Q_h are the harmonic powers of the ith inverter.

The droop controller concept can be used for power sharing purposes. Moreover, the VI-based approaches can be added to the system loops to improve the droop controller functions. However, additional studies are required to understand more completely the key tenets of the harmonic droop controller. The existing harmonic droop controllers have led to microgrid stability and robustness issues, so far lacking in the literature.

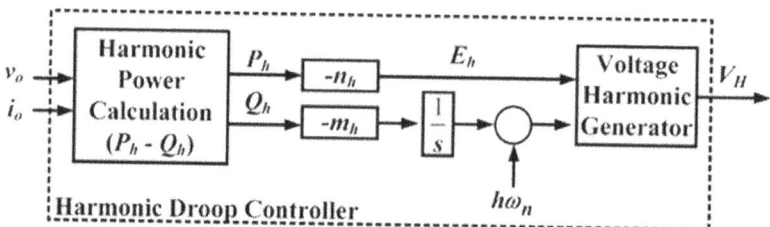

FIGURE 2.10   Harmonic droop controller.

2.5 HARMONIC MITIGATION METHODS AT SECONDARY

As shown in Figures 2.1 and 2.11, the second level of the smart microgrid hierarchical control needs communication links to get data from the microgrid measurement devices and inject information to the local controller of each DG. In this section, the harmonic compensation methods used at these two levels of the microgrid are discussed. The main drawbacks of the methods used on the secondary controller are the delays and faults of the communication links and time-consuming.

The data used at the secondary level, as shown in Figures 2.1 and 2.11 with red lines, mostly are output current and voltage of each DG and the voltage at PCC. These inputs of the secondary controller are measured and transmitted through the low bandwidth communication links. The literature review shows that the secondary controller was applied to the microgrid in order to achieve five main tasks, as shown in Figure 2.11 [3]. These objectives will be explained in the following.

There have been several research studies in which the harmonic components are extracted at the secondary level of the smart microgrid by using harmonic extraction methods such as second order generalized integrator (SOGI), MSOGI, Fourier-based methods, etc. Then, the extracted harmonics are transmitted to the local controller of each DG to use in harmonic compensator tools such as PR controllers, VI, adaptive VI, etc. [24,25].

The HD can be mitigated in microgrid by using harmonic compensator approaches on the secondary controller of the smart microgrid. In these methods, the output voltage or current of the DG is captured and transmitted to the secondary control. The compensated current or voltage signal then is injected to the primary level to use in the local control loops. The VI loop and PR controllers are sometimes used in the primary controller to make sure that the signals are harmonic free [26–28].

In [29,30], a virtual admittance calculator is used in the secondary controller to determine the amplitude of the virtual admittance in order to use in an adaptive virtual admittance control loop in the primary controller. The virtual admittance

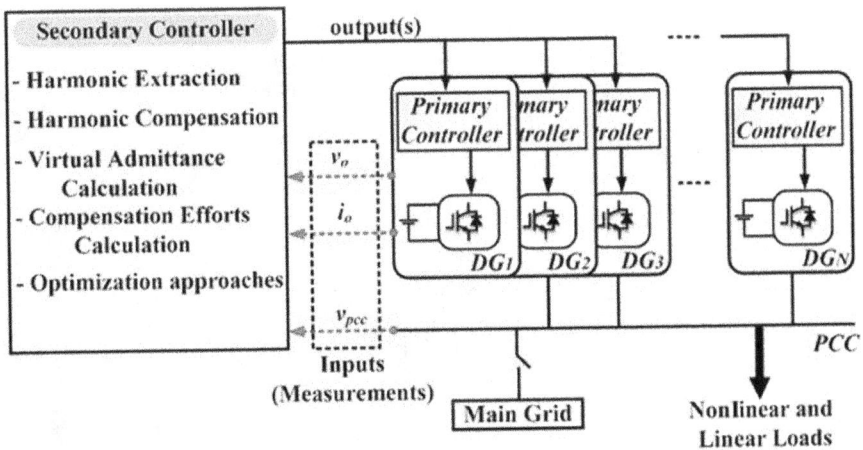

FIGURE 2.11 Secondary controller in microgrids.

is calculated based on each inverter rated power, inverter and line impedance, and loads. Therefore, the proposed virtual admittance calculation would be sensitive to the system parameters and configuration.

Another approach used at the secondary level for selective harmonic cancellation at the critical bus is to calculate compensation effort at the secondary controller and then send this value to the primary controller of each DG [31,32]. As shown in Figure 2.12, the compensation effort (C_t^h) can be calculated on the secondary controller of the smart microgrid based on (10). The input and output of the proposed secondary controller are the measured voltage at PCC and compensation effort value, respectively. The harmonic and fundamental components of v_{pcc} are extracted by using MSOGI-frequency locked loop (FLL) and then through the LPFs, and based on the following equations, the compensation effort is calculated:

$$C_t^h = v_{dq}^h \left(k_{p,h} + \frac{k_{i,h}}{s} \right)(\text{THD}_{\text{ref}} - \text{THD}) \left(\frac{3 \times \text{HD}_h}{\sum_{h=5,7,11} \text{HD}_k} \right) \left(\frac{S_{o,i}}{\sum_{n=1}^{N} S_{o,n}} \right) \quad (2.10)$$

where $k_{i,h}$, $k_{p,h}$, THD, and THD_{ref} are the PI controllers integral and proportional gains for each selected harmonic, calculated and reference THDs, respectively. S_o is the rated power of DG. The HD for each harmonic and THD are obtained as follows:

$$\text{HD}_h = \sqrt{\frac{\left(v_d^h \right)^2 + \left(v_q^h \right)^2}{\left(v_d^f \right)^2 + \left(v_q^f \right)^2}}, \quad \text{THD} = \sqrt{\sum_{k=5,7,11} \text{HD}_k^2} \quad (2.11)$$

Finally, the calculated compensation signal is transmitted to the local controller, and after multiplying by the harmonic frequency, it is added to the output voltage of the VI loop at the primary level.

The last approach utilized at the secondary level of the smart microgrid, which is proposed in [33], is an optimization-based method for harmonic reduction purposes. In this approach, the particle swarm optimization (PSO) is utilized on the secondary controller to optimize the injected voltage to the primary controller in order to minimize selected harmonic components, THD of the critical bus, and also average THD in multi-bus smart microgrid. The voltage harmonic amplitudes of all buses in the multi-bus microgrid are measured and transferred through the communication links to the secondary controller. The optimized angle and amplitude of the voltage

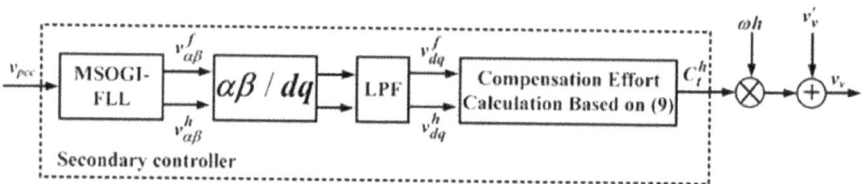

FIGURE 2.12   Input, output, and the compensation effort calculation method for the secondary controller.

are obtained based on the PSO algorithm and injected to the primary controller, in which it is added to the droop controller output voltage.

The five main objectives of the secondary level shown in Figure 2.11, which are used in the secondary controller as a tool for harmonic compensation applications, are briefly discussed in this section. Although the use of the secondary controller for mitigating selected harmonics and THD of the PCC are addressed in the previous studies, a number of challenges regarding the harmonic mitigation goals remain to be addressed:

1. All approaches have relied on the communication links. However, the failure and delays in these links have not been well analysed in the literature.
2. The robustness of the proposed harmonic mitigation approaches against cyber-attacks, which is a crucial issue in today's microgrids, is not addressed.
3. The individual harmonics and THD reduction in a multi-bus multi-inverter microgrid, which can be a major issue in recent years due to the increase demand and growing network, are not well considered so far.

2.6   THE PROPOSED HARMONIC MITIGATION PROCEDURE

Due to an increasing need for delivering high-quality power, and also growing number of DG units with nonlinear loads connected to the system, which have led to propose new harmonic mitigation methods, it is of interest to have a clear procedure for applying these proposed methods to the smart microgrids. Therefore, the six-step guideline for applying the harmonic mitigation method is proposed in this section. The proposed flowchart is shown in Figure 2.13. The six steps of the proposed flowchart are discussed as follows:

Step 1: Determine microgrid configuration and parameters. The first step is to configure the studied microgrid and determine microgrid parameters such as fundamental and switching frequencies, line impedances, rated power of the microgrid and converters, etc. The microgrid structure based on its components like inverter, transmission lines, loads, etc. would be different. Moreover, the microgrid operation modes, which are islanded and grid-connected modes, should be defined in this step.

Step 2: Filter design consideration. In this step, a type of passive filter will be chosen, which can be LC, LCL, etc. The filter parameters can be obtained based on each DG parameter.

Step 3: Determine inner controllers' parameters. As discussed in previous sections, the inner controllers consist of voltage and current controllers. The inner controllers commonly used in literature are PI, PR, and proportional controllers. The parameters of these controllers based on opened and closed loop analysis can be calculated. The block diagram shown in Figure 2.8 can be used to obtain the inner controller parameters. Moreover, the harmonic compensation parameters used in inner controller can be determined in this step.

Step 4: Determine droop coefficients. The droop coefficients, which are shown in Figure 2.9 for a conventional one, should be designed in this step.

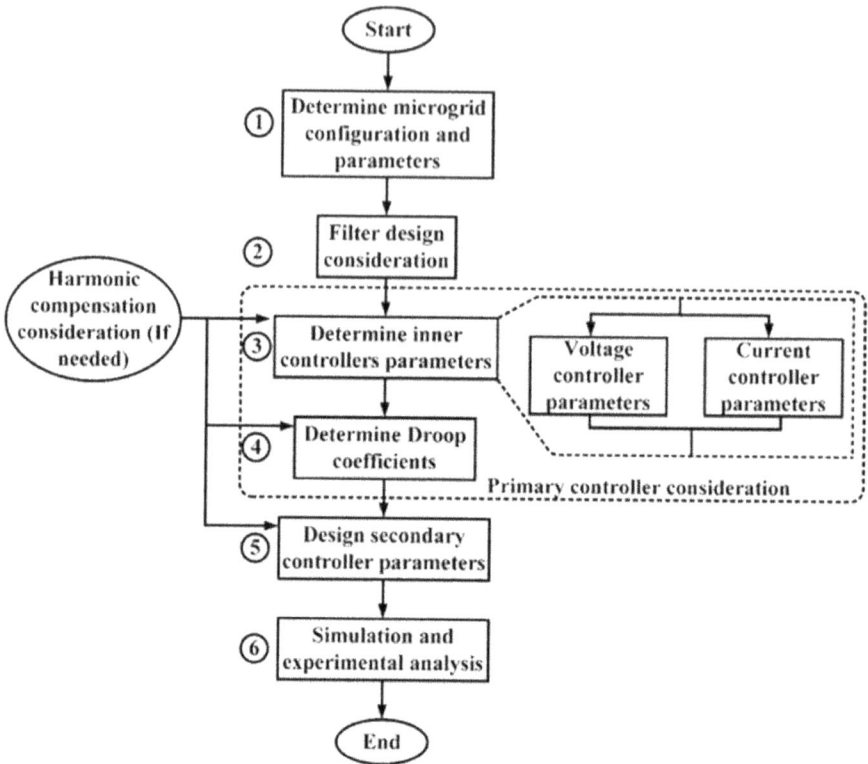

FIGURE 2.13   The proposed flowchart to apply harmonic mitigation methods to microgrids.

The stability analysis has been used to determine the proper droop parameters. If a VI (as shown in Figure 2.8) or/and harmonic droop controller (as
shown in Figure 2.10) is also used, their parameters should be determined
in this step.

Step 5: Design secondary controller parameters. If there is another control
level (secondary or tertiary level) in microgrid hierarchical control for the
proposed harmonic mitigation method, the proper parameters should be
obtained in this step.

Step 6: Simulation and experimental analysis. Finally, after careful designing
of the microgrid and its controller's parameters, the simulation and experimental results can be achieved.

2.7   SIMULATION RESULTS

In this section, the simulation results for two case studies are made in MATLAB®/
Simulink® by applying two different harmonic mitigation methods, which are discussed in previous sections. The PR controller and PSO-based method are used on
the primary and secondary controllers of the studied microgrid hierarchical control,
respectively. This section aims to show how to use the proposed harmonic mitigation

flowchart, which is shown in Figure 2.13, and to present the results for voltage harmonic reduction by using PR controller and PSO-based method in uninterruptible power supply (UPS) and multi-bus microgrids, respectively.

2.7.1 PR Controller Used at Primary Level

As it is discussed in the previous section and based on the proposed flowchart in Figure 2.13, the first step in applying harmonic mitigation methods is to determine microgrid structure and basic parameters of the system. The local controllers, which are current and voltage controllers, and the power configuration of the UPS system are shown in Figure 2.14. A 3 kW inverter is connected to the load through the LC filter in order to supply the demanding power. The nonlinear load is three-phase diode rectifier. The inverter current (i_s), output voltage (v_o), and output current (i_o) are measured and used at the local controller to generate the proper signals for three-phase inverter by using sinusoidal pulse width modulation (SPWM). The system parameters are shown in Table 2.3.

Several methods are reported in the literature to properly determine power filter parameters, which should be done in step two of the proposed flowchart. In Refs. [34,35], the following equations are used to obtain the LC filter parameters:

$$C < \lambda_c \frac{P_{inv}}{\omega_n} V_o^2, L < \lambda_l \frac{V_o}{\omega_n I_s} \tag{2.12}$$

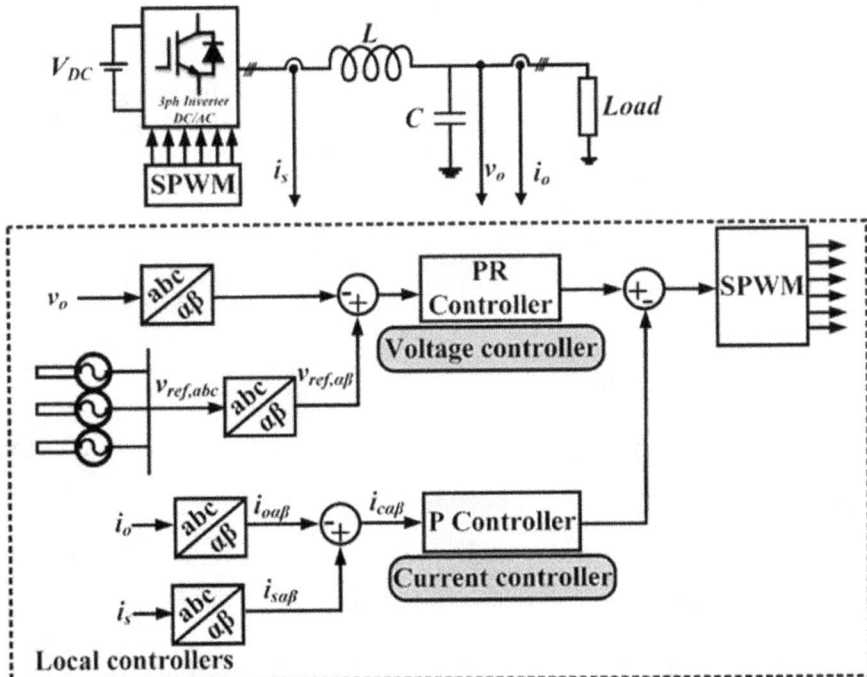

FIGURE 2.14 Local controllers and power structure of a UPS.

TABLE 2.3

System Parameters

Parameter	Value
DC link voltage V_{dc}	700 V
Fundamental frequency f	50 Hz
Switching frequency f_s	10 kHz
Linear load	$R = 50\,\Omega$
Nonlinear load	$R = 14\,\Omega$
Rated voltage (rms)	230 V

where P_{inv}, ω_n, V_o, and I_s are rated power of the inverter, fundamental frequency, rms value of the rated voltage, and rms value of the inverter current, respectively. In practical applications, the value of the capacitor voltage ripple λ_c and inductance current ripple λ_l is usually recommended to be about 5%. Therefore, the values of L and C are considered as 5 mH and 9 µF in this case study, respectively.

In step three, the parameters of the current and voltage controllers should be designed. Based on the opened and closed loops analysis [36,37], the following values are used for inner controller parameters:

$$k_{pv} = 15, \; k_{rn} = 50, \; k_{5n} = 1{,}000, \; k_{7n} = 700, \; k_{11n} = 1{,}000, \; k_{pi} = 10, \; \omega_c = \sqrt{2} \;\; (2.13)$$

It should be noted that the non-ideal PR controller is utilized in the voltage controller to compensate voltage harmonics. The values of harmonic orders 5, 7, and 11, and THD of the output voltage are shown in Figure 2.15. The results show that the harmonic components and THD of the output voltage are reduced by using PR controller, and the results have met the harmonic standards discussed in Section 2.3.

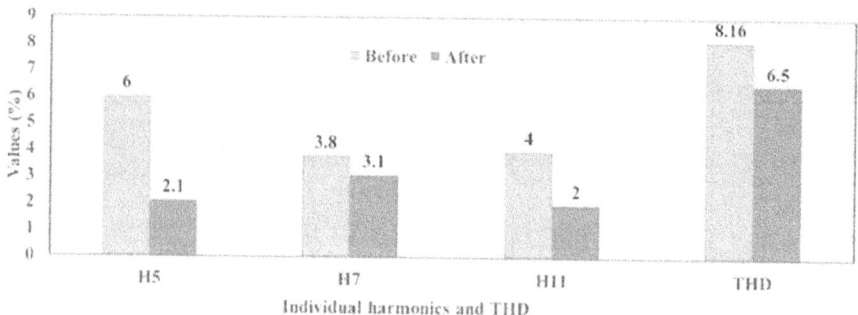

FIGURE 2.15   Harmonic mitigation results for UPS by using PR controller.

2.7.2 PSO-Based Method Used at Secondary Level

In [33], the authors suggested to use PSO optimization algorithm in the secondary controller of the microgrid to find the best voltage amplitude and angle for each DG and then inject the optimized voltage to the primary controller. The main goal of this approach is to minimize harmonic components (5 and 7) and THD of the critical bus in order to meet the harmonic standard requirement.

The microgrid structure with primary and secondary controllers is shown in Figure 2.16. The microgrid includes six buses, in which the loads are connected and two DGs. The nonlinear loads are three-phase diode rectifiers. In the proposed approach, the harmonic voltages of all buses in the microgrid are measured and transmitted (red dashed lines) to the secondary controller. The microgrid parameters are the same as the ones in Table 2.3. Hence, the passive filter and inner controller parameters will be the same as the ones in the previous case study.

The conventional droop controller, as shown in Figure 2.9, is used to share power in the studied microgrid. The droop coefficients are shown in Table 2.4. The PSO optimization algorithm finds the best solution based on the following equations:

$$v_i^{k+1} = av_i^k + c_1 r_1 \left(P_{\text{best},i}^k - x_i^k \right) + c_2 r_2 \left(G_{\text{best}} - x_i^k \right), \; x_i^{k+1} = x_i^k + v_i^{k+1} \qquad (2.14)$$

where x and v are the position and velocity of the ith particle, respectively. The best position of each particle and the best particle position between all particles are $P_{\text{best},i}^k$ and G_{best}, respectively. The values for c_1, c_2, and a, which are the variables of the PSO algorithm are shown in Table 2.4. r_1, r_2, and k are the random numbers between 0 and 1, and iteration number, respectively.

The simulation results for reduction in harmonic components (5 and 7) and THD of the critical bus, which is bus number six in Figure 2.16, are shown in Figure 2.17. The results show that the HD in multi-bus microgrid can be effectively reduced by using the proposed optimization-based method, and the results that have met the harmonic standards are discussed in Section 2.3.

2.8 CONCLUSION AND FUTURE WORKS

The overall contribution of this chapter is summarized briefly in this section. Moreover, a number of challenges and future trends of the harmonic control methods are presented in the following.

The penetration of microgrids including DG units, which produce power from the renewable energy resources by using power electronic devices, loads (linear and

---

TABLE 2.4
Droop Coefficients and PSO Parameters

G_P	G_Q	c_1	c_2	a
5×10^{-5}	3×10^{-3}	2	1	0.99

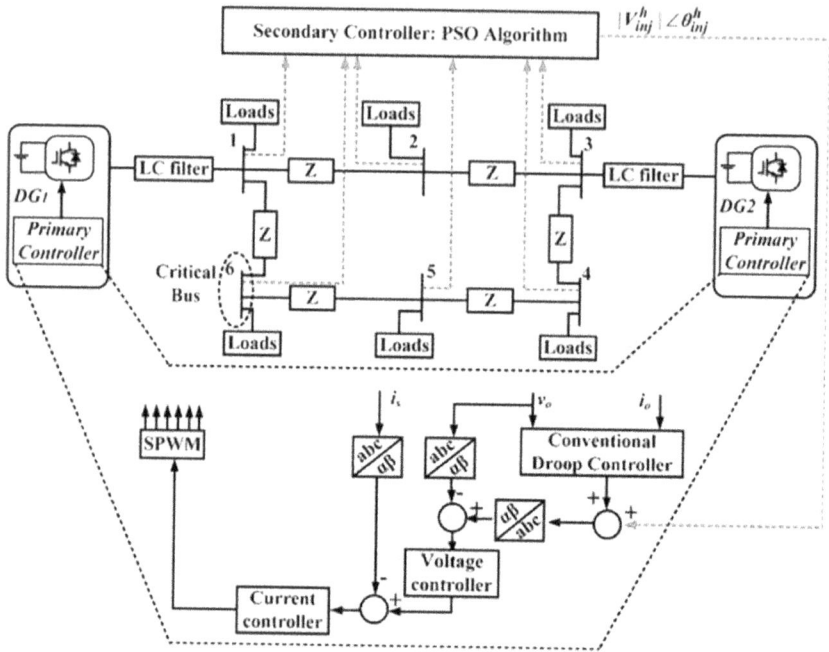

FIGURE 2.16 Multi-bus microgrid structure, primary controllers, and secondary input and outputs.

nonlinear), communication links (if secondary and tertiary levels exist), etc., has increased in the recent decades. Moreover, the users and companies are increasingly demanding for the power with high reliability and high power quality. Therefore, a critical issue here is to reduce HD in microgrids to improve the system operation and meet the increasing demand. Hence, the overall aim of this chapter is to give a general look about basic concept of the harmonic mitigation methods used in literature [34–37].

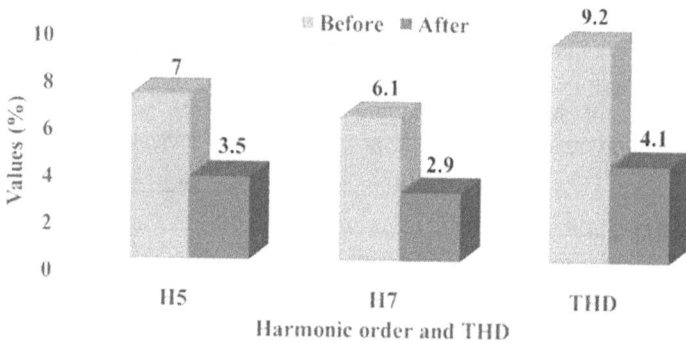

FIGURE 2.17 Reduction in harmonic components and THD of the critical bus by using the PSO-based method.

First, the parts in microgrids, which have the main contribution in generating harmonics, are introduced and explained. Then, the main practical harmonic standards, which are IEEE 519, IEC 61000-2-2, and EN 50160, are introduced and discussed. It is revealed that with the presence of the power electronic devices and nonlinear loads polluting the microgrid with harmonics, the microgrid harmonic standards should be improved. On the other hand, the proposed harmonic mitigation methods should meet the microgrid harmonic standards.

The harmonic compensation methods are summarized and addressed based on their application in hierarchical control of the microgrid. The methods utilized on the primary controller are categorized into three groups: PR controllers, VI scheme, and droop-based methods. The basic concept of each method along with their application for harmonic reduction purposes is summarized and discussed. Moreover, the main approaches used on the secondary controller of the smart microgrid are presented. Although these methods can be applied in the smart microgrid hierarchical control to decrease harmonic components and THD levels, further questions and challenges remain to be solved in microgrids from the harmonic perspective.

Furthermore, a flowchart is proposed to be used as a guidance for applying harmonic compensation methods into the microgrids. The simulation results for two approaches, which are PR controller and optimization-based method, applied based on the proposed flowchart confirm the potential of these methods to reduce HD in microgrids.

Only a few works in literature have considered the microgrid as a multi-bus multi-inverter network. Therefore, it is suggested to use the harmonic mitigation methods in the multi-bus multi-DG microgrids and try to reduce HD with different goals such as reduction in individual harmonics, reduction in THD of the critical bus, and decreasing THD of all buses in such microgrid.

It is also observed that it is necessary to have an effective control modelling containing harmonic analysis and power converter equations along with the hierarchical control of the microgrid. Moreover, the application of the optimization approaches, predictive control-based methods, and artificial intelligence is not well considered in the literature for microgrid harmonic applications. It is concluded that the microgrids still need a reliable, robust, and fast harmonic compensation approach, which can be used in the hierarchical control and have an efficient performance under different situations such as changing from the grid-tied mode to islanded mode or vice versa, plug and play of the converters, cyber-attacks, etc.

REFERENCES

1. M. A. Jirdehi, V. S. Tabar, S. Ghassemzadeh, and S. Tohidi, "Different aspects of microgrid management: A comprehensive review," *Journal of Energy Storage*, vol. 30, p. 101457, 2020.
2. K. Rajesh, S. Dash, and R. Rajagopal, "Load frequency control of microgrid: A technical review," in Drück H., Mathur J., Panthalookaran V., Sreekumar V. (eds), *Green Buildings and Sustainable Engineering*, pp. 115–138, Singapore: Springer, 2020.
3. B. Adineh, R. Kaypour, P. Davari, and F. Blaabjerg, "Review of harmonic mitigation methods in microgrid: From a hierarchical control perspective," *IEEE Journal of Emerging and Selected Topics in Power Electronics*, 2020. doi:10.1109/JESTPE.2020.3001971.

4. O. P. Mahela, B. Khan, H. H. Alhelou, and P. Siano, "Power quality assessment and event detection in distribution network with wind energy penetration using stockwell transform and fuzzy clustering," *IEEE Transactions on Industrial Informatics*, vol. 16, no. 11, pp. 6922–6932, Nov. 2020.

5. O. P. Mahela, B. Khan, H. Haes Alhelou, and S. Tanwar, "Assessment of power quality in the utility grid integrated with wind energy generation," *IET Power Electronics*, vol. 13, no. 13, pp. 2917–2925, 14 Oct. 2020.

6. O. P. Mahela et al., "Recognition of power quality issues associated with grid integrated solar photovoltaic plant in experimental framework," *IEEE Systems Journal*, doi:10.1109/JSYST.2020.3027203.

7. O. P. Mahela, A. G. Shaik, B. Khan, R. Mahla, and H. H. Alhelou, "Recognition of complex power quality disturbances using S-transform based ruled decision tree," *IEEE Access*, vol. 8, pp. 173530–173547, 2020.

8. O. P. Mahela, B. Khan, H. H. Alhelou, S. Tanwar, and S. Padmanaban, "Harmonic mitigation and power quality improvement in utility grid with solar energy penetration using distribution static compensator," *IET Power Electron*, vol. 14, pp. 912–922, 2021.

9. R. K. Pachauri et al., "Impact of partial shading on various PV array configurations and different modeling approaches: A comprehensive review," *IEEE Access*, vol. 8, pp. 181375–181403, 2020.

10. R. K. Pachauri, O. P. Mahela, B. Khan, A. Kumar, S. Agarwal, H. H. Alhelou, and J. Bai, "Development of Arduino assisted data acquisition system for solar photovoltaic array characterization under partial shading conditions," *Computers & Electrical Engineering*, vol. 92, p. 107175, 2021.

11. R. K. Pachauri et al., "Shade dispersion methodologies for performance improvement of classical total cross-tied photovoltaic array configuration under partial shading conditions," *IET Renewable Power Generation*, vol. 15, pp. 1796–1811, 2021.

12. B. Khan, G. Agnihotri, P. Rathore, A. Mishra, and G. Naidu, "A cooperative game theory approach for usage and reliability margin cost allocation under contingent restructured market," *International Review of Electrical Engineering*, vol. 9, no. 4, pp. 854–862, 2014.

13. B. Khan, G. Agnihotri, and A. S. Mishra, "An approach for transmission loss and cost allocation by loss allocation index and co-operative game theory," *Journal of The Institution of Engineers (India): Series B*, vol. 97, pp. 41–46, 2016.

14. X. Wang, F. Blaabjerg, and W. Wu, "Modeling and analysis of harmonic stability in an ac power electronics- based power system," *IEEE Transactions on Power Electronics*, vol. 29, no. 12, pp. 6421–6432, 2014.

15. H. Laaksonen and K. Kauhaniemi, "Voltage and current THD in microgrid with different dg unit and load configurations," in *CIRED Seminar 2008: SmartGrids for Distribution*, pp. 1–4, IET, 2008.

16. R. Teodorescu, F. Blaabjerg, M. Liserre, and P. C. Loh, "Proportional-resonant controllers and filters for grid-connected voltage-source converters," *IEE Proceedings-Electric Power Applications*, vol. 153, no. 5, pp. 750–762, 2006.

17. A. Micallef, M. Apap, C. Spiteri-Staines, and J. M. Guerrero, "Mitigation of harmonics in grid-connected and islanded microgrids via virtual admittances and impedances," *IEEE Transactions on Smart Grid*, vol. 8, no. 2, pp. 651–661, 2015.

18. U. B. Tayab, M. A. B. Roslan, L. J. Hwai, and M. Kashif, "A review of droop control techniques for microgrid," *Renewable and Sustainable Energy Reviews*, vol. 76, pp. 717–727, 2017.

19. Q.-C. Zhong, "Harmonic droop controller to reduce the voltage harmonics of inverters," *IEEE Transactions on Industrial Electronics*, vol. 60, no. 3, pp. 936–945, 2012.

20. H. Dong, S. Yuan, Z. Han, Z. Cai, G. Jia, and Y. Ge, "A comprehensive strategy for accurate reactive power distribution, stability improvement, and harmonic suppression of multi-inverter-based micro-grid," *Energies*, vol. 11, no. 4, p. 745, 2018.

21. T.-L. Lee and P.-T. Cheng, "Design of a new cooperative harmonic filtering strategy for distributed generation interface converters in an islanding network," *IEEE Transactions on Power Electronics*, vol. 22, no. 5, pp. 1919–1927, 2007.

22. P.-T. Cheng, C.-A. Chen, T.-L. Lee, and S.-Y. Kuo, "A cooperative imbalance compensation method for distributed-generation interface converters," *IEEE Transactions on Industry Applications*, vol. 45, no. 2, pp. 805–815, 2009.

23. H. Moussa, A. Shahin, J.-P. Martin, B. Nahid-Mobarakeh, S. Pierfederici, and N. Moubayed, "Harmonic power sharing with voltage distortion compensation of droop controlled islanded microgrids," *IEEE Transactions on Smart Grid*, vol. 9, no. 5, pp. 5335–5347, 2018.

24. Y. Han, P. Shen, X. Zhao, and J. M. Guerrero, "An enhanced power sharing scheme for voltage unbalance and harmonics compensation in an islanded ac microgrid," *IEEE Transactions on Energy Conversion*, vol. 31, no. 3, pp. 1037–1050, 2016.

25. J. He, Y. W. Li, J. M. Guerrero, F. Blaabjerg, and J. C. Vasquez, "An islanding microgrid power sharing approach using enhanced virtual impedance control scheme," *IEEE Transactions on Power Electronics*, vol. 28, no. 11, pp. 5272–5282, 2013.

26. M. Savaghebi, A. Jalilian, J. C. Vasquez, and J. M. Guerrero, "Secondary control for voltage quality enhancement in microgrids," *IEEE Transactions on Smart Grid*, vol. 3, no. 4, pp. 1893–1902, 2012.

27. J. Zhou, S. Kim, H. Zhang, Q. Sun, and R. Han, "Consensus-based distributed control for accurate reactive, harmonic, and imbalance power sharing in microgrids," *IEEE Transactions on Smart Grid*, vol. 9, no. 4, pp. 2453–2467, 2016.

28. A. Das, A. Shukla, A. Shyam, S. Anand, J. M. Guerrero, and S. R. Sahoo, "A distributed-controlled harmonic virtual impedance loop for ac microgrids," *IEEE Transactions on Industrial Electronics*, 2020. doi:10.1109/TIE.2020.2987290.

29. C. Blanco, F. Tardelli, D. Reigosa, P. Zanchetta, and F. Briz, "Design of a cooperative voltage harmonic compensation strategy for islanded microgrids combining virtual admittance and repetitive controller," *IEEE Transactions on Industry Applications*, vol. 55, no. 1, pp. 680–688, 2018.

30. C. Blanco, D. Reigosa, J. C. Vasquez, J. M. Guerrero, and F. Briz, "Virtual admittance loop for voltage harmonic compensation in microgrids," *IEEE Transactions on Industry Applications*, vol. 52, no. 4, pp. 3348–3356, 2016.

31. M. H. Andishgar, E. Gholipour, and R.-A. Hooshmand, "Voltage quality enhancement in islanded microgrids with multi-voltage quality requirements at different buses," *IET Generation, Transmission & Distribution*, vol. 12, no. 9, pp. 2173–2180, 2018.

32. M. H. Andishgar, E. Gholipour, and R.-A. Hooshmand, "Improved secondary control for optimal total harmonic distortion compensation of parallel connected DGs in islanded microgrids," *IET Smart Grid*, vol. 2, no. 1, pp. 115–122, 2019.

33. R. Keypour, B. Adineh, M. H. Khooban, and F. Blaabjerg, "A new population-based optimization method for online minimization of voltage harmonics in islanded microgrids," *IEEE Transactions on Circuits and Systems II: Express Briefs*, vol. 67, no. 6, pp. 1084–1088, 2020.

34. T. I. Incorporated, "Three-level, three-phase sic ac-to-dc converter reference design," in *Technical report, Texas Instruments Incorporated*, 05 Nov. 2018.

35. X. Ruan, X. Wang, D. Pan, D. Yang, W. Li, and C. Bao, *Control Techniques for LCL-Type Grid-Connected Inverters*. Singapore: Springer, 2018.

36. Q.-C. Zhong, "Robust droop controller for accurate proportional load sharing among inverters operated in parallel," *IEEE Transactions on Industrial Electronics*, vol. 60, no. 4, pp. 1281–1290, 2011.

37. W. Yao, M. Chen, J. Matas, J. M. Guerrero, and Z.-M. Qian, "Design and analysis of the droop control method for parallel inverters considering the impact of the complex impedance on the power sharing," *IEEE Transactions on Industrial Electronics*, vol. 58, no. 2, pp. 576–588, 2010.

3 Energy Management in Deregulated Power Market with Integration of Microgrid

Vasundhara Mahajan and Ritu Jain

CONTENTS

3.1 INTRODUCTION OF TRADITIONAL AND DEREGULATED POWER SYSTEM

3.1.1 TRADITIONAL POWER SYSTEM

Earlier, the power sector was vertically integrated, with a single utility overseeing generation, transmission, and distribution. There is a monopoly in the power sector, so the cost of energy is decided by the governing organization, and the rules and regulations decided by it are followed by all other entities. Due to monopoly in the

DOI: 10.1201/9781003278030-3

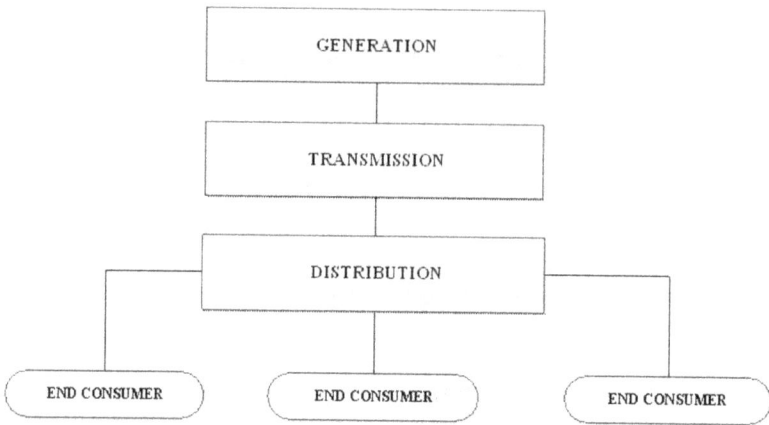

FIGURE 3.1 Traditional power system.

power sector, the cost of energy is increasing rapidly with the expansion of the power system. Therefore, it has become necessary that the power sector be restructured and privatized in order to reduce the cost of energy [1].

The main issues in developing countries' move toward deregulation are continuously increasing demand and inefficient management. On the other hand, in developed countries, the main objective of reregulating their power sectors is to supply energy to consumers at cheaper prices [2]. The structure of the traditional power sector is shown in Figure 3.1.

3.1.2 DEREGULATED POWER SYSTEM

The word "deregulation" means changing the monopoly rules and regulations of the regulated industry in order to improve the way electric utilities do business and how consumers can buy power at a cheaper rate [3]. The two main objectives of a deregulated power system are to decrease the cost of energy supplied to the consumer and to maximize the social benefit. The social benefit is the difference between the revenue collected from the DISCOS and the cost paid to GENCOS for their energy. Social benefit is the criteria for measuring the performance of the market [4–9]. It is used by the ISO to manage and control the transmission system and to install new transmission lines [10–14]. The deregulated model of the power system is shown in Figure 3.2.

3.1.3 NEED OF DEREGULATION

The need for deregulation is to introduce competition in the power sector by allowing private bodies to enter the market. Competition is required to control the continuously rising price of electricity. Deregulation allows its customers to buy energy at lower prices. It gives the customer a choice to have energy from the desired sellers. It helps in maintaining the power sector efficiently. It helps the ISO

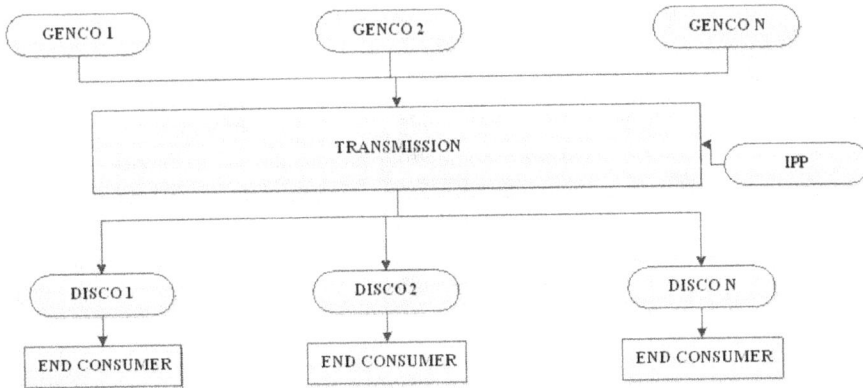

FIGURE 3.2 Deregulated power system.

to maintain the reliable and efficient operation of the power system. It helps the ISO gain maximum social benefit by allowing different investors to bid in the market. It increases employment opportunities. It encourages new participants to enter the power market and invest at different locations and times to enhance economic development [15].

3.1.4 STEPS TOWARD DEREGULATION

There are mainly two steps to moving from the regulated power sector to the reregulated or deregulated power sector [15].

1. Restructuring
2. Privatization

Restructuring: It means separating the functions of generation, transmission, and distribution. It is of two types:

a. Vertical unbundling means separating the power sector vertically into generation, transmission, and distribution.
b. Horizontal unbundling means separating the generation and distribution sectors horizontally. For economic and feasibility reasons, the transmission sector is not horizontally unbundled.

Privatization: Privatization means the selling of utility assets by the government to private industries. In a deregulated power system, the generation and distribution sectors are privatized. There is a monopoly in the transmission sector because it is infeasible and uneconomical to build a separate transmission system for a particular body. Thus, there is no competition in transmission, but there is TOA (transmission open access), which means that any entity that does not own the transmission system has the right to use it.

3.2 INTRODUCTION OF MICROGRID

3.2.1 Microgrid Definition

The power grid is the heart of the power system, so it is necessary that it goes through various innovations. The utilities are spending large amounts of money and time to make the system more intelligent, vulnerable, and reliable so that it can sustain natural and man-made disasters. One such way to provide self-healing to the power system is by allocating small and self-governing entities named microgrids (MG) [16]. A microgrid is a group of micro sources, loads and batteries that represent themselves as a single entity which can reciprocate the control signals sent by the central control centre [17]. MG is the low-voltage intelligent distributed network which is composed of micro sources or distributed generations (DGs), energy storage devices (ESDs), and loads. It can operate either in grid-connected mode or in islanded mode. From the grid's side, it is termed as the controlled entity which can be operated as the aggregated load and a micro source for power and ancillary services. From the consumer's side, it is termed the low-voltage distribution system [18]. Allocations of MGs are encouraged in the power system because they enhance reliability, improve power quality and reduce carbon emissions due to the incorporation of renewable energy sources [19].

3.2.2 Architecture of Microgrid

The microgrid architecture comprises a few components [20–23]:

Micro-Source Controller (MSC): Each micro source, such as DGs based on renewable energy sources, fuel cells, and micro-turbines, has a MSC, which has the capability to control and monitor the active and reactive power generated by each source.

Load Controller (LC): Each controllable load in MG has a LC which is responsible for load management. It receives instructions from MGCC for load management during emergency conditions by load shedding or by demand response.

Microgrid Central Controller (MGCC): It serves as a link between the MSC, the LC, and the distribution system operator (DSO). Its main responsibility is to minimize the generation cost and maximize the profit of MG. It sends control signals to MSC and LC to optimize their set points in such a way that the security and economics of the system are not affected. In a decentralized power system, each MG has its own MGCC, which coordinates with the DSO.

Distribution System Operator (DSC): It is the interface between the distribution companies and the independent system operator (ISO). It is responsible for managing and operating the resources in the distribution system.

Independent System Operator (ISO): It is responsible for managing and controlling the whole power system. It has to ensure the secure and reliable operation of the power system. It is an impartial authority with no personal vested interest or benefit in the monetary business [2]. Thus, it does not own

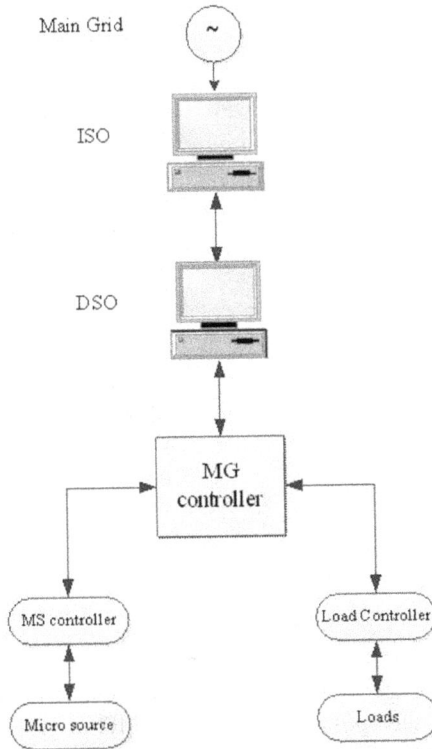

FIGURE 3.3 Architecture of microgrid.

any generation company, but to maintain reliable and secure operations; it procures various emergency services like spinning reserve. Figure 3.3 shows the architecture of MG. The typical layout of the MG, having different energy sources and loads along with interconnection with the main grid, is shown in Figure 3.4.

3.2.3 MODES OF OPERATION

By optimizing the renewable energy sources, microgrids have the ability to self-heal, thus improving the reliability of the distribution network by minimizing the chances of load shedding, increasing power quality, reducing carbon emissions and reducing prices by optimally scheduling the renewable energy sources and providing energy supply to remote areas [24–28]. The main feature of MG is that it can be operated in both grid-connected and islanded modes. The decision on the mode of operation is taken by MGCC considering the economic and security constraints. The MG is usually operated in a grid-connected mode for the economic operation of the power system, but it should have sufficient capacity to supply its load during emergency conditions under an islanded mode of operation. The MG is isolated from the main grid through switches at the point of common coupling [29,30]. In grid-connected

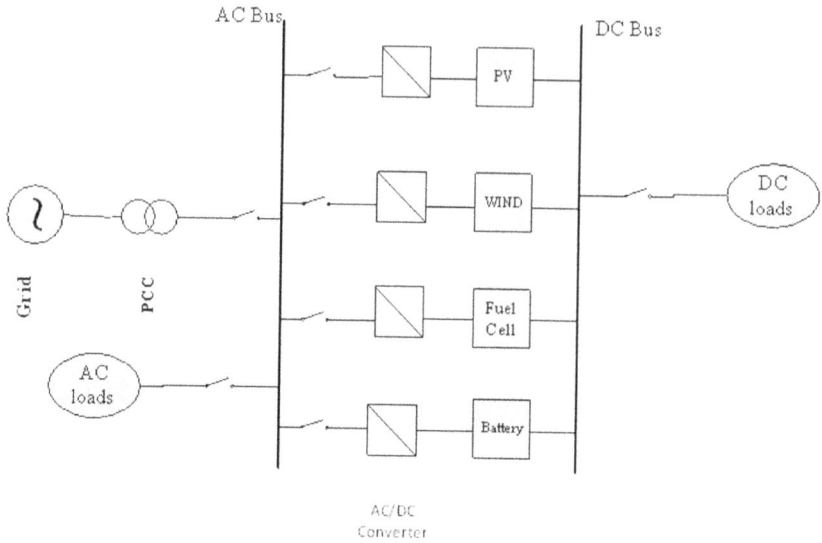

FIGURE 3.4   Layout of microgrid.

mode, the MG acts as a reserve for the main grid when there is a sudden increase in load demand, variable generation, or an inaccurate load forecast. There is a large literature available on the optimal scheduling of MG in grid-connected mode, but very little literature is available on the islanded operation of MG. The optimal energy management of MG can be performed in a centralized or decentralized manner. In the centralized method, all the information regarding the available generation and load is collected for centralized operation and control [29,31,32]. However in the decentralized method, every entity is considered as an agent that is free to take decisions [33,34].

3.3   UNCERTAINTY MODELLING OF WIND AND SOLAR

3.3.1   MODELLING OF WIND POWER

Wind power is the most commonly used renewable energy source, and it has been widely developed in recent years. The various advantages of wind energy are zero pollution, short gestation periods, and low capital costs. The power output of a wind turbine generator (WTG) depends on the wind speed [35]. As the wind speed increases, the power output of the wind energy increases approximately as the cube of the wind speed shown in the equation below.

$$P_w = \frac{1}{2} \times \rho \times A \times v^3 \qquad (3.1)$$

where P_{WT} is the power generated from WTG, ρ is the density of air in kg/m^3, A is the area of blades in m^2 and v is the wind speed in m/s. Thus, the power generated from WTG is defined as in the equation below [36].

$$P_{WT} = \begin{cases} 0, v < v_{in}, v > v_{out} \\ P_r \times \dfrac{(v_i - v_{in})}{v_r - v_{in}}, v_{in} < v_i < v_r \\ P_r, v_r < v < v_{out} \end{cases} \tag{3.2}$$

where P_r is the rated power output of WTG, v_{in} is the cut-in speed of wind in m/s, v_r is the rated wind speed and v_{out} is the cut-out wind speed. Wind speed is intermittent in nature and it follows a weibull distribution [37]. Thus, to model the stochastic nature of wind speed WPDF is used. The WPDF for wind speed v m/s is given as;

$$f(v) = \frac{k}{c}\left(\frac{v}{c}\right)^{k-1} e^{\left\{-\left(\frac{v}{c}\right)^k\right\}} \tag{3.3}$$

where c is the scale factor and k is the shape factor. For Rayleigh PDF, the value of k is 2. This is the preferred PDF as it has periods of both low and high wind speed. Hourly mean wind speed and standard deviation of wind are used as the input data to create the PDF for wind speed.

3.3.2 Uncertainty Modelling of Solar Power

Solar irradiation follows the bimodal distribution, which is a combination of two unimodal distributions. The beta distribution is used to model the solar irradiance [38], which is shown in the Eqs. (3.4) and (3.5).

$$f_{pv}(ir) = \begin{cases} \left(\dfrac{\Gamma(A+B)}{\Gamma(A)\Gamma(B)} \times ir^{A-1} \times (1-ir)^{(B-1)}\right), & \text{for } 0 \le ir \le 1, A \ge 0, B \ge 0 \\ 0, & \text{otherwise} \end{cases} \tag{3.4}$$

$$B = (1-\mu) \times \left(\frac{\mu \times (1+\mu)}{\sigma^2} - 1\right) \tag{3.5}$$

$$A = \frac{\mu \times B}{1-\mu}$$

Here, μ and σ are the mean and standard deviation of solar irradiation respectively. The solar power generated is shown in Eq. (3.6). Here, η is the efficiency of solar panel.

$$P_{pv}(ir) = \eta^{pv} \times R^{pv} \times ir \tag{3.6}$$

3.4 PROBLEM FORMULATION AND SYSTEM DATA

3.4.1 Objective Function

The aim of this paper is to minimize the total generation cost, including the cost of energy from conventional generators in the main grid and the cost of energy from

DG in the MG, by efficiently utilizing the RES [39]: the objective function is subjected to inequality and equality constraints, as shown in [39]. The objective function is shown below.

$$\text{Cost} = \sum_{t=1}^{24}\left[\left(\sum_{x=1}^{N_{CG}} a \times P_{x,t}^2 + b \times P_{x,t} + c\right) + \left(\sum_{y=1}^{N_{DG}} d * P_{y,t} + e\right) + \left(\sum_{z=1}^{N_C} f \times P_z^2 + g \times P_z + h\right)\right]$$

(3.7)

Here, t is the number of hours, N_{CG} is the number of conventional generators in main grid, N_{DG} is the number of renewable-based DG in the MG and N_c is the number of diesel generators in the MG. The first term of the objective function represents the generation cost of a conventional generator. The second term represents the generation cost of renewable-based DG in MG. The third term represents the generation cost of the thermal generator located in MG. The cost function of conventional generators on the main grid and thermal generators in MG are considered quadratic in nature, whereas the cost function of renewable-based DG is considered linear.

3.4.2 System Data

The IEEE 30 bus system is considered the main grid in this study, while the 6 bus test system is considered the MG. The IEEE 30 bus system is composed of 6 generators, 41 transmission lines, and 21 loads [40,41]. Whereas MG has three generators, of which two are renewable-based DG and one is a thermal generator, with 3 loads and 11 transmission lines. The 6 bus system data is taken from [39]. The data for solar irradiation and wind speed is taken from [42,43]. The wind and solar power available for a day in MG are shown in Figures 3.5 and 3.6, respectively. The load data for MG is taken from [24] and is shown in Figure 3.7. Figure 3.8 shows the load data for the main grid, taken from [44] and normalized according to the system data. The proposed algorithm is shown in Figure 3.9.

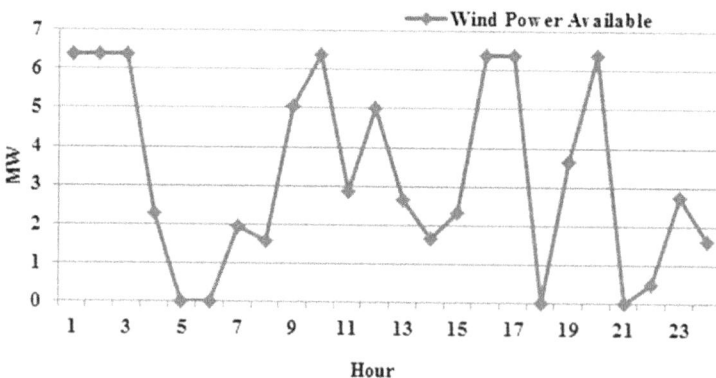

FIGURE 3.5   Wind power available.

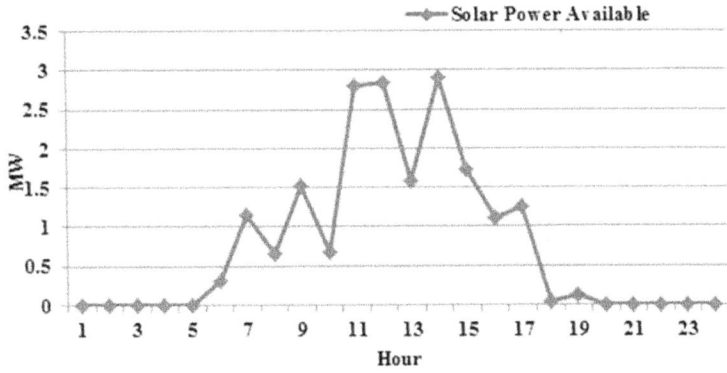

FIGURE 3.6 Solar power available.

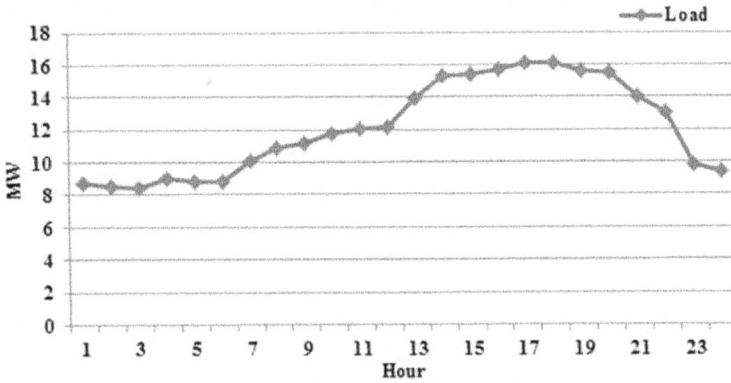

FIGURE 3.7 Load data of microgrid.

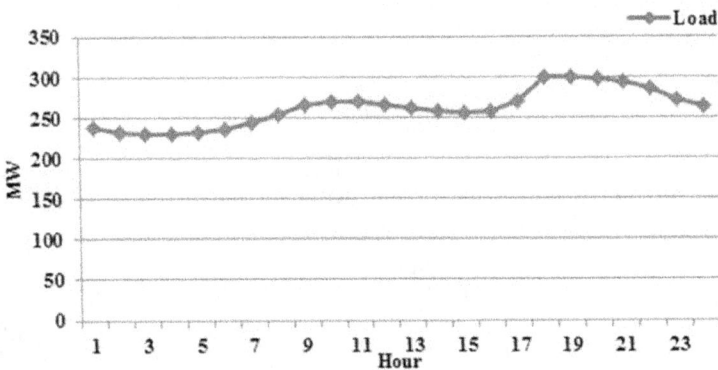

FIGURE 3.8 Load data of main grid.

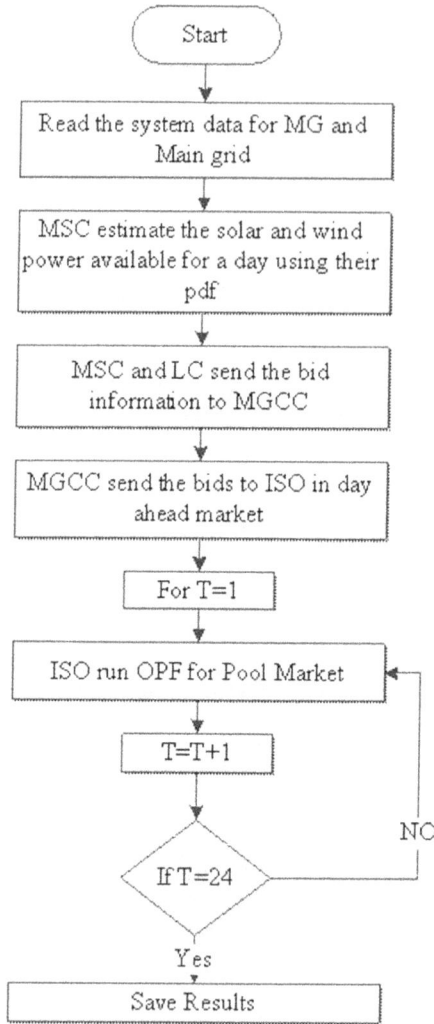

```
                        ┌─────────────┐
                        │    Start    │
                        └──────┬──────┘
                               ▼
              ┌────────────────────────────────┐
              │ Read the system data for MG and │
              │            Main grid            │
              └────────────────┬───────────────┘
                               ▼
              ┌────────────────────────────────┐
              │  MSC estimate the solar and wind │
              │ power available for a day using their │
              │               pdf               │
              └────────────────┬───────────────┘
                               ▼
              ┌────────────────────────────────┐
              │   MSC and LC send the bid       │
              │    information to MGCC          │
              └────────────────┬───────────────┘
                               ▼
              ┌────────────────────────────────┐
              │  MGCC send the bids to ISO in day │
              │          ahead market           │
              └────────────────┬───────────────┘
                               ▼
                        ┌─────────────┐
                        │   For T=1   │
                        └──────┬──────┘
                               ▼
              ┌────────────────────────────────┐
              │   ISO run OPF for Pool Market   │◄──┐
              └────────────────┬───────────────┘   │
                               ▼                    │
                        ┌─────────────┐             │
                        │   T=T+1     │             │
                        └──────┬──────┘             │
                               ▼                    │  NO
                          ◇─────────◇               │
                         ◇  If T=24  ◇──────────────┘
                          ◇─────────◇
                               │
                              Yes
                               ▼
              ┌────────────────────────────────┐
              │        Save Results            │
              └────────────────────────────────┘
```

FIGURE 3.9   Proposed algorithm.

3.5   RESULTS AND DISCUSSION

In this study, the IEEE 30 bus system is treated as a main grid that is interconnected with a MG. The MG is composed of six buses, each with a wind-based plant, a solar-based plant, and a thermal generator. The wind plant has a maximum generation capacity of 6.4 MW. It consists of eight turbines of 800 kW. The solar plant has a maximum generation capacity of 5 MW. The thermal generator in the MG has generation limits of 0–10 MW. It acts as a support to the MG when there is unavailability of power by RES. The generation from PV and wind is estimated by its historical mean solar irradiance, mean wind speed and standard deviation by using its probability density function (pdf). For forecasting the wind speed, Weibull pdf is used.

Whereas, for solar irradiance Beta pdf is used. After estimating the power output and load, the MSC and LC will send the information about their bid quantity and price to the MGCC. The MGCC will send the day-ahead bidding to ISO in pool market. After having the biddings from all the market participants in day-ahead, the ISO performs optimization and clears the market. Now the information regarding day-ahead market clearance is displayed in Open Access Same-time Information System (OASIS). The ISO always have some reserves available in case of emergency during real-time market operations. If there are any defaulters in real-time market operations then they are penalized according to the norms decided by system operator. Figure 3.10 shows the scheduled dispatch of wind, solar and thermal plant of MG after market clearance. From Figure 3.10, it is clear that MG is unable to satisfy all its internal loads during few hours of a day so it will purchase rest of the power from main grid. Figure 3.11 shows the surplus and deficit power of the MG for the scheduled day.

Figure 3.11 shows that MG has a power deficiency during hours 4–6, 14, 15, 18, 19, 21, and 22. Thus, it has to purchase power from the main grid during that interval to satisfy its internal loads, whereas it has surplus power during hours 1–3, 7–13, 16, 17, 20, 23, and 24, and this power is delivered to the main grid. The + sign denotes a power surplus, while the − sign denotes a power deficit. The total generation cost and power loss in the system are shown in Table 3.1. Table 3.2 shows the nodal prices for all the buses in MG where GB represents the Generator Bus and LB represents the Load Bus in MG. It shows that during the peak load period of MG, i.e., hour 18, the nodal prices of its buses are high. This is due to the high congestion in the system. However, during the minimum load period, i.e., hour 3, the nodal prices for the buses of MG are the minimum. The pricing mechanism chosen in this work is based on the nodal price. The generators have paid the generator bus nodal price, whereas the load has to pay the load bus nodal price.

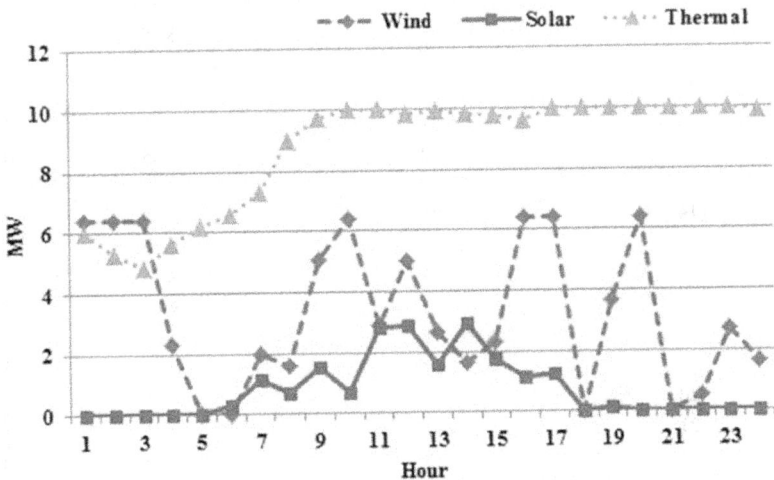

FIGURE 3.10 Scheduled generation of MG.

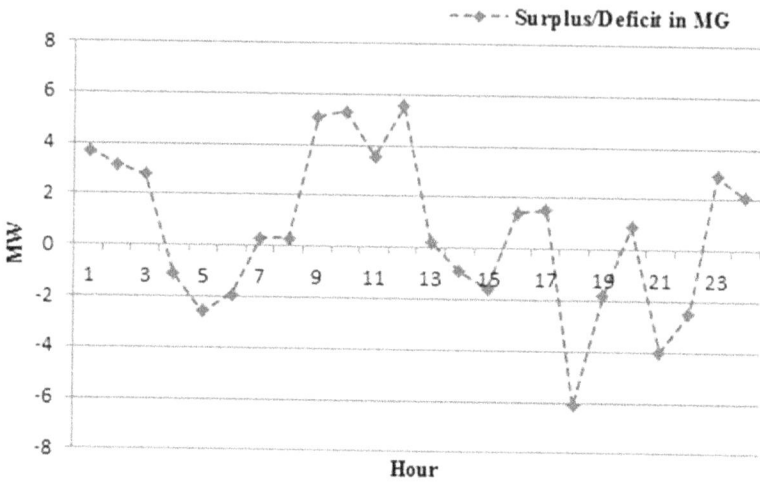

FIGURE 3.11 Surplus/deficit in MG.

TABLE 3.1
Total Generation Cost and Power Loss in the System

Hour	Gencost $/MW	Power Loss in MW
1	660.00	6.69
2	640.43	6.34
3	629.08	6.13
4	640.17	6.41
5	651.75	6.67
6	662.81	6.87
7	686.70	7.25
8	727.79	8.05
9	759.56	8.55
10	780.16	8.85
11	785.37	8.94
12	766.90	8.65
13	763.71	8.62
14	753.91	8.45
15	749.22	8.38
16	748.53	8.30
17	791.95	8.98
18	916.99	10.94
19	909.97	10.81
20	891.88	10.51
21	888.81	10.55
22	853.39	10.02
23	788.17	9.04
24	762.68	8.68

TABLE 3.2

Nodal Price at Generator Bus and Load Bus of MG

Hour	GB1	GB2	GB3	LB1	LB2	LB3
1	3.31	3.31	3.30	3.31	3.31	3.31
2	3.27	3.27	3.26	3.27	3.27	3.27
3	3.24	3.25	3.24	3.25	3.25	3.25
4	3.28	3.28	3.28	3.29	3.29	3.28
5	3.31	3.32	3.31	3.32	3.32	3.32
6	3.34	3.34	3.33	3.34	3.34	3.33
7	3.37	3.37	3.37	3.38	3.38	3.37
8	3.46	3.46	3.45	3.46	3.46	3.46
9	3.50	3.50	3.49	3.51	3.50	3.49
10	3.53	3.53	3.52	3.53	3.53	3.52
11	3.54	3.53	3.52	3.54	3.54	3.53
12	3.51	3.50	3.49	3.51	3.51	3.50
13	3.52	3.51	3.51	3.52	3.52	3.51
14	3.50	3.50	3.49	3.51	3.51	3.50
15	3.50	3.50	3.49	3.51	3.50	3.50
16	3.48	3.49	3.48	3.49	3.49	3.49
17	3.55	3.55	3.54	3.56	3.56	3.55
18	3.72	3.73	3.72	3.74	3.74	3.73
19	3.71	3.71	3.70	3.72	3.72	3.71
20	3.68	3.68	3.67	3.69	3.69	3.68
21	3.69	3.69	3.68	3.70	3.70	3.69
22	3.64	3.65	3.64	3.65	3.65	3.64
23	3.55	3.55	3.53	3.55	3.55	3.54
24	3.52	3.51	3.50	3.52	3.52	3.51

3.6 CONCLUSION

In this study, the energy management of the main grid in interconnection with a microgrid is shown. The mode of operation for MG is the grid-connected mode. The main objective is to minimize the generation cost of the system. It is also explained that during a power deficit, the MG will take assistance from the main grid to supply its internal loads, and during surplus availability of resources, the MG can support the main grid. The probability distribution function approach is used to accurately estimate the RES available in MG. For estimating the wind speed, the Weibull probability distribution function is used, and for solar irradiance, the beta distribution is used. In this study, the day-ahead market operations are performed. This study can be further extended to include real-time market operations considering the islanded mode of MG.

REFERENCES

1. L. L. Lai, *Power System Restructuring and Deregulation: Trading, Performance and Information Technology*. John Wiley & Sons, 2001. Hoboken, New Jersey, United States.

2. K. Bhattacharya, M. Bollen, and J. E. Daalder, *Operation of Restructured Power Systems.* Springer Science & Business Media, 2012. Springer, Boston, MA, United States.
3. Y. R. Sood, N. P. Padhy, and H. Gupta, "Wheeling of power under deregulated environment of power system-a bibliographical survey," *IEEE Transactions on Power systems,* vol. 17, pp. 870–878, 2002.
4. M. Shahidehpour and M. Alomoush, *Restructured Electrical Power Systems: Operation: Trading, and Volatility.* CRC Press, 2001. Springer, Boston, MA, United States.
5. B. Khan, G. Agnihotri, S. E. Mubeen, and G. Naidu, "A TCSC incorporated power flow model for embedded transmission usage and loss allocation," *AASRI Procedia,* vol. 7, pp. 45–50, 2014.
6. B. Khan, G. Agnihotri, G. Gupta, and P. Rathore, "A power flow tracing based method for transmission usage, loss & reliability margin allocation," *AASRI Procedia,* vol. 7, pp. 94–100, 2014.
7. B. Khan, G. Agnihotri, P. Rathore, A. Mishra, and G. Naidu, "A cooperative game theory approach for usage and reliability margin cost allocation under contingent restructured market," *International Review of Electrical Engineering,* vol. 9, no. 4, pp. 854–862, 2014.
8. B. Khan and G. Agnihotri, "A comprehensive review of embedded transmission pricing methods based on power flow tracing techniques," *Chinese Journal of Engineering,* vol. 2013, Article ID 501587, 13 pages, 2013.
9. B. Khan, G. Agnihotri, and A.S. Mishra, "An approach for transmission loss and cost allocation by loss allocation index and co-operative game theory," *Journal of the Institution of Engineers (India): Series B,* vol. 97, pp. 41–46, 2016.
10. B. Khan and P. Singh, "Optimal power flow techniques under characterization of conventional and renewable energy sources: A comprehensive analysis," *Journal of Engineering,* vol. 2017, Article ID 9539506, 16 pages, 2017.
11. P. Singh and B. Khan, "Smart microgrid energy management using a novel artificial shark optimization," *Complexity,* vol. 2017, Article ID 2158926, 22 pages, 2017.
12. T. Molla, B. Khan, B. Moges, H. H. Alhelou, R. Zamani, and P. Siano, "Integrated optimization of smart home appliances with cost-effective energy management system," *CSEE Journal of Power and Energy Systems,* vol. 5, no. 2, pp. 249–258, June 2019.
13. Z. Tang, Y. Lin, M. Vosoogh, N. Parsa, A. Baziar, and B. Khan, "Securing microgrid optimal energy management using deep generative model," *IEEE Access,* vol. 9, pp. 63377–63387, 2021.
14. S. P. Bihari et al., "A comprehensive review of microgrid control mechanism and impact assessment for hybrid renewable energy integration," *IEEE Access.* doi:10.1109/ ACCESS.2021.3090266.
15. M. Shahidehpour, H. Yamin, and Z. Li, *Market Operations in Electric Power Systems: Forecasting, Scheduling, and Risk Management.* John Wiley & Sons, 2003. Springer, Boston, MA, United States.
16. B. Lasseter, "Microgrids [distributed power generation]," in *2001 IEEE Power Engineering Society Winter Meeting. Conference Proceedings (Cat. No. 01CH37194),* 2001, pp. 146–149.
17. R. H. Lasseter, "Microgrids," in *2002 IEEE Power Engineering Society Winter Meeting. Conference Proceedings (Cat. No. 02CH37309),* 2002, pp. 305–308.
18. R. Lasseter, A. Akhil, C. Marnay, J. Stephens, J. Dagle, R. Guttromson, A. Meliopoulous, R. Yinger, and J. Eto, "The CERTS microgrid concept," *White Paper for Transmission Reliability Program, Office of Power Technologies, US Department of Energy,* vol. 2, p. 30, 2002.
19. T. Logenthiran and D. Srinivasan, "Short term generation scheduling of a microgrid," in *TENCON 2009-2009 IEEE Region 10 Conference,* 2009, pp. 1–6.

20. N. Hatziargyriou, *Microgrids: Architectures and Control*. John Wiley & Sons, 2014. Springer, Boston, MA, United States.
21. M. W. Khan, J. Wang, L. Xiong, and S. Huang, "Architecture of a Microgrid and Optimal Energy Management System", in book Multi Agent Systems - Strategies and Applications, Edited by Ricardo López - Ruiz. IntechOpen, London, United Kingdom, 2020.
22. J. Kaur, Y. R. Sood, and R. Shrivastava, "Optimal resource utilization in a multi-microgrid network for Tamil Nadu state in India," *IETE Journal of Research*, pp. 1–11, 2019.
23. J. Kaur, Y. R. Sood, and R. Shrivastava, "A two-layer optimization approach for renewable energy management of green microgrid in deregulated power sector," *Journal of Renewable and Sustainable Energy*, vol. 9, p. 065905, 2017.
24. A. Khodaei, "Microgrid optimal scheduling with multi-period islanding constraints," *IEEE Transactions on Power Systems*, vol. 29, pp. 1383–1392, 2013.
25. M. Shahidehpour, "Role of smart microgrid in a perfect power system," in *IEEE PES General Meeting*, 2010, p. 1.
26. B. Kroposki, R. Lasseter, T. Ise, S. Morozumi, S. Papathanassiou, and N. Hatziargyriou, "Making microgrids work," *IEEE Power and Energy Magazine*, vol. 6, pp. 40–53, 2008.
27. I.-S. Bae and J.-O. Kim, "Reliability evaluation of customers in a microgrid," *IEEE Transactions on Power Systems*, vol. 23, pp. 1416–1422, 2008.
28. S. Kennedy and M. M. Marden, "Reliability of islanded microgrids with stochastic generation and prioritized load," in *2009 IEEE Bucharest PowerTech*, 2009, pp. 1–7.
29. A. G. Tsikalakis and N. D. Hatziargyriou, "Centralized control for optimizing microgrids operation," in *2011 IEEE Power and Energy Society General Meeting*, 2011, pp. 1–8.
30. C. Hou, X. Hu, and D. Hui, "Hierarchical control techniques applied in micro-grid," in *2010 International Conference on Power System Technology*, 2010, pp. 1–5.
31. N. Hatziargyriou, G. Contaxis, M. Matos, J. P. Lopes, G. Kariniotakis, D. Mayer, J. Halliday, G. Dutton, P. Dokopoulos, and A. Bakirtzis, "Energy management and control of island power systems with increased penetration from renewable sources," in *2002 IEEE Power Engineering Society Winter Meeting. Conference Proceedings (Cat. No. 02CH37309)*, 2002, pp. 335–339.
32. T. Logenthiran, D. Srinivasan, and A. M. Khambadkone, "Multi-agent system for energy resource scheduling of integrated microgrids in a distributed system," *Electric Power Systems Research*, vol. 81, pp. 138–148, 2011.
33. T. Logenthiran, D. Srinivasan, and D. Wong, "Multi-agent coordination for DER in MicroGrid," in *2008 IEEE International Conference on Sustainable Energy Technologies*, 2008, pp. 77–82.
34. J. Oyarzabal, J. Jimeno, J. Ruela, A. Engler, and C. Hardt, "Agent based micro grid management system," in *2005 International Conference on Future Power Systems*, 2005, p. 6.
35. M. G. Masters, *Renewable and Efficient Electric Power Systems*. Hoboken, NJ: Wiley-Interscience, John Wiley & Sons, Inc, vol. 75, p. 76, 2004. Springer, Boston, MA, United States.
36. N. Gupta, "A review on the inclusion of wind generation in power system studies," *Renewable and Sustainable Energy Reviews*, vol. 59, pp. 530–543, 2016.
37. S. S. Reddy, "Optimal scheduling of thermal-wind-solar power system with storage," *Renewable Energy*, vol. 101, pp. 1357–1368, 2017.
38. M. Mazidi, A. Zakariazadeh, S. Jadid, and P. Siano, "Integrated scheduling of renewable generation and demand response programs in a microgrid," *Energy Conversion and Management*, vol. 86, pp. 1118–1127, 2014.

39. A. J. Wood, B. F. Wollenberg, and G. B. Sheblé, *Power Generation, Operation, and Control*. John Wiley & Sons, 2013. Springer, Boston, MA, United States.

40. O. Alsac and B. Stott, "Optimal load flow with steady-state security," *IEEE Transactions on Power Apparatus and Systems*, vol. 3, pp. 745–751, 1974.

41. R. D. Zimmerman, C. E. Murillo-Sánchez, and D. Gan, "MATPOWER: A MATLAB power system simulation package," *Manual, Power Systems Engineering Research Center, Ithaca NY*, vol. 1, pp. 10–17, 1997.

42. N. D. Catalog. Available: https://data.nrel.gov/.

43. V. K. Prajapati and V. Mahajan, "Demand response based congestion management of power system with uncertain renewable resources," *International Journal of Ambient Energy*, pp. 1–14, 2019. doi:10.1080/01430750.2019.1630307.

44. P. ISO. Available: https://www.pjm.com/markets-and-operations/energy.aspx.

4 Business Models for Different Future Electricity Market Players

Hosna Khajeh, Hooman Firoozi,
Hannu Laaksonen, and Miadreza Shafie-khah

CONTENTS

4.1 INTRODUCTION

In traditional centralized power systems, power was produced by only bulk generators and delivered to the customers through transmission and distribution networks. In fact, the flow of power was totally unidirectional. In this way, generating

DOI: 10.1201/9781003278030-4

companies, system operators, and large-scale retailers and customers were introduced as the main players of electricity markets. Regarding the conventional trading structure, small-scale electricity consumers were not able to participate in the markets. They were submissive ratepayers who were not subjected to the variation of the market prices. In this way, flexible capacities of small customers are not utilized in energy markets.

However, today's energy system is experiencing revolutionary changes due to the environmental crisis such as global warming. Environmental problems have led to an increase in the penetration of renewable resources in all levels of the power system. To this end, distribution network located customers are increasingly equipped with renewable resources (such as PV panels) which have produced a bidirectional flow of power. In this way, business models are key drivers incentivizing consumers to change to proactive consumers or "prosumers". Prosumers are end-users with the production capability who can change their consumption and production according to external signals (such as electricity prices) [1].

The other reason accelerating revolution of the power system is the recent development in ICT technology facilitates bidirectional communication, enabling prosumers and consumers to change their consumption power according to the system's needs [2]. Recently, new projects and market designs are developed aiming to focus on bringing customers at the heart of energy systems [3–5]. For this purpose, novel business models should be defined in order to shift the value proposition of the existing trading based on the power grid's requirements [6–15]. They should attract more and more small-scale customers to play the role of prosumers. In addition to energy, the introduced business models should seek to exploit the maximum flexibility potential of demand-side resources so as to deal with problems related to the intermittency of renewable resources in the power grid [16].

In this regard, this chapter firstly introduces the existing and newly emerged players participating in different markets and trading structures in a smart grid environment. These market players are prosumers, aggregators, virtual power plants (VPP), community managers, local market operators, and system operators including distribution system operators (DSO) as well as transmission system operators (TSO). Afterward, different business models and trading structures are introduced in the section. The business models and trading structures that are assessed in this chapter are local electricity markets, peer-to-peer (P2P) trading, aggregator-based trading model, and flexibility local markets. In local markets, prosumers are assumed to trade locally under the supervision of a local market operator. In P2P trading, prosumers are able to choose their trading partners and trade energy bilaterally. In an aggregator-based model, however, prosumers negotiate contracts with an aggregator to provide the power system with energy. In each model, market participants are specified and their roles and responsibilities are defined as well. Finally, the chapter introduces a new local market structure for trading flexibility services. It proposes a structure in which small-scale resources will be able to provide both local and system-wide flexibility and ancillary services. This business model is a driver unlocking the flexibility potential of small-scale prosumers and consumers so as to participate in increasing the reliability and security of the local and system-wide power networks.

4.2 PLAYERS IN FUTURE ELECTRICITY MARKETS

4.2.1 PROSUMERS

The new technological advances provide the ultra-fast bidirectional flow of information between end-users and utilities. In light of the new technology, end-users who were previously non-active consumers and submissive ratepayers are increasingly equipped with energy resources such as solar panels and smart meters. These consumers are now able to produce their own power as well as managing their consumption and production through the use of smart meters. They can also sell their surplus to the grid (through retailers and aggregators) and make profits accordingly. These emerging consumers who actively control their production and consumption are called proactive consumers or "prosumers" [2].

During different time slots, a prosumer may play the role of either a seller or a buyer. This role depends mainly on the prices of electricity (both selling and buying prices) as well as the prosumer's net power profile. In addition to renewable energy resources, a prosumer may have various flexible energy resources (FERs) including electrical storage (batteries and electric vehicles) and controllable appliances. FERs help the prosumer to better control their production and consumption. By scheduling the FERs, a prosumer is able to maximize its profits taking into account the external signals such as electricity prices.

Energy management systems (EMS) are responsible for obtaining the optimal operation of FERs by considering the related operating constraints. The EMS is equipped with sensors and actuators for monitoring and controlling energy consumption and the production of prosumers [17]. This system should also satisfy the constraints imposed by the owner's convenience level. The other responsibility of EMS can be obtaining the optimal bidding strategy for the prosumer based on the data received from the local or retail electricity prices. The optimal scheduling of the FERs should be updated if the data on the prices update constantly (in the case that it participates in local markets). The EMS also needs to take into account the real-time preference of the owner. Accordingly, there are just a few appliances whose control will not interfere with the prosumer preference. The controllable appliances need to be inherently flexible appliances in terms of their working time and/or working power [18]. For some appliances, the EMS is able to change their operating power as well as their operating time. For example, EVs which can be charged with different power rates are in this group. In contrast, just working time of some appliances is allowed to be controlled. Washing machines and dishwashers are two examples.

A prosumer may adopt different strategies with different objective functions. They may aim to be self-sufficient. In this way, the prosumers try to be islanded, utilizing its flexible resources to capture the unbalanced power resulted from its renewable energy resources. Hence, islanded prosumers need to be highly flexible [19]. Hence, storage-based resources are required to make the prosumer highly flexible. On the other hand, a prosumer may decide to operate in the grid-connected mode. It means that it is able to trade electricity with the grid. The EMS decides on the amount of the imported and exported power based on the prosumer's objective function. Normally, it is more profitable for prosumers to consume their renewable production and sell the production surplus to the grid. However, in general, the energy policy and motivation programs play an important role in the decision making of the EMS. The prosumer may choose to provide flexibility services to the grid providing that there exist decent motivation schemes.

Some prosumers may choose to be aggregated with other prosumers and join a local community. In this regard, the independent prosumers cooperate with each other to enhance their own community. The community may aim to maximize its total profits, maximize its self-sufficiency, or minimize the environmental impacts of consuming and producing electricity. In the local community, a hierarchical control and management scheme is adopted. Therefore, each prosumer operates autonomously using its own local EMS. On top of the control layer, a central EMS coordinates the operation of its prosumers so as to achieve the community's objective function [19]. Figure 4.1 shows the structure of a prosumer in islanded mode and in a community-based structure.

Other prosumers may choose to negotiate a contract with an aggregator. In this regard, they give permission to the aggregator to provide a specified amount of energy or flexibility in specified time slots. They receive benefits if they fulfil their commitment. Thus, the aggregator plays the role of a broker between the prosumers and the utility (or wholesale markets if the aggregator participates in the markets). In the aggregator-based model, the objective function of prosumer's EMS is to schedule the FERs aiming to follow the predefined contract and the signals received by the aggregator.

4.2.2 ENERGY COMMUNITIES

An energy community consists of a group of prosumers and/or consumers who voluntarily join the community with a specific goal. Energy communities can have various kinds of energy resources such as an energy storage system, photovoltaic arrays, and wind turbines which can be regarded as shared assets, meaning that they belong

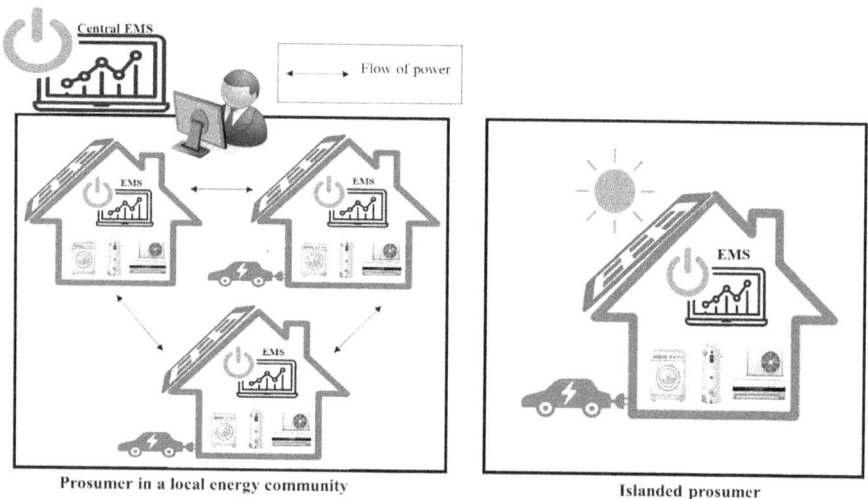

Prosumer in a local energy community | Islanded prosumer

FIGURE 4.1 Structures of a prosumer in an islanded mode and in a community.

to all of the members. Forming the energy community can follow different objectives such as maximizing the community's total profits, minimizing its costs, or increasing the self-sufficiency of the members.

Different members can constitute different types of communities. In this regard, the community can be a residential energy community, an industrial energy community, etc. The short physical distance between the members of the community can form a local energy community. In this kind of community, the local production aims to satisfy the local demand. Moreover, anyone from the neighbouring area can voluntarily register as a member of this community and being supplied locally. In addition, the total income and costs of the community will be shared between the members [20]. Thus, the capital costs of the shared resources can be split between the members while they can all benefit the profits obtained from these resources. Regarding the management of the community, a non-profit manager who can be selected from the existing members of the community can be nominated and take the responsibility of community management as well as the related monetary and technical considerations [21].

An energy community can be considered as an independent player who may participate in wholesale or local markets. For instance, it can sell its production surplus to wholesale markets and/or provide ancillary services to the market operators. In this way, the main aim of the community is to maximize the total profits of its members by participating in the market.

4.2.3 AGGREGATORS, VPPS, AND COMMUNITY MANAGERS

In most power systems, small prosumers and consumers are not allowed or are not able to participate in the wholesale markets. In these markets, there are minimum capacities that each player needs to submit to the market. Thus, it prevents these players from taking part in wholesale energy and ancillary service markets. In addition, in the wholesale markets where small players are allowed to participate, the prosumers are not motivated enough to take part and compete against those large-scale players with huge capacities. Another reason is that a prosumer that is fully capable of providing flexibility services may lack the needed information. For instance, small consumers and prosumers often lack information on time slots in which system peaks will happen, the prices of provision different kinds of flexibility services, the available up-to-date technologies that can help them to manage their consumption and production, and the cost-benefit analysis regarding deploying these technologies. Besides, prosumers may not have the ability and knowledge of forecasting the prices and build profitable bidding strategies, accordingly. In this situation, an aggregator acts as a broker, aggregating prosumers to reach the permissible capacity utilizing a communication interface [22]. An aggregator may be responsible for intervening to fill information gaps between the system operators (TSO and DSO) and different agents.

An aggregator may be in charge of aggregating specific FERs. For instance, it can be an EV aggregator [23], a distributed energy resources (DER) aggregator [24], or an energy storage aggregator [25]. It also may aggregate prosumers/or consumers [26]. To a large extent, the aggregator can aggregate several microgrids or demand-response resources [27].

The main aims of aggregating several resources are participating in wholesale markets and/or contributing to the provision of flexibility services for the grid. The aggregator may decide to take part in electricity and/or ancillary service markets. For this purpose, it executes a contract with each participant for controlling its FERs or for selling a specified amount of energy and flexibility for some time slots. Some aggregators may offer dynamic tariffs or incentive programs to their aggregated prosumers and consumers. Hence, in the former situation, they control their resources indirectly.

Aggregators mainly run optimization problems in order to optimally schedule and coordinate their resources. They can be separated from the utility or they can be associated with and established by it. It should be noted that the scheduling and control actions taken on the aggregator's resources may affect the load flows, transformers, and line thermal capacities of distribution systems. Hence, distribution network-related constraints should be taken into account regarding the area which is controlled by an aggregator.

A VPP can be also regarded as an aggregated distributed energy unit. A VPP combines the capabilities of a number of DER so as to increase power generation and enable them to trade energy with open markets [28]. A VPP can manage its internal DERs and controllable loads in an efficient way, making traditional fuel-based and renewable-based energy to operate together in harmony. Moreover, the efficiency and reliability of the system can be improved by the efficient integration of DER [29]. Furthermore, a VPP can increase synergy and interactivity by coordinating several DERs and EMS located in different areas. As the resources can be located in various areas, a VPP is able to manage resources to participate in providing voltage-based ancillary services as well as energy.

A community manager can be selected from one of the members of the community whose main goal should be in line with the objective of forming the community. In other words, the main goal for which the members gather together should be taken into account by the community manager. Thus, a community manager can be considered as one kind of an aggregator. The community may have some shared resources such as energy storage or renewable-based resources that should be scheduled by the manager of the community. In this light, the community manager needs to schedule these resources with the target to satisfy the total objective of the community. If the members gather together as a community aiming to increase their sufficiency, the manager should schedule the shared resources with the objective of satisfying the community's demand in every time slot. In this regard, the balance constraints and community's network should be taken into consideration by the manager of the community.

4.2.4 Market Operators

The main responsibility of a market operator as an autonomous entity is to match bids and offers of the electricity sellers and buyers. It can have different objectives based on the type of the market. For example, the market operator may aim to maximize the social welfare of the players, minimize the costs of generation units, increase the liquidity of the market, decrease carbon footprint regarding producing electricity, or minimize energy losses in the networks.

If the objective of the market is defined to be maximizing the social welfare of players, selling offers should be matched with buying offers in order to reach the pre-defined objective. In other words, the market operator accepts bids and offers such that the total selling and buying quantities match while the social welfare of the whole players is maximized. The market operator should also consider the imbalances of the market as the demand should be satisfied in every time slot.

Different pricing mechanisms can be deployed by the market operator. If it adopts "pay as bid" approach, the accepted sellers simply receive revenue according to the prices that they have previously offered through their offering curves. Similarly, buyers should pay for the energy according to the bids that they have submitted. The advantage of this pricing approach is that this mechanism is simple and intuitive for the different market players [30]. However, it cannot incentivize market participants enough to participate in the market competitively. In comparison, the market operator may deploy "uniform pricing" approach through which the sellers and buyers receive the same prices, corresponding to the intersection of the aggregated selling offers with the aggregated buying bids. This approach sounds to be pretty fair especially for trading electricity.

Moreover, the market operator should take into account the constraints related to the networks in its optimization problem. In this way, a mathematical model of distribution and transmission networks should be presented. The topology of the network and characteristic of the transformers and lines need to be considered in the formulation. The topology of networks describes the network graph which may vary according to the on-off status of switches.

4.2.5 TRANSMISSION SYSTEM OPERATORS (TSO)

The main responsibilities of TSOs in power systems are to maintain the frequency of the power system by keeping the balance between the generation and demand as well as regulating voltage and congestion management of transmission networks [31]. The frequency of the power system is controlled using different reserve products. For example, in Finland, the primary reserve is called FCR which requires to automatically react to the real-time frequency deviation in a constant way. FCR reserve itself is split up into two types which include frequency containment reserve for normal condition (FCR-N) and frequency containment reserve for disturbance condition (FCR-D) [32]. FCR-N is deployed all the time when operating the power system, while FCR-D is procured when a large deviation of frequency happens [33]. Hence, FCR-D is not utilized constantly in all of the time slots. The other type of reserve is frequency restoration reserve which is categorized into automatic frequency restoration reserve (aFRR) and manual frequency restoration reserve (mFRR). These reserves are considered secondary and tertiary reserves as well. Automatic FRR is an automatically centralized reserve which is activated according to a power unbalance signal. The unbalance signal is calculated based on the frequency deviation in the Nordic synchronized area. In comparison, mFRR is deployed manually in some situations such as power outages, power-constrained violations regarding cross-border connections, and also unexpected sustained activation of aFRR [34]. In addition to the above-mentioned services, a new type of reserve market has been introduced recently which is entitled "fast frequency reserve". This reserve is responsible for capturing rapid frequency fluctuations in low inertia situations.

The online states of the power system determine the actions related to the congestion management of the transmission networks. In this regard, TSOs utilize Flexible ramping products along with ancillary service products in order to avoid congestion in transmission networks [34]. In addition, different active power and reactive power products may be deployed by the TSO to regulate the voltage of the buses.

In order to better operate the power system, cooperation among TSOs is required. For instance, a study proves that optimal cooperation between TSOs will enhance flexibility in transmission networks which could in turn facilitate the higher penetration of renewable-based power into the system [35].

In general, as previously mentioned, TSOs utilize different types of FERs to enhance the flexibility of transmission networks. These flexible resources can be located in low-voltage, medium-voltage, or high-voltage networks [36]. In addition to generators which are large-scale flexibility providers, flexible resources connected to distribution networks have considerable potential for providing TSO-level flexibility services such as frequency services. Thus, a new structure for ancillary service markets is needed in order to exploit the potent flexibility of distribution network located resources (Figure 4.2).

4.2.6 DISTRIBUTION SYSTEM OPERATOR (DSO)

DSOs' main responsibility is to operate the distribution networks. A DSO should ensure that voltages of nodes remain in the predefined levels and it manages the power flowing through the network in a way that it does not create congestion and violate the thermal limits of the lines. Thus, congestion management of different feeders and also reactive power (Q)–voltage (U) management (QU- or Volt/Var) which is deployed for distribution network voltage control are also other vital responsibilities of DSOs [37]. However, with the growing number of renewable-based distributed energy resources, smart metering, and smart grid technologies, the role of DSOs is going to undergo revolutionary changes.

DSOs in smart networks are burdened with various additional responsibilities. For example, they need to handle and collect the mass data that are received from the smart meters and manage and use them for the forecasting and risk management

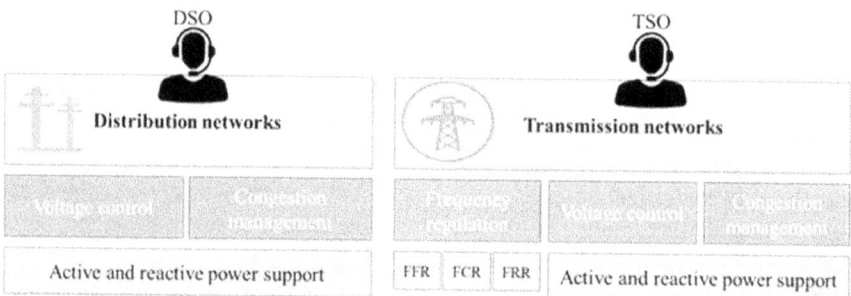

FIGURE 4.2 An overview of flexibility services adopted by system operators.

as well as operation and planning of the distribution networks [36]. In addition, they may also need to manage local and microgrid markets in the future distribution system. Moreover, voltage control and congestion management of feeders are becoming more challenging owing to the high penetration of intermittent renewable-based power generation as well as the growing number of EVs with uncertain charging behaviour in distribution networks. As a result, DSOs require to deploy more flexibility services in the future in order to address the new challenges associated with power variability and uncertainty.

For this purpose, the DSO should deploy both active power and reactive power flexibility of the distribution network located resources. These resources include storage-based resources, controllable appliances, and inverters of inverter-based DGs which can assist in enhancing the feeder transfer capability by absorbing or injecting the required amount of reactive power. As these FERs can be located in various nodes and locations, they can be enormously helpful for DSOs. In this way, the DSO can utilize distributed voltage control for voltage regulation purposes. According to the studies conducted by [38] distributed voltage control is one of the most efficient control actions which can be adopted by DSOs due to its low communication burden as well as its high control quality. In the distributed voltage control, the distribution system should be divided into different subsystems. Each subsystem has its own control areas. Each control area will work autonomously, aiming to optimize its local network [39].

4.3   BUSINESS MODELS

Traditionally, there exist some wholesale markets for trading energy and ancillary services between different large-scale players. The only players who can participate in these markets are large-scale generating companies such as huge fuel-based generators, aggregators, as well as some retailers who submit their required demand. TSOs also play the role of buyers in ancillary service markets. Regarding the traditional structure, small-scale resources such as prosumers cannot actively contribute to providing energy and flexibility services. For this reason, new business models and market structures have been defined in the recent studies, trying to involve more distribution network located customers and FERs as much as possible. In this section, these market structures are introduced and the participants and the trading architecture are discussed as well.

4.3.1   LOCAL ELECTRICITY MARKETS

In the last decade, a rise in the production of distributed renewable energy was reported significantly. The high deployment of renewable energy sources can alleviate the problem of global warming and environmental issues. In addition, the growing amount of demand in the power system has led the policymaker to design different incentive programs so as to motivate consumers to turn into prosumers. As a consequence, a concept of local energy markets was brought forward. The local energy market seeks to enable any prosumer to utilize the capacity of its energy resources and flexible resources at its highest potential.

Local energy markets can gather together different types of energy resources at different levels of the power system. In this way, it may create a competitive environment in which players can compete against the same-sized competitors. It means that, for example, a distribution network located prosumer who has a capacity in a range of several kilowatts competes with the distribution network located prosumers with the same range capacities. In addition, local markets help to overcome challenges related to geographical limitations and lack of network infrastructure. In the local market environment, prosumers and consumers are able to trade throughout the local network (for example an islanded microgrid) in order to effectively utilize their available capacities. This helps to reduce the energy losses in the networks and lower the cost of energy since all of the flexible potentials of the end-users can be exploited [40].

Various market platforms can be used so as to connect various stakeholders and players in the local market. In addition, the local market can have different architectures. It can deploy different pricing mechanisms as well as a different market settlement which can affect the motivation of the players. Although several structures are available, the local market should follow two main goals. First, it should optimize the utilization of local energy resources and integrate distributed energy resources. Second, it needs to assure that there exists no discrimination and create a competitive environment for trading energy.

4.3.1.1 Market Participants

Different parties can be involved in local electricity markets. Prosumers and consumers are the main players of these markets, playing the role of both sellers and buyers. It means that prosumers submit selling offers for their surplus production, while consumers bid to supply their demand. It should be noted that during some time slots, a prosumer may turn into a consumer since the amount of its demand exceeds the production.

The local market operator is responsible for matching selling offers with buying bids considering the offered prices and the required quantities. The local market operator can be a DSO. In other words, a DSO may operate several local markets in its territory. In situations in which the local market operator is an independent entity (i.e. the local market operator is not a DSO), it should be constantly in touch with the DSO to check whether the trading power does not violate the network constraints.

4.3.1.2 Local Market Architectures and Clearing Process

A local market may form like an islanded microgrid, meaning that it acts as an autonomous network. This kind of local market aims at satisfying the local demand by utilizing local production. In this way, the local market should have enough production capacities and also FERs to meet the local demand. Additionally, the production surplus of this market may be curtailed during time slots in which the production exceeds the demand. In comparison, a local market operator can trade the surplus power with the upstream grid in order to satisfy the balance-related constraint. Figure 4.3 describes the architectures of the above-mentioned two local markets. The second structure is more likely to exploit the maximum potential of local players since the production surplus does not need to be curtailed and remains unused.

A local market can have a pool-based architecture. In the pool-based architecture, players submit their bids and offers to the local market operator. The operator clears the offers and bids with regard to the main objective of the market. This structure is

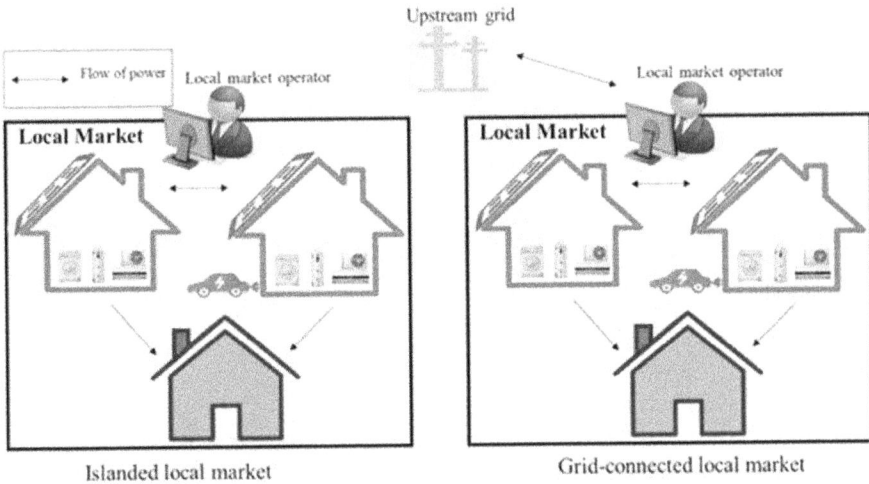

FIGURE 4.3   Islanded and grid-connected structures of a local market.

similar to the current structure of the pool-based wholesale markets. In this regard, in each time slot, the bids of demand are aggregated according to their prices in descending order. The offers of production are also aggregated according to their offered prices in ascending order. The intersection of the aggregated bids and offers determines the local market price in the studied time slot. The remaining production and demand will be traded with the upstream grid. Figure 4.4 depicts the clearing approach of the pool-based local market. In contrast, a local market may have a peer-to-peer structure in which players may be given an opportunity to select their trading peers. This structure will be fully discussed in the following subsection.

4.3.2   PEER-TO-PEER (P2P) ELECTRICITY TRADING

In a P2P trading structure, prosumers and consumers can trade electricity bilaterally. This trading structure aims to increase energy democracy and incentivize prosumers and consumers to be more active in energy sectors. However, it may have an adverse effect on residential participants while they need to be constantly active for meeting their required energy or for selling their production.

    Considering the consumers' preferences in choosing their sources of energy is one of the drives which can promote P2P trading. These days, public awareness of the problem associated with global warming and climate change is rising. As a result, energy-related sustainability becomes more and more popular. Hence, P2P trading can pave the way for empowering consumers through providing them with their required information, the option to choose their source of energy, and the ability to manage their consumption and production which can in turn lead to making profits or decreasing their energy costs. In addition to this, a consumer and prosumer may negotiate a long-term contract bilaterally with the satisfaction of both parties. In this regard, some may prefer to choose the neighbours as their trading partners to empower their local community.

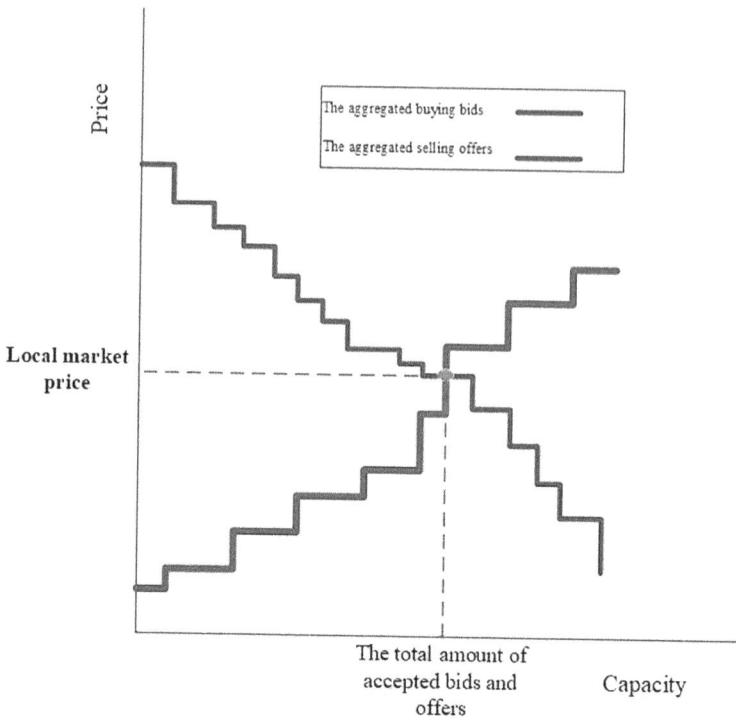

FIGURE 4.4 The clearing approach of a pool-based local market.

P2P trading can be a suitable trading option for the network with weak interconnections. For example, in rural areas, this kind of trading may be beneficial as it
decreases the costs of network reinforcement while incentivizing end-users to be
equipped with renewable resources and become self-sufficient and making profits by
selling their surplus production.

Nevertheless, the unsupervised P2P can result in some issues related to the reliability
and security of the system. First, renewable-based power is extremely uncertain and
volatile in nature. Thus, the seller who is equipped with this kind of resource needs
to utilize FERs (such as storage) as well in order to fulfil its promises on supplying a
specified demand. In addition, the unsupervised trading power may endanger network-
related constraints. Some trading can result in congestions in the distribution network,
and some may violate the voltage threshold in distribution networks. Hence, the amount
and direction of trading power should be checked beforehand by the DSO (or any other
entity who is in touch with the DSO) in order not to violate the network constraints.

4.3.2.1 Trading Participants

In P2P trading, a seller can trade with several buyers and vice versa. Sellers are
prosumers or local communities which have production surplus. Buyers consist of
consumers or local communities with unsupplied demand. As stated before, the
DSO (or any other entity receiving signals from the DSO) should supervise the P2P
trading between the local peers. In this regard, the peers need to give the DSO the

information about their upcoming trading, sellers, and buyers. Then, the DSO should check whether the trading does not endanger the security of the local network.

4.3.2.2 P2P Trading Architecture

Authors in [41] have categorized the P2P trading structure into three different architectures for participants of distribution networks. These architectures are entitled "full P2P trading", "community-based P2P trading", and "hybrid P2P trading". In the full P2P trading architecture, small prosumers and consumers can trade energy with each other, while in a community-based one, a prosumer or consumer should join a community. Then, communities can trade energy in a P2P trading structure. In a hybrid architecture, prosumers and consumers as well as communities can perform P2P energy trading. These architectures are depicted in Figure 4.5.

In order to deal with the imbalances resulted from the P2P trading of end-users, consumers and prosumers may decide to trade bilaterally with their neighbours while trading their surplus with the upstream grid. In this way, the peers can decide on their trading partners in the first stage. Then, if the demand of a consumer does not fully satisfy with the chosen peers, it can be supplied by the upstream grid. Similarly, if a prosumer has a surplus production after providing the promised demand, it is able to sell it to the upstream grid. Thus, this structure can fully cover the imbalances stemming from intermittent characteristics of renewable resources and exploit all of the potentials of the local resources. This structure is also illustrated in Figure 4.6.

4.3.3 AGGREGATOR-BASED ELECTRICITY TRADING

The number of aggregators who are participating in energy markets has been rising in the last decade [42]. Some European companies such Voltalis in France and REstore in Belgium are currently playing the role of an aggregator and participate in different energy markets [43]. Not only do they participate in wholesale energy markets by aggregating consumers and prosumers, they also provide some services for end-users such as consultancy tools as well as optimization and monitoring-related equipment so as to have more control over them. The aggregators then analyse the profiles of the end-users and build a bidding strategy accordingly. In order to achieve a better income, a cooperative relationship between the aggregators and their resources is required [44].

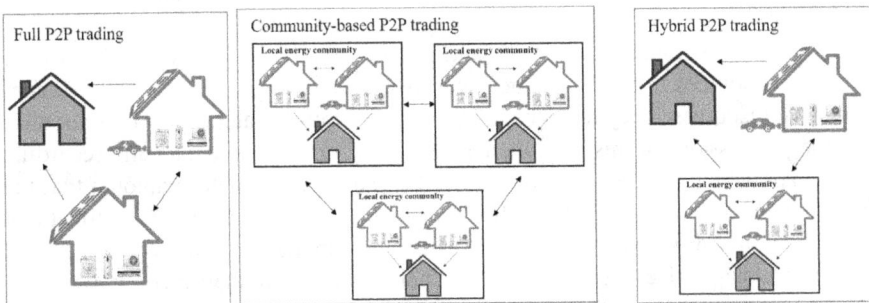

FIGURE 4.5 Different architectures of P2P trading.

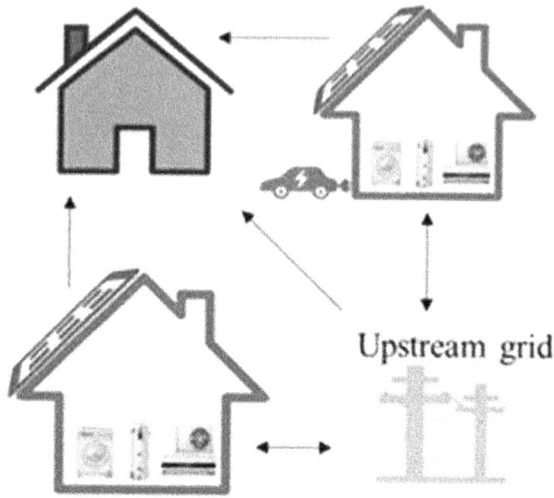

FIGURE 4.6 A structure of P2P trading in which players can trade their surplus with the upstream grid.

4.3.3.1 Participants

In this structure, the aggregator purchases the energy or flexibility from the prosumers and consumers. In this way, the aggregator may define the groups of prosumers and/or consumers as well as appropriate tariffs for each group. The tariff can be determined according to the time of the day and also the services that the prosumers will provide. In addition, the aggregator may schedule DGs and FERs of consumers/prosumers and build a bidding strategy in accordance with the available forecasts obtained from the data which are received from smart meters. The aggregator may also take into account the uncertainties of renewables and errors of forecast when scheduling the resources [44]. Note that if the aggregator is only responsible for aggregating different kinds of DGs, it may call a VPP. Thus, the main role and responsibility of a VPP are similar to those of an aggregator.

From the viewpoint of consumers and prosumers, they may submit a list of their available resources to the aggregator and give permission to the aggregator to control these resources. They may negotiate contracts with the aggregator to grant him/her the permission. In this way, their FERs are controlled automatically regarding the signals which are received from the aggregator. The monetary amount that the prosumers receive and consumers should pay is in accordance with the contract that they have signed with the aggregator. In another architecture, the prosumers and consumers play active roles and schedule their resources on their own. In each time slot, they try to optimize their schedule according to the signals receiving from the aggregator as well as their own profits.

4.3.3.2 Aggregator-Based Trading Architecture

The aggregator-based trading structure mainly consists of two-level hierarchical architecture [45]. In the higher level, an aggregator aims to minimize different terms of cost, including the incentives that should pay to the prosumers, the penalty costs incurred by not fulfilling the promises (which may be resulted from the uncertainty of renewables and error of forecast), and a term aiming at a fair distribution of profits/ costs among the aggregated prosumers and consumers [45]. It should be noted that the aggregator needs to define the amount of exchanged active and reactive power of prosumers/consumers as well as the related incentives, beforehand.

At the lower level, consumers/prosumers receive signals regarding the power reference pattern which was set by the aggregator and schedule their resources. Figure 4.7 illustrates this architecture.

In the architecture in which aggregator has full permission for controlling the resources, the scheduling problem turns into a single-level optimization problem. In other words, the lower level optimization which is performed by the prosumers and consumers is omitted from the problem.

4.3.4 LOCAL FLEXIBILITY MARKETS

Consumers and prosumers are able to provide both system-wide (TSO-level) and local (DSO-level) flexibility services using their FERs. As stated before, electric vehicles, different kinds of storage, as well as controllable appliances are examples of consumers' and prosumers' flexible resources. In terms of system-wide services, the active power flexibility of end-users can be aggregated and submitted to different ancillary services markets. When it comes to local services, local flexibility services can be procured using the active power flexibility of prosumers and consumers from their FERs as well as reactive power support from inverters of DGs. Local flexibility services are those services helping DSOs to operate the network effectively. Both services associated with voltage regulations and congestion management can be regarded as local services.

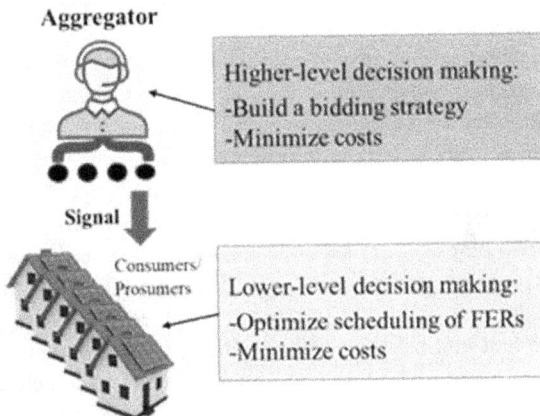

FIGURE 4.7 A two-level hierarchical architecture for the aggregator-based trading.

In conventional power systems, prosumers and consumers do not play roles in providing system-wide and local services. However, recent studies are analysing the active participation of small-scale resources in the provision of flexibility services. In this regard, local flexibility markets can assist in motivating consumers and prosumers to actively contribute to the provision of these services.

4.3.4.1   Roles of Different Participants

In local flexibility markets, system operators including TSOs and DSOs play the roles of buyers while consumers and prosumers are sellers of the local market. Moreover, the presence of a local flexibility market operator is needed to settle the market and match selling offers with the buying bids. It should also determine the prices of flexibility taking into account the type of flexibility that the seller can offer. It is noticeable that providing system-wide flexibility services by the local flexibility market must not violate the constraints of the distribution network. Thus, the local flexibility market operator should be constantly in touch with the TSO to ensure that providing flexibility services does not endanger local network security.

4.3.4.2   Local Flexibility Market Architecture

The trading architecture of the local flexibility market can be pool-based or in a form of P2P trading. In the pool-based architecture, all of the sellers and buyers submit their offers and bids, and the local flexibility market operator matches the selling offers with the bidding bids taking into account the network constraints. In comparison, the operators may negotiate bilaterally with the prosumers and consumers and determine the price accordingly. P2P architecture can be helpful for providing DSO-level flexibility services. In this regard, for example, a DSO can contact directly with the nodes related to feeders which are going to be congested.

4.4   SUMMARY AND CONCLUSION

In the last decade, small consumers become so motivated to change their roles from submissive ratepayers to proactive consumers (prosumers) who can make profits by managing their consumption and production. The reasons behind this revolutionary change are the newly emerged advances in technology, the advent of smart meters, and the global awareness of environmental issues. In this regard, a growing number of small-scale end-users are equipped with renewable energy resources and storage-based resources as well as smart meters. However, the existing wholesale markets and trading structure are not able to follow this revolution. Hence, the existing markets fail to accommodate the prosumers' need. In addition, they cannot motivate consumers and prosumers to play active roles in providing the required network flexibility.

In order to exploit the maximum energy and flexibility potential of end-users, new business model and trading structures are needed. Accordingly, this chapter analyses four different trading structures in which different prosumers and consumers can sell and buy energy. The first structure is related to local electricity markets, with the aims of incentivizing small end-users to compete with each other in a local environment. In this way, local consumption can be satisfied by the local production which in turn may decrease network losses. Another studied structure is peer-to-peer trading aiming to respect the players' preferences in choosing their trading partners. Peer-to-peer structure can promote energy democracy in the local level of power systems while motivating small-scale end-users to be more active. In comparison, an aggregator-based structure introduces a hierarchical architecture which can be beneficial for consumers and prosumers who are not willing to be constantly active. In this situation, the aggregator may provide the aggregated prosumers and consumers with the required equipment and facilities so as to control them. Finally, the chapter proposes a local flexibility market seeking to satisfy the system operators' needs. In this market, TSOs and DSOs can buy their needed flexibility from prosumers and consumers. Similar to local electricity markets, the architecture of the local flexibility market can be either pool-based or peer-to-peer-based. As a result, the introduced local markets and trading structure can be included in the future electricity markets.

REFERENCES

1. M. Gough, S. F. Santos, M. Javadi, R. Castro, and J. P. S. Catalão, "Prosumer flexibility: A comprehensive state-of-the-art review and scientometric analysis," *Energies*, vol. 13, no. 11, p. 2710, 2020.
2. H. Khajeh, M. Shafie-khah, and H. Laaksonen, "Blockchain-based demand response using prosumer scheduling," in *Blockchain-based Smart Grids*, Elsevier, 2020, pp. 131–144.
3. M. He, F. Zhang, Y. Huang, J. Chen, J. Wang, and R. Wang, "A distributed demand side energy management algorithm for smart grid," *Energies*, vol. 12, no. 3, p. 426, 2019.
4. E. M. Craparo and J. G. Sprague, "Integrated supply-and demand-side energy management for expeditionary environmental control," *Applied Energy*, vol. 233, pp. 352–366, 2019.
5. J. W. Forbes Jr, "System and method for generating and providing dispatchable operating reserve energy capacity through use of active load management," *Google Patents*, May-2019.
6. B. Khan, G. Agnihotri, S. E. Mubeen, and G. Naidu, "A TCSC incorporated power flow model for embedded transmission usage and loss allocation," *AASRI Procedia*, vol. 7, pp. 45–50, 2014.
7. B. Khan, G. Agnihotri, G. Gupta, and P. Rathore, "A power flow tracing based method for transmission usage, loss & reliability margin allocation," *AASRI Procedia*, vol. 7, pp. 94–100, 2014.
8. B. Khan, G. Agnihotri, P. Rathore, A. Mishra, and G. Naidu, "A cooperative game theory approach for usage and reliability margin cost allocation under contingent restructured market," *International Review of Electrical Engineering*, vol. 9, no. 4, pp. 854–862, 2014.
9. B. Khan and G. Agnihotri, "A comprehensive review of embedded transmission pricing methods based on power flow tracing techniques," *Chinese Journal of Engineering*, vol. 2013, Article ID 501587, 13 pages.
10. B. Khan, G. Agnihotri, and A. S. Mishra, "An approach for transmission loss and cost allocation by loss allocation index and co-operative game theory," *Journal of The Institution of Engineers (India): Series B*, vol. 97, pp. 41–46, 2016.
11. B. Khan and G. Agnihotri, "A novel transmission loss allocation method based on transmission usage," in *2012 IEEE Fifth Power India Conference*, 2012, pp. 1–3.
12. P. Rathore, G. Agnihotri, B. Khan, and G. Naidu, "Transmission usage and cost allocation using shapley value and tracing method: A comparison," *Electrical and Electronics Engineering: An International Journal (ELELIJ)*, vol. 3, pp. 11–29, 2014.

13. B. Khan and G. Agnihotri, "An approach for transmission usage & loss allocation by graph theory," *WSEAS Transactions on Power Systems*, vol. 9, pp. 44–53, 2014.

14. S. Khare, B. Khan, and G. Agnihotri, "A shapley value approach for transmission usage cost allocation under contingent restructured market," in *2015 International Conference on Futuristic Trends on Computational Analysis and Knowledge Management (ABLAZE)*, Noida, 2015, pp. 170–173.

15. B. Khan, G. Agnihotri, and G. Gupta, "A multipurpose matrices methodology for transmission usage, loss and reliability margin allocation in restructured environment," *Electrical & Computer Engineering: An International Journal*, vol. 2, no. 3, p. 11, September 2013.

16. S. Talari, H. Khajeh, M. Shafie-khah, B. Hayes, H. Laaksonen, and J. P. S. Catalão, "The role of various market participants in blockchain business model," in *Blockchain-based Smart Grids*, Elsevier, 2020, pp. 75–102.

17. G. Pau, M. Collotta, A. Ruano, and J. Qin, Smart home energy management. *Energies*, vol. 10, no. 3, p.:382, 2017.

18. Z. Zhang, R. Li, and F. Li, "A novel peer-to-peer local electricity market for joint trading of energy and uncertainty," *IEEE Transactions on Smart Grid*, vol. 11, pp. 1205–1215, 2019.

19. A. C. Luna, N. L. Diaz, M. Graells, J. C. Vasquez, and J. M. Guerrero, "Cooperative energy management for a cluster of households prosumers," *IEEE Transactions on Consumer Electronics*, vol. 62, no. 3, pp. 235–242, 2016.

20. A. R. Servent, *The European Parliament*. Macmillan International Higher Education, 2017.

21. N. Verkade and J. Höffken, Collective energy practices: A practice-based approach to civic energy communities and the energy system. *Sustainability*, vol. 11, no. 11, 3230, 2019. https://doi.org/10.3390/su11113230.

22. K. T. Ponds, A. Arefi, A. Sayigh, and G. Ledwich, "Aggregator of demand response for renewable integration and customer engagement: Strengths, weaknesses, opportunities, and threats," *Energies*, vol. 11, no. 9, p. 2391, 2018.

23. M. G. Vayá and G. Andersson, "Optimal bidding strategy of a plug-in electric vehicle aggregator in day-ahead electricity markets under uncertainty," *IEEE Transactions on Power Systems*, vol. 30, no. 5, pp. 2375–2385, 2014.

24. M. Di Somma, G. Graditi, and P. Siano, "Optimal bidding strategy for a DER aggregator in the day-ahead market in the presence of demand flexibility," *IEEE Transactions on Industrial Electronics*, vol. 66, no. 2, pp. 1509–1519, 2018.

25. J. E. Contreras-Ocana, M. A. Ortega-Vazquez, and B. Zhang, "Participation of an energy storage aggregator in electricity markets," *IEEE Transactions on Smart Grid*, vol. 10, no. 2, pp. 1171–1183, 2017.

26. J. Iria, F. Soares, and M. Matos, "Optimal supply and demand bidding strategy for an aggregator of small prosumers," *Applied Energy*, vol. 213, pp. 658–669, 2018.

27. H. Khajeh, A. A. Foroud, and H. Firoozi, "Robust bidding strategies and scheduling of a price-maker microgrid aggregator participating in a pool-based electricity market," *IET Generation, Transmission & Distribution*, vol. 13, no. 4, pp. 468–477, Feb. 2019, doi:10.1049/iet-gtd.2018.5061.

28. S. Yu, F. Fang, Y. Liu, and J. Liu, "Uncertainties of virtual power plant: Problems and countermeasures," *Applied Energy*, vol. 239, pp. 454–470, 2019.

29. M. Braun and P. Strauss, "A review on aggregation approaches of controllable distributed energy units in electrical power systems," *International Journal of Distributed Energy Resources*, vol. 4, no. 4, pp. 297–319, 2008.

30. A. Akbari-Dibavar, B. Mohammadi-Ivatloo, and K. Zare, "Electricity market pricing: Uniform pricing vs. pay-as-bid pricing," in *Electricity Markets*, Springer, 2020, pp. 19–35.

31. D. Tchoubraev and D. Wiczynski, "Swiss TSO integrated operational planning, optimization and ancillary services system," in *2015 IEEE Eindhoven PowerTech*, 2015, pp. 1–6.
32. P. H. Divshali and C. Evens, "Stochastic bidding strategy for electrical vehicle charging stations to participate in frequency containment reserves markets," *IET Generation, Transmission & Distribution*, vol. 14, no. 13, pp. 2566–2572, 2020.
33. P. H. Divshali and C. Evens, "Optimum operation of battery storage system in frequency containment reserves markets," *IEEE Transactions on Smart Grid*, vol. 11, pp. 4906–4915, 2020.
34. G. De Zotti, S. A. Pourmousavi, H. Madsen, and N. K. Poulsen, "Ancillary services 4.0: A top-to-bottom control-based approach for solving ancillary services problems in smart grids," *IEEE Access*, vol. 6, pp. 11694–11706, 2018.
35. P. G. Thakurta, J. Maeght, R. Belmans, and D. Van Hertem, "Increasing transmission grid flexibility by TSO coordination to integrate more wind energy sources while maintaining system security," *IEEE Transactions on Sustainable Energy*, vol. 6, no. 3, pp. 1122–1130, 2014.
36. H. Khajeh, H. Laaksonen, A. S. Gazafroudi, and M. Shafie-khah, "Towards flexibility trading at TSO-DSO-customer levels: A review," *Energies*, vol. 13, no. 1, p. 165, 2020.
37. C. Zhang, D. Yi, N. C. Nordentoft, P. Pinson, and J. Østergaard, "FLECH: A Danish market solution for DSO congestion management through DER flexibility services," *Journal of Modern Power Systems and Clean Energ*, vol. 2, no. 2, pp. 126–133, 2014.
38. A. Abessi, V. Vahidinasab, and M. S. Ghazizadeh, "Centralized support distributed voltage control by using end-users as reactive power support," *IEEE Transactions on Smart Grid*, vol. 7, no. 1, pp. 178–188, 2015.
39. A. Abessi, A. Zakariazadeh, V. Vahidinasab, M. S. Ghazizadeh, and K. Mehran, "End-user participation in a collaborative distributed voltage control and demand response programme," *IET Generation, Transmission & Distribution*, vol. 12, no. 12, pp. 3079–3085, 2018.
40. Í. Munné-Collado, E. Bullich-Massagué, M. Aragüés-Peñalba, and P. Olivella-Rosell, "Local and Micro Power Markets," *Micro Local Power Mark*, 2019.
41. T. Sousa, T. Soares, P. Pinson, F. Moret, T. Baroche, and E. Sorin, "Peer-to-peer and community-based markets: A comprehensive review," *Renewable & Sustainable Energy Reviews*, vol. 104, pp. 367–378, 2019.
42. B. Shen, G. Ghatikar, Z. Lei, J. Li, G. Wikler, and P. Martin, "The role of regulatory reforms, market changes, and technology development to make demand response a viable resource in meeting energy challenges," *Applied Energy*, vol. 130, pp. 814–823, 2014.
43. A. M. Carreiro, H. M. Jorge, and C. H. Antunes, "Energy management systems aggregators: A literature survey," *Renewable & Sustainable Energy Reviews*, vol. 73, pp. 1160–1172, 2017.
44. P. Faria, J. Spínola, and Z. Vale, "Reschedule of distributed energy resources by an aggregator for market participation," *Energies*, vol. 11, no. 4, p. 713, 2018.
45. G. Ferro, R. Minciardi, L. Parodi, M. Robba, and M. Rossi, "Optimal control of multiple microgrids and buildings by an aggregator," *Energies*, vol. 13, no. 5, p. 1058, 2020.

5 Distributed Generation, Storage and Active Network Management

Chethan Parthasarathy, Hosna Khajeh,
Hooman Firoozi, Hannu Laaksonen,
and Hossein Hafezi

CONTENTS

DOI: 10.1201/9781003278030-5

5.1 INTRODUCTION

Power systems are changing due to global drivers such as climate change and environmental issues as well as increasing dependency on electricity. Therefore, there are needs for (i) large-scale integration of renewable, low-emission (CO_2) energy sources in high-, medium- and low-voltage (HV, MV and LV) networks, (ii) improving energy efficiency of the whole energy system, and (iii) enhance electricity supply reliability. In Figure 5.1 the main impacts (A)–(D) of these changes on power systems are presented.

   Previously, distribution grid-connected renewable energy sources (RES) and other distributed generation (DG) units were usually required to be disconnected during faults and disturbances. Recent technological advances along with social acceptance have led to an increase in the number of DG units in the electricity distribution networks. Due to constantly increasing number of connected DG units, HV grid stability has become a central issue, and it could get worse if, for example, a large share of the DG units is disconnected during HV grid faults or frequency disturbances. Therefore, supply reliability and quality with the integration of RES are tried to be ensured by setting stricter grid code requirements for the RES and other DG units connected at different voltage levels (HV, MV and LV). These grid codes define, for example, the required HV grid stability supporting functionalities like voltage and frequency fault-ride-through (FRT) requirements of the DG units during HV grid faults and frequency disturbances. These DG unit grid code requirements have been made mainly from the HV grid stability point of view, and less attention has been paid on distribution network effects (like protection and islanding detection). However, in the future, this is not enough. In addition, there is a need for flexibility from distributed energy resources (DER), because DER (controllable generation units, energy storages, controllable loads and electric vehicles) at different voltage levels (HV, MV and LV) has potential to

a. Provide different local (corresponding voltage level) and system-wide (whole power system /HV network) technical flexibility services by active (P) and reactive (Q) power control which could, in addition to grid codes and regulations, be realized by future technical ancillary service/flexibility markets
b. Simultaneously improve energy efficiency, i.e. reduce the demand for distribution network capacity (coordinated voltage control and congestion management), reduce losses and increase reliability of electricity supply to the customers (intended island/microgrid) operation

But this potential of DG units and full integration of RES cannot be realized without active management of the distribution grids and flexibilities connected in distribution grids. For example, different types of energy storage systems (ESSs) have been deployed extensively as flexible energy resources aiming to

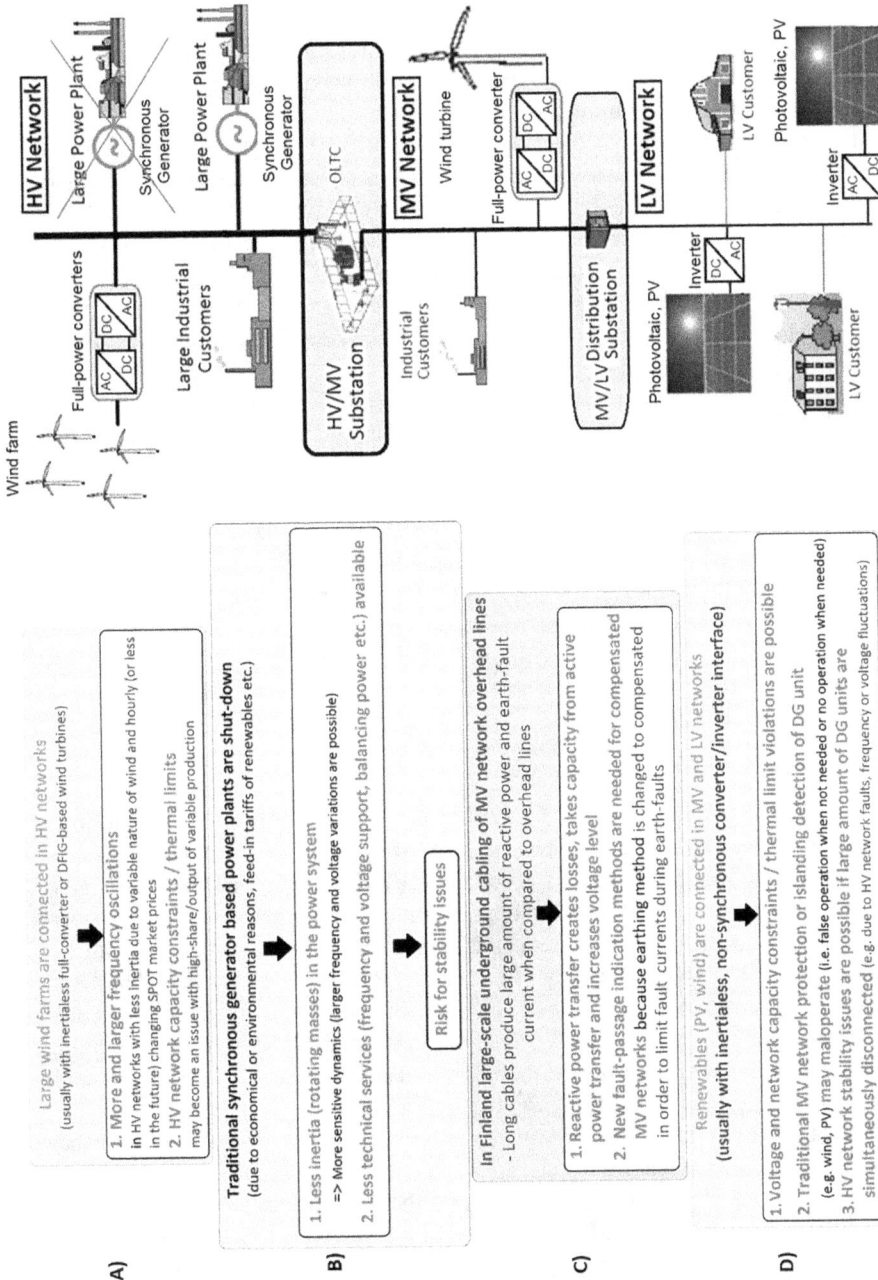

FIGURE 5.1 Main impacts of ongoing changes on power systems.

increase the flexibility at different levels of the power system including local (distribution network) and system-wide (transmission network) levels. In order to effectively integrate DGs and ESSs and exploit their maximum flexibility potential, appropriate management and control structures are needed. The recent advances in ICT have enabled active participation of these resources, accommodate all DG generation and energy storage options and also facilitate optimal management of these resources.

The most efficient way to meet energy demand with increasing integration of RESs in the distribution networks is to incorporate innovative solutions, technologies and grid architectures. Developing economically viable yet innovative grid architectures becomes essential with the increased role of non-dispatchable and DGs based on RES. Such solutions are realized by active control of the DGs as part of Active Network Management (ANM) schemes. Smart grids provide the platform to implement ANM schemes, where the DGs are interconnected and inter-communicable in real time to work in tandem to supply energy demands.

In this chapter, an overview of DG, i.e. their classification and role in ANM of smart grid operations and adjoining control methodologies, will be addressed in detail. ESSs play a crucial role as flexible energy resources for ANM in smart grids. Their technology types and various applications they tend to use will be explained. Battery energy storage systems (BESSs), especially Lithium-ion batteries with their current technology and economic maturity, are considered as a viable option for stationary grid applications. Design and control of Li-ion battery integration by means of power electronic converters for ANM in medium voltage (MV) distribution system, along with managing other flexibility services of DGs, are the scope of this chapter.

5.2 DG SOURCES

DGs have made a significant contribution to producing energy in the last decade. Most energy players and utilities have found that utilizing DGs can decrease their net costs, as the marginal costs of producing electricity by renewable-based DGs are very low, near to zero [1]. System operators including transmission system operators (TSOs) and distribution system operators (DSOs) can also benefit from the large installation of DGs since they offer benefits for the power systems. The DGs' advantages from the operators' viewpoints can be supporting network voltage and frequency, reducing network losses, reducing transformers loading stress, promoting system reliability, as well as providing environmental benefits [2]. In addition, end-users who are equipped with DGs are also able to take advantage of the economic benefits of installing DGs. In this way, not only they can be self-sufficient, but they can also sell their production surplus and make profits. DGs mainly assist in satisfying distribution network located demand, and they are located in distribution networks (low voltage (LV) and MV levels) [3–12]. Photovoltaic (PV) panels and wind turbines are currently the most popular renewable-based DERs that are located in distribution networks. Micro-CHP (combined heat and power) and fuel cells are the other common DGs which can be adopted in distribution networks. The following sections give more information about these DGs.

5.2.1 PV PANEL

There exist various methods estimating the active power produced by a PV module. In one of the first researches, the power output of a PV module was proposed to be estimated based on the cell temperature as well as solar irradiance. The cell temperature is in turn dependent on the ambient temperature as stated in Eq. (5.1) [13,14].

$$\theta^{\text{cell}} = \theta^{\text{ambient}} + G\left(\frac{\theta^{\text{NOCT}} - 20}{800}\right) \tag{5.1}$$

The output power captured by a PV cell is estimated accordingly, utilizing Eq. (5.2).

$$P^{pv} = P^{\text{STC}} G\left(1 + c\left(\theta^{\text{cell}} - 25\right)\right) \tag{5.2}$$

where, in the above equations, c indicates the power-temperature coefficient. θ^{cell} is the PV cell temperature, θ^{ambient} denotes ambient temperature while θ^{NOCT} refers to the temperature associated with the nominal
operation of the cell all in [°C]. In addition, P^{pv} and P^{STC} are the output power of the PV cell and the power produced under the standard test condition, respectively. Finally, the solar irradiance is shown by G [13].
The output power of PV can be also calculated according to its simplified equivalent circuit illustrated in Figure 5.2 [15]. Thus, the following equations are adopted in order to calculate the produced PV power equipped with the boost converter [15]:

$$I^{pv} = I^{ph} - I^{sa}\left(e^{\frac{q(V+IR^s)}{nkT}} - 1\right) - \frac{V + IR^s}{R^{sh}} \tag{5.3}$$

$$P^{pv} = \eta^{\text{boost}} I^{pv} V^{pv} \tag{5.4}$$

where in Eq. (5.3), I^{ph} denotes the photocurrent and I^{sa} refers to the saturation current of the diode. Moreover, the series and shunt resistances are represented by R^s and R^{sh}, respectively while n indicates the factor related to the diode ideality, k is Boltzmann's constant, q refers to the electron charge and T is a parameter expressing the absolute temperature in Kelvin. In Eq. (5.4), P^{pv} is obtained utilizing I^{ph}, the open circuit voltage, V^{pv}, and the efficiency of the boost converter, η^{boost}.

Simplified electrical model of a solar cell

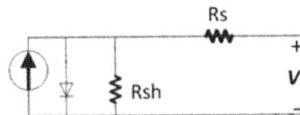

FIGURE 5.2 An electrical single diode model for a PV cell.

It is worth mentioning that the power produced by a single PV cell is so small. However, PV modules can be connected in parallel and series in various topologies in order to generate more power.

In general, a PV system that is connected to the grid may consist of a boost DC–DC converter, a Maximum Power Point Tracking (MPPT) controller, a voltage source inverter and some other equipment. The main responsibility of the boost converter is to balance the system while the inverters convert the output DC power of the PV system to the AC power. In order to ensure the efficient operation of the PV panel, the point in which the output power of the PV panels reaches its maximum value should be found. In this regard, an MPPT controller is deployed so as to track the MPP of the panel [16,17].

5.2.2 WIND TURBINE

The active power produced by a wind turbine is dependent on some factors such as the area and location where the wind turbine's rotor blades are spinning (swept area), the wind speed as well as the air mass density. In addition, the output power of the wind turbine is restricted by a coefficient of power denoted by c_p. If the coefficient of power equals its optimal value, the maximum wind turbine output is obtained. This value can be calculated by Eq. (5.5) [18].

$$P^{WT} = 0.5c_p{}^{opt}(\gamma,\beta)A^s\rho^{air}(WS)^3 \tag{5.5}$$

where P^{WT} is the maximum output of wind power, $c_p{}^{opt}(\gamma,\beta)$ is the coefficient of wind power which is a function of speed ration (γ) and blade pitch angle (β), the parameter A^s indicates the swept area in which the rotor spins, ρ^{air} is the air mass density and finally WS is the wind speed.

In the following model, wind power is considered a non-linear function of the wind speed. There also exists another model aiming to explain the linearized relationship between wind power output and the wind power speed. The linear model can be adopted especially for scheduling several energy resources. This model is expressed by Eq. (5.6) [19].

$$P^{WT} = \begin{cases} 0 & WS < WS^{\text{cut-in}} \\ \dfrac{P^r}{WS^r - WS^{\text{cut-in}}}WS + P^r\left(1 - \dfrac{WS^r}{WS^r - WS^{\text{cut-in}}}\right) & WS^{\text{cut-in}} \le WS < WS^r \\ p^r & WS^r \le WS < WS^{\text{cut-off}} \\ 0 & WS \ge WS^{\text{cut-off}} \end{cases} \tag{5.6}$$

where $WS, WS^r, WS^{\text{cut-in}}, WS^{\text{cut-off}}$ are wind speed, rated wind speed, cut-in wind speed and cut-off wind speed, respectively. Additionally, P^r denotes the rated power of the wind turbine.

5.2.3 Micro-CHP

A CHP is utilized in order to combine the heat with electricity production. A micro-CHP is regarded as a decentralized small-scale CHP located at the customer level of the electrical network. The micro-CHP is able to simultaneously produce heat and power which increases the efficiency of the system. The maximum capacity of the micro-CHP is usually below 15 kW. The energy efficiency of the CHP unit can be assumed to be constant. However, in practice, the CHP unit's efficiency varies with dynamic operation due to the variation of the output power of the micro-CHP [20]. Moreover, ramping constraints need to be applied in energy scheduling problems since the CHP requires some time to reach the steady-state after its set-point changes [20].

5.2.4 Fuel Cells

A fuel cell can produce electricity by converting the chemical energy originating from hydrogen and oxygen into electricity. Fuel cells can be also located at customer levels and utilized as DGs. In the solid-oxide fuel cell, anode supplies hydrogen and catalytically splits it into a number of protons and electrons. The electrons are then flowing towards the positive side, i.e. the cathode by flowing through the external circuit. The oxygen then reacts with the protons and also the electrons flowing in the circuit, forming water formula [21]. The solid-oxide fuel cell can operate in parallel with PV panels, meaning that it can be integrated with solar power. In the nighttime, when PV panels cannot produce electricity, the fuel cell can be deployed to supply the demand.

5.3 CHALLENGES IN DG IMPLEMENTATION

The challenges associated with DGs can be different according to the type of DG, the amount of intermittent power injected from renewable-based DGs, the type of distribution network, as well as the location of DGs. However, DGs give birth to some new technical challenges in the power system. DGs are mainly installed in the vicinity of residential loads. It results in the bidirectional flow of power in distribution networks.

The connection of DGs to the distribution network exerts significant impacts on voltage profiles and also on the network power flow. These effects can be positive or negative. The positive effects include improving the reliability of supply and reducing losses of power system by bringing the generation closer to consumption. The negative impact is increasing the voltage magnitude at nodes with DGs which may violate the maximum permissible value in the moments with high generation. Accordingly, the voltage control is the most serious challenge, and voltage regulation of the distribution network needs more advanced strategy [22]. Moreover, the connection of DGs to the distribution networks exacerbates the challenges related to the traditional Volt-Var control equipment. Traditional and expensive Volt-Var control actions are significantly delayed in order to react to the fast fluctuations resulted from the output power of renewable-based DGs [23]. Besides, the voltage regulation and control devices in the traditional distribution networks are mostly designed to operate without DGs. In this light, the network voltage magnitudes are assumed to decrease along the distribution feeder starting from the substation to the customers. However, with the presence of DGs, the mentioned assumption is no longer valid [24].

In addition to regulation problems, a large standalone DG (like wind turbine) can result in power quality issues, especially in a weak and rural distribution network during the time in which the DG is switched on and off [22]. According to [25], integration of DGs can affect power quality. Voltage dips are a significant event that can occur due to failures of the DG.

In terms of protection, the connection of a large number of DGs on the distribution network's feeders has significant influence on the applicable protection practices. DGs can have a major contribution to the short-circuit currents. This issue may, for example, result in unexpected operation of fault indicators which are deployed to locate fault position. Besides, some additional factors should be considered when installing a number of DGs in distribution networks [26]. These factors include the protection of the installed DGs from internal faults, the protection issues associated with the faulted distribution network from the fault currents produced by a DG, and the issues related to islanding detection (anti-islanding/loss-of-mains) as the high DG installation may increase islanding in distribution networks. Also, management of these DGs in the distribution systems needs to be efficiently controlled. Hence, integrating vast amounts of DGs raises multiple challenges related to the distribution system's control, operation and protection. Such challenges are mitigated by innovative distribution grid management architectures reinforcing the need for rapidly controllable flexible energy sources. The following section shall introduce such control architectures in detail.

5.4 MANAGEMENT-RELATED SOLUTIONS

There are several studies implying that active management of distribution networks and also active management of DGs can help to increase the hosting capacity and accommodate higher number of DGs connected to the current networks [27]. Figure 5.3 presents an overview of the management-related solutions.

Future ANM methods can enable active control and facilitate the deployment of DGs' available flexibilities during both operation modes including grid-connected and islanded [28]. However, previous studies state that the benefits brought by

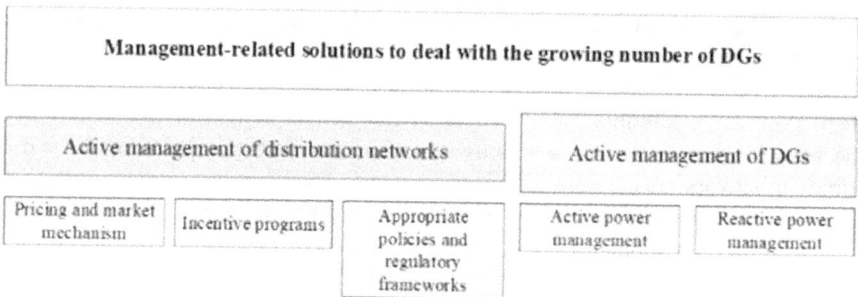

FIGURE 5.3 An overview on the management-related solutions of the growing number of DGs in distribution networks.

utilizing DGs are expected to exceed the cost of the active management implementation [26], there exist some uncertainties related to the cost of active management in distribution networks.

From the economic side, the appropriate price control mechanism is needed in order to recover the expenses of applying ANM. Additionally, developing active distribution networks requires new commercial arrangements. In this way, different incentive schemes and market mechanism should be employed to effectively integrate DGs and motivate the owners of these resources to actively participate in different programs. On the other hand, the lack of policies and well-defined regulatory frameworks in the traditional design of distribution networks limits the high utilization of DGs. Hence, supporting DG integration requires appropriate policies applied to the distribution networks.

As previously stated, the growing number of DGs in distribution networks can have both negative and positive impacts on the power system. The active management of DGs not only is able to decrease the negative impacts but can also enhance the flexibility of the network. In light of this, the following subsections deal with the management of active power and reactive power of DGs seeking to help the system operators and enhance the flexibility of the networks.

5.4.1 DG Inverters' Reactive Power Management

Inverter-based DGs can be regarded as excellent alternatives to resolve the issue related to the rapid response and control of the voltage variation resulted from DGs. Power flowing in feeders is restricted by the line branch's thermal capacity and also by bus voltages along the feeder. When the flow of one of the branches reaches its maximum thermal capacity or one of the bus voltages of the feeder approaches the upper or lower limits, no further power is able to flow the feeder. In this way, DG inverters can assist in increasing the feeder transfer capability by absorbing/injecting the specified amount of reactive power [29].

Inverter-based DGs are equipped with power electronic devices so as to procure the required reactive power in less than 50 ms. This will avoid fast voltage fluctuations stemming from the transient cloud passing [30]. With the help of this feature, inverter-based DGs can be independent on the control actions of traditional distribution system such as deploying capacitor banks, static Var compensators as well as on-load tap changers. Furthermore, DG inverters produce a fast response and also provide more flexible reactive power support. Note that, unlike shunt capacitors, the inverters are able to both absorb and inject reactive power to assist in controlling voltage.

In order to highly exploit the flexible capacities of DGs for operating the distribution networks, the reactive power capacity of the DG inverter should be highly utilized. For example, a PV inverter can operate to its full capacity during daytime when there exists active power injection. However, both reactive power absorption and injection can be used during evening time when there exists no PV power.

The reactive power of DG's inverters is generally limited by the nominal value of the active power output of the DGs. However, the capacity of inverters may be over-sized to provide surplus reactive power support as well as maintaining the fully active power capability. The maximum reactive power should satisfy the following constraint [31].

$$|Q^{DG}| \le \sqrt{\left((1+OF)S^{DG,r}\right)^2 - \left(P^{DG,r}\right)^2} \tag{5.7}$$

In Eq. (5.7), *OF* is the over-sizing factor of DG inverter (per unit) in comparison with the normal values of DG units [31].

5.4.2 DGs Active Power Management

The appropriate management of DGs' active power can be seen as potent DSO-level flexible resources, reducing the need for DSO operational actions which include grid reinforcement and reconfiguration [32]. DGs may be curtailed in a dynamic or static way [33]. The curtailment of DGs is performed in order to increase system flexibility and provide the system with downward flexibility. When the system needs downward flexibility, it has surplus production which violates the balance constraint of the power system. Hence, it should curtail.

In the static curtailment, the system operator imposes a predefined threshold related to the injection of active power produced by renewable-based DGs, whereas in dynamic curtailment, the injection of active power is under the full control and may be curtailed due to the network constraints.

Curtailment resulted from network constraints can be performed either voluntary or involuntary [34]. In voluntary cases, an ex-ante agreement was reached between the DG owner and the network operator which specifies the amount of curtailment as well as the possible compensation. It should be noted that the DG owners are required to sign the contract voluntarily. The DG owners may also participate in flexibility markets. However, the existing flexibility markets for balancing purposes are designed for large-scale flexibility products. Hence, the DG owners should firstly be aggregated and then participate in selling downward flexibility services. Involuntary curtailment due to network constraints is resulted from an obligation for the network operators including DSOs and TSOs. However, this kind of curtailment may decrease future investment in renewable-based energy resources.

The interactions between DG owners and the DSO have been covered by some research such as IMPROGRES project [34]. This project states that the location of DGs highly depends on regulation and policies designed by the DSO. In order to find the optimal location of DGs, the appropriate cost mechanism should be determined for the DSO. For example, the DG owners may encounter several curtailments providing that they invest in the locations with high network reinforcement costs. However, curtailment with appropriate compensation can avoid overinvestment in the grid and also motivate renewable DG investors to find the best location in which the reinforcement costs of the grid are the lowest. This may lead to increasing the capability of the grid to host a large amount of renewable-based DG capacities.

5.4.3 Distribution Network Management with DGs

The increasing number of DGs in distribution networks leads to power systems restructuring the existing management and control architectures. Power systems have been traditionally managed in a centralized way. They consist of generation, transmission, and distribution levels. In a generation level, generators are centrally dispatched through the centralized pool-based markets. The transmission system is then responsible for transmitting the electrical high-voltage (HV) level power to lower-level systems. Transmission networks are also centrally managed by TSOs. The distribution system delivers electrical power to final customers and end-users. These networks are controlled and operated by DSOs. However, the dramatic growth of DG in distribution networks creates considerable challenges [35]. The traditional centralized architecture fails to exploit potent flexibility capacities of new distribution network located resources such as DGs since in the centralized paradigm, these resources are not able to actively participate in energy and flexibility provision. Accordingly, the power system needs to adopt new control and management methods in order to adapt to the changes associated with the growing number of DGs in distribution networks.

5.4.4 Hierarchical Architecture for Distribution Network

The deployment of hierarchical management aims to facilitate the integration of DGs at different levels of the power system. This architecture contains different levels of management. In each level, the related elements are controlled and managed through the external signals receiving from the upstream layers. In other words, the controller in each layer has a certain degree of autonomy which should set its functionalities in accordance with the upstream layers.

Regarding the LV level of distribution networks, at the first hierarchical level, the microgrid is managed and controlled by its control centre named a microgrid control centre (MGCC). The centre is located on the LV side of the MV/LV secondary substation and has some operational functionalities associated with the management and control of the DGs in the MG. The MGCC acts as an interface between the DSO and the MG. At a second hierarchical level, each DG unit is managed and controlled locally through a micro source controller (MC). Loads can be also controlled locally through a load controller (LC) [35]. Figure 5.4 describes this kind of control and management structure.

At MV levels of the system, the MGs, the DG units located at MV levels, and MV loads are taken as active cells. The cells are given a certain degree of autonomy. In this regard, the decision making of each cell follows a hierarchical structure. In other words, a central controller is in charge of collecting data from different control units as well as establishing rules for the downstream control units [35]. With regard to the hierarchical structure, consider an aggregator who is responsible for aggregating DGs in the distribution system. In this way, the aggregator negotiates a contract with the owner in order to control their DGs. The aggregator directly controls DGs by sending its control signal to the DGs. The aggregator, itself, receives the control signals from the upper entity, which can be the DSO. This management method can be regarded as a hierarchical management and control method.

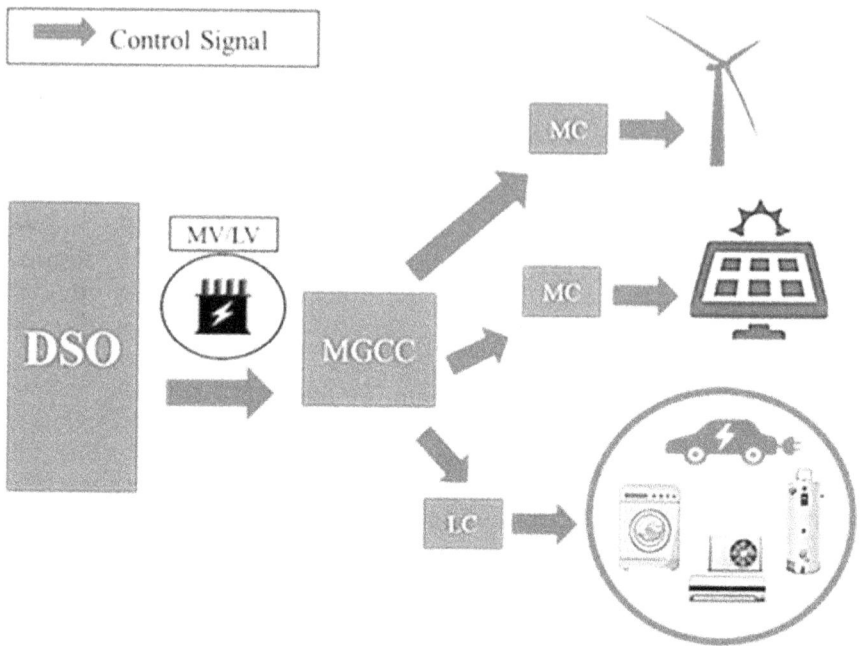

FIGURE 5.4   Hierarchical control of DGs located in a microgrid.

5.4.5   Decentralized Architecture for Distribution Network

Unlike a hierarchical approach, each controller in a decentralized management method has full control over its elements. The recent advances in artificial intelligence enable each controller to act as an independent agent and form the multi-agent system. In this regard, various agents in the distribution grid can communicate with each other aiming to optimize the global objective function. However, each agent has its own objective function. In the decentralized approach, different grid components can be counted as independent agents. For example, flexible loads, electric vehicles, switchers, on-load tap changing and DGs can have their own objective and autonomous management [35].

For example, consider a DG owner acting as an autonomous agent. It decides for controlling the DG autonomously with the target to maximize its profits by participating in the local market. It controls its resources using an energy management system (EMS). On the other hand, the local market operator (LMO), which can be the DSO or receives signals from the DSO, is responsible for finding the optimal operating points for players based on their offers and bids.

The aim of the local market operator can be maximizing the social welfare of all of the players with respect to the network constraints imposed by the DSO. Therefore, the DG owners have their own autonomous objectives while they follow a community-based objective by participating in the local market. Figure 5.5 illustrates the decentralized management method of DGs in the local market environment. The DGs may also participate in flexibility local markets by providing active power and reactive power support for the grid.

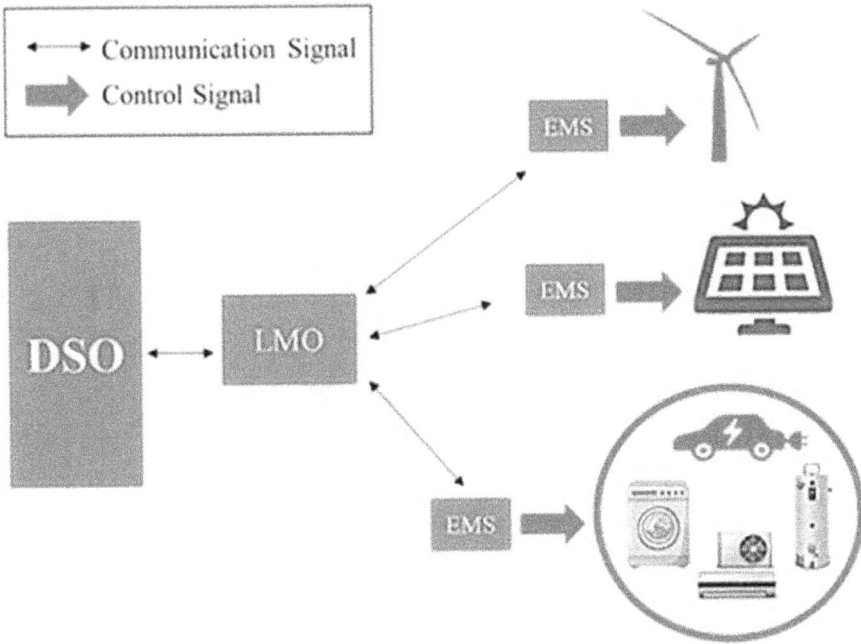

FIGURE 5.5   Decentralized control and management of DGs in a local market.

An overview of various control architectures provides a clear view on utilization of FESs to mitigate various challenges posed by the integration of DGs in the distribution system. The following section provides a detailed account on utilizing ESSs as FESs in the distribution system, with focus on their classification and range of applications they are capable of tending in smart grid applications.

5.4.6   FUTURE DISTRIBUTION NETWORK MANAGEMENT ARCHITECTURE – POTENTIAL GENERAL APPROACH

In the future distribution grid zones with flexibilities (Figure 5.6), i.e. FlexZones, could be seen as building-blocks of a smart, flexible and resilient distribution grid as described in Figure 5.6. FlexZone could be also called as an active cell, zone with DER or local energy community. Also, for example, one utility grid connected MV or LV microgrid could create one FlexZone (Figure 5.6a). FlexZone approach could also create a basis to implement new business and market models for flexibilities (flexibility service markets). As illustrated in Figure 5.6, there can be different levels of FlexZones, and higher level FlexZone can consist of multiple lower level FlexZones like grid-connected nested microgrids (i.e. one/multiple smaller LV microgrids inside larger MV microgrid).

In the future different local (DSO) ANM functionalities could be realized by (de)centralized, hierarchical and coordinated management solutions at HV/MV, MV/LV substations with dedicated management units, i.e. FlexZone Units (FZUs),

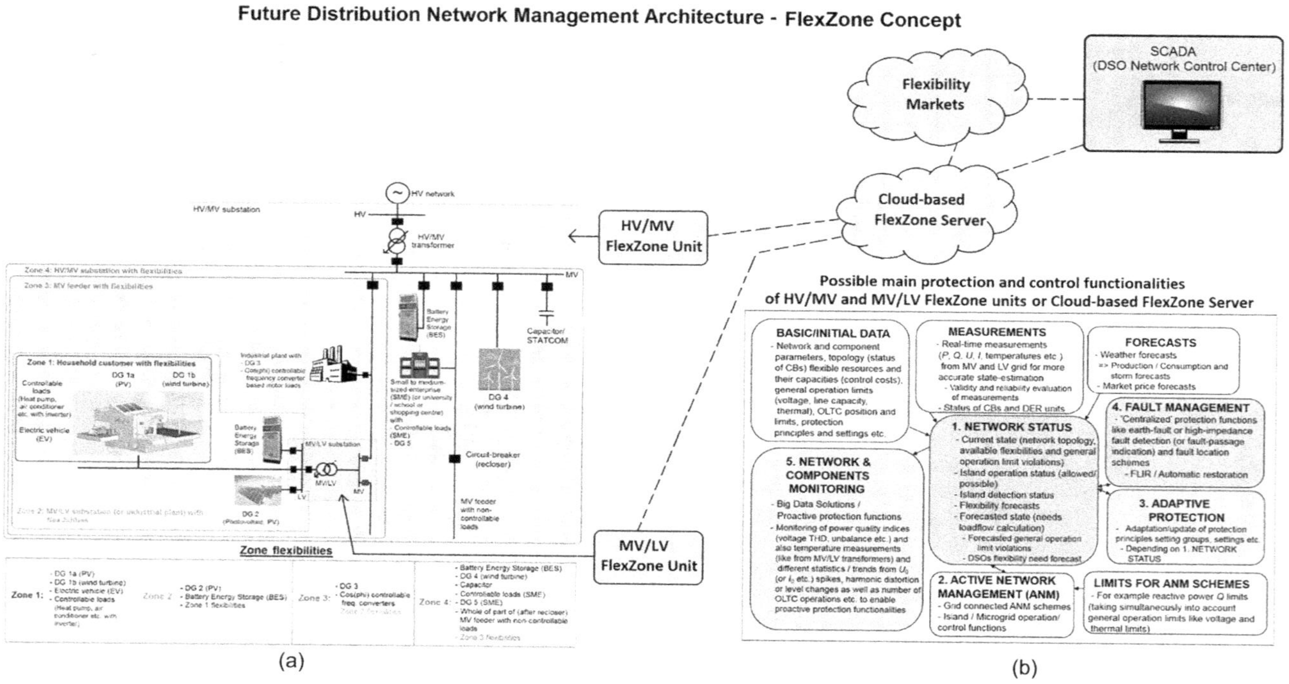

FIGURE 5.6 Potential future distribution network management architecture (FlexZone Concept) and (a) Different levels of FlexZones with possible flexibilities, (b) Possible main protection and control functionalities of HV/MV and MV/LV management units.

because it is more feasible to distribute also the needed processing and calculation capacity closer to the actual measurement points and controlled flexibilities. With this kind of distributed data processing approach, it is possible to avoid too extensive 'raw data' transfer and reduce the risk related to loss of one central management unit or communication. However, still fast, secure and reliable communication between different devices will play an essential role in future smart grids with flexibilities to enable the needed management and protection functionalities. HV/MV and MV/LV FZUs could include, in addition to different ANM functionalities, also other functionalities like protection/fault indication & fault location, islanding detection & logic, status monitoring, predictive protection, available flexibilities, flexibility forecasts and historian from flexibilities control/use (Figure 5.6b).

ANM may simultaneously have an effect on protection settings if, for example, the network topology is changed. On the other hand, active management of flexibilities could be used to enable correct and reliable operation of certain islanding detection methods or, for example, due to earth-fault, the grid topology may be changed, and it may have an effect on ANM functionalities such as voltage control or losses minimization. Therefore, dependencies between ANM and protection functionalities require careful planning and development to enable creation of future-proof solutions for the Smart Grids.

In the future, one alternative could also be that some of the less critical/high-speed communication-dependent DSO FZU functionalities (Figure 5.6) like, for example, monitoring or predictive protection-related big data solutions, flexibility forecasts, some ANM schemes, etc. would be alternatively located in cloud servers. This approach could enable more flexible and scalable solutions when only most communication and time-critical protection and islanding detection applications would remain at actual HV/MV or MV/LV FZUs. Communication reliability and cyber security will play a more and more important role in the future grid protection and management solutions and, for example, potential short data packet loss should not cause false operation of FZU functionalities (Figure 5.6). In possible cloud server–based applications the role of redundant back-up schemes, like hot-standby or hot-hot schemes, becomes also significant.

5.5 ESSs

Integration of RES has been progressing at a faster pace at all the voltage levels in the power systems, particularly in the LV and MV distribution grids. RESs are typically intermittent and low inertia generation sources leading to large voltage and frequency instabilities in the distribution systems compared to traditional centralized power systems. Flexible energy sources are capable of providing stability in the modern power grids with higher RES penetration. ESSs play key roles with their ability to provide multiple flexibility services in smart grids spread over different time ranges. In this section, various ESS technologies capable of acting as flexible energy sources will be explained along with their application ranges for smart grid applications.

5.5.1 ESS Technologies for Stationary Grid Applications

Energy storage technologies for stationary grid applications are primarily classified based on the nature of energy conversion. From the literature [36–38], a brief outline of the classification of energy storage types is defined as below. Also, Figure 5.7 provides details on the technologies applicable for grid scale applications.

1. **Mechanical Energy Storage Systems:** Store and convert electrical energy in various forms of kinetic and/or potential energy
2. **Electrical Energy Storage Systems:** Store the electrical energy by transforming electrical or magnetic fields with the aid of capacitors and superconducting magnets
3. **Electrochemical Energy Storage Systems:** Chemical energy of active materials is converted into electrical energy during the discharge phases and vice versa while charging. Simultaneous Redox reactions are responsible for the energy conversion
4. **Chemical Energy Storage Systems:** These systems store energy in the chemical bonds of atoms and molecules and released by electron transfer to generate electrical energy
5. **Thermal energy storage systems:** Store energy in the form of heat or ice, and converted to electrical energy when required

5.5.2 Application of ESSs Smart Grids

ESSs play a key role in smart grids by bridging gaps in power generation and demand, especially in the modern power grids where higher amount of RES are integrated in the MV and LV distribution systems. ESSs are capable of tending multiple applications and services in the distribution systems and are mainly classified based on the duration of their usage. Based on the available literature [39–42], three major categories of applications are observed based on their local and system-wide requirements. Table 5.1 provides details on their classification based on their points of usage, i.e. generation, transmission & distribution and end users.

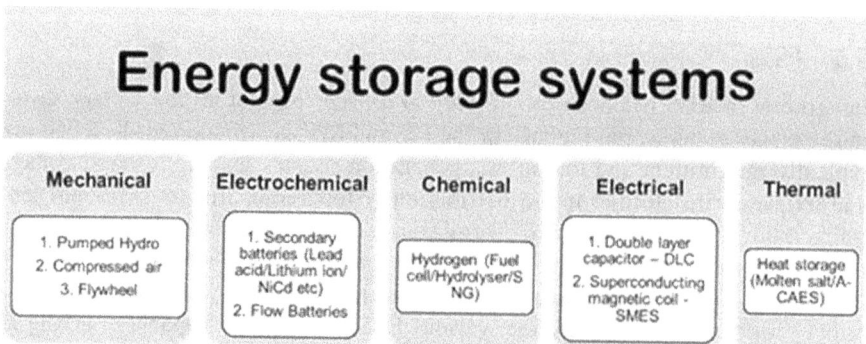

FIGURE 5.7   Classification of electrical ESSs.

TABLE 5.1
Classification of ESS Applications

Generation	Transmission and Distribution	End Users
• Energy arbitrage	• Update deferral • Reduce circuit and line overload	Reduce demand charges

• Ancillary services
  • Frequency regulation
  • Spinning reserves
  • Supplemental reserves
  • Ramping
	• Grid resiliency
	  • Outage migration
	  • Backup power
		• Optimize retail rates

• Capacity
  • Peak energy
  • Flexibility
	• Voltage support/power quality	• Power quality/UPS

• Reliability
  • Voltage support/reactive power
  • Black start
  • Frequency response
	• Congestion relief	• Onsite renewables

Based on the duration of energy dispatch, they are classified into three major categories [43],

1. **Reserve and Response Ancillary Services:** Power quality services where dispatch varies between micro-seconds to minutes
   a. Supply interruptions
   b. Voltage sags or dips
   c. Voltage swell
   d. Harmonic distortion

2. **Transmission and Distribution Grid Support:** Dispatch time typically varies from few seconds to hours supporting ancillary services in T & D grids to operate as specified by grid codes.
   a. Grid frequency support
   b. Voltage control
   c. Spinning reserve
   d. Congestion relief

3. **Energy Management:** Application where duration of dispatch varies from several hours to days
   a. Energy arbitrage
   b. Load levelling
   c. Peak shaving
   d. Black start
   e. Non-spinning reserve

Figure 5.8 provides a pictorial representation of the power ratings and the overall discharge duration required from energy storage technologies for various applications. It defines the requirements or specifications to select various energy storage technologies for particular applications in the smart grid. Based on the requirements defined in Figure 5.8, suitable ESS technologies that can be utilized for various grid applications are depicted in Figure 5.9. The power/energy requirements for a particular application must match with the characteristics of the ESS technology to be deployed. It is also important to have the detailed load curves, while conduction feasibility analysis of ESS technologies considering the fact that the energy & power densities, cost and life-cycle characteristics vary from each ESSs. Based on mix and match of various technologies, it is possible to develop hybrid ESS solutions to cater to particular load requirements in the smart grid applications.

From Figure 5.9 it is evident that the Lithium-ion (Li-ion)-based BESSs are capable of tending applications in all the time ranges, i.e. reserve and response applications, T & D grid support and energy management. Also, their superior energy and power densities, shelf and cycle life and low self-discharge make them an ideal candidate for flexibility applications in smart grids. Hence, Li-ion batteries are modelled to cater to various power system applications by the authors of this chapter. Based on [44], integration of lithium-ion batteries in the MV distribution system by means of power converters shall be explained, which is followed by a case study, to verify the operations of the developed converter controls for battery integration.

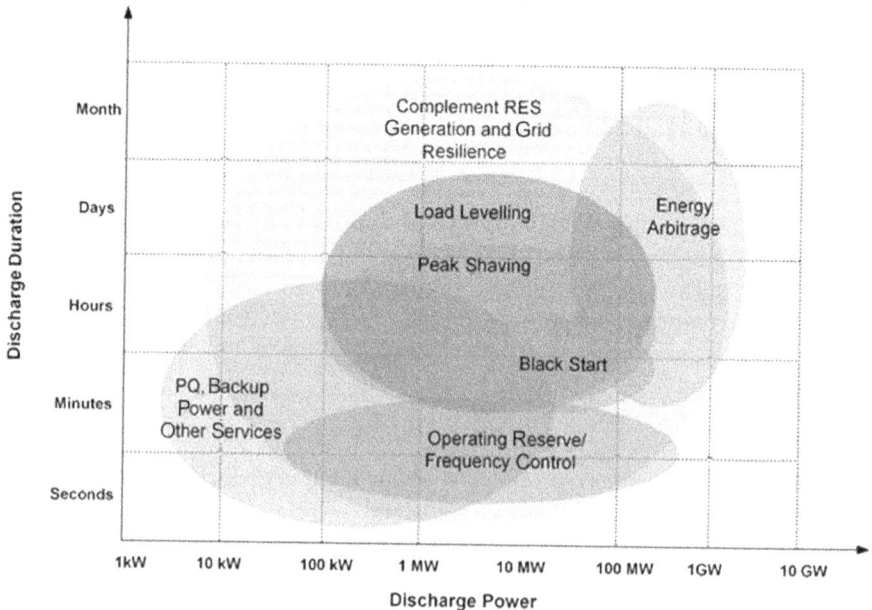

FIGURE 5.8 ESS requirements for grid applications.

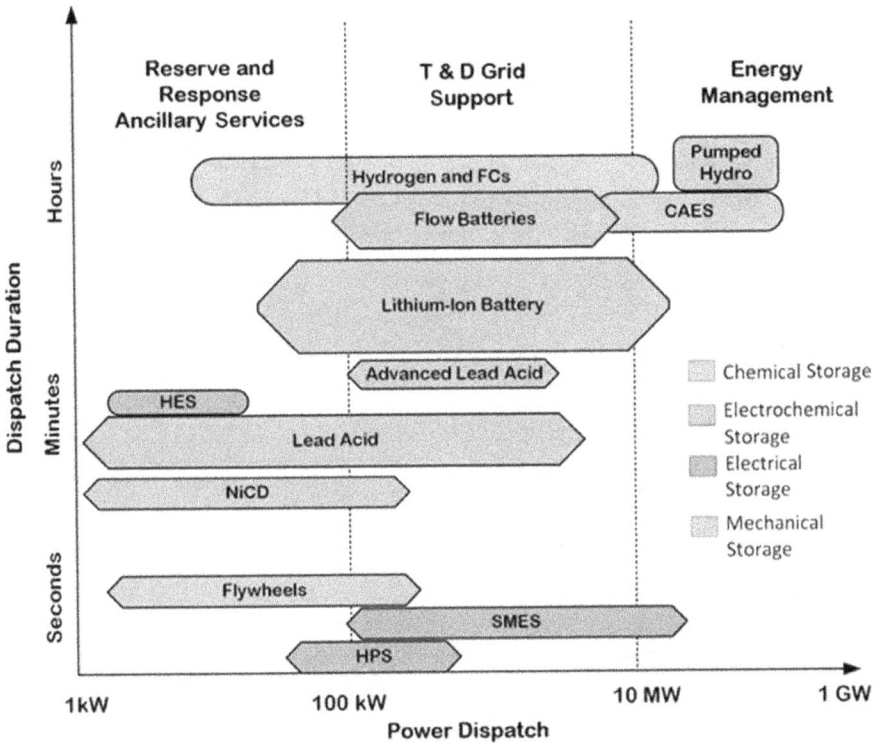

FIGURE 5.9 ESS capability for land-based applications.

5.6 INTEGRATION OF BATTERY ESSs

BESSs are typically inverter-based FESs capable of providing both active and reactive power management. Integration of these BESSs follows the same methodology despite being used in centralized or decentralized control architectures. Hence, in this section, design and integration methodology of Lithium-ion BESSs to the MV distribution system is presented.

5.6.1 POWER ELECTRONIC CONVERTERS

Power electronic (PE) converters provide the vital technology for integration of BESS to the power grids. Maintaining various grid code requirements in the modern distribution systems is satisfied by the PE converters, as most RESs are connected to the grid through PE. Simultaneously PE interface controls BESS in active (P) and reactive (Q) power flow modes, keeping in view the current and voltage variations across the battery pack which affects BESS health, performance and lifetime [45]. This section briefly reviews the available and widely used PE topologies for large-scale BESSs discussing the main parts of the system in detail. The used configuration and its modelling aspects to integrate BESS to the studied real-life smart grid pilot will be addressed in the next section.

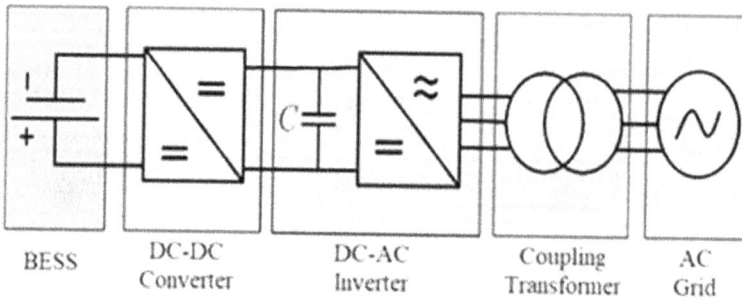

FIGURE 5.10 Typical PE units for BESS grid integration.

Figure 5.10 shows the elements of PE units for a typical BESS grid integration which including three main parts:

- DC-DC converter
- DC-AC inverter
- Coupling transformer

In order to design an optimized BESS system, selection and design of each part need to be investigated in detail. Different parameters affect selection and design of PE units such as power ratio, energy density, response time (speed of system) and grid interconnection requirements.

5.6.1.1 DC-DC Converter

The presence of DC-DC converter increases system performance and flexibility, but it also adds losses and costs. Therefore, it is worth considering to connect BESS directly to DC-AC inverter DC bus. BESSs with and without DC-DC converter have been used in recent applications.

BESS without DC-DC Converter

PE system design without DC-DC conversion stage could be a cost-effective solution in one sense, but it can add complexity to the DC-AC inverter system. Some issues can be linked to this solution. The BESS voltage varies depending on State of Charge (SoC), and this introduces DC bus voltage variation which will affect the DC-AC inverter operation performance. Also, safety problems can be associated with this solution since over-voltages or over-currents can happen at BESS side, and their management is not easy by only grid side DC-AC inverter. Another possible drawback could be low-order harmonic injection to the battery, and this also can affect BESS health and lifetime [46].

BESS with DC-DC Converter

Adding a DC-DC conversion stage eliminates above-mentioned low-order current harmonics flowing in the battery since the DC-DC converter isolates the BESS from inverter DC bus. With DC-DC converter, BESS can be designed to lower voltage level (less cells in series). Moreover, it can be used as protective and current limiting device which will increase the safety and controllability of the system.

Simple bidirectional Buck-Boost DC-DC converter is widely used due to its simplicity when a lower voltage level is used for BESS side and higher voltage level at the DC bus of the grid side DC-AC inverter. However, Buck-Boost converters have their own intrinsic limits for levelling up/down the voltage [47]. This issue can be dealt with using a full-bridge bidirectional DC-DC converter which also increases the costs and losses (due to increased number of the switching modules).

For high power applications, advanced solutions such as isolated DC-DC converters [48] and isolated dual active bridge (DAB) converters have been proposed [49] which include high frequency isolation transformer (20 kHz). This transformer can eliminate the need for coupling transformer, and since it is working in high frequency, it can decrease the size and weight of the system considerably. These are interesting solutions, but the technology is in research stage, and at the moment no commercial solutions are available.

5.6.1.2 DC-AC Inverter

DC-AC inverter is a necessary part of the PE system in the BESS grid integration applications. The main task of it is to convert the DC voltage to the AC voltage and guarantee fulfilment of grid code requirements [50]. Two-level inverters are widely used in different applications, including also BESS integration (with centralized BESS at DC bus), due to the maturity of the technology and availability of the commercial solutions. Recently three-level, five-level and, more general, multi-level inverters have also been developed and used in a wide range of applications.

More specifically, cascaded H-bridge (CHB) converters are used for BESS integration where the BESS can be equally distributed among sub-modules (SMs) in the form of smaller battery packs. However, advanced control algorithms are required in order to ensure balanced SoC for all battery packs.

Modular Multilevel Converter (MMC) is another developing solution in which both centralized BESS at DC bus and distributed BESSs among SMs have been proposed [51]. However, distributed approach can be preferred where the benefits of cascaded structure can be better utilized. With centralized BESS at DC bus, some positive features of the MMC structure will be lost [52,53].

Apart from technology readiness, using CHC and MMC solutions are linked with a few issues [45]:

- Need for an extra battery management system (BMS) to ensure balanced SoC among the battery modules
- One of the main advantages using CHB or MMC solutions is to eliminate the need for coupling transformer (otherwise these solutions cannot be economically justified), but in BESS application unbalanced SoC can cause DC current injection to the grid (this should be limited to 0.5% according to [50])
- Complexity of the control and system cost.

5.6.1.3 Coupling Transformer

Conventional grid connected PE for BESS system consists of a coupling transformer. Transformer-based solution is favoured so far [45], and for large-scale BESS system it could be the preferred solution providing galvanic isolation between grid and ESS. However, several transformer-less solutions have been investigated as well. The advantages of using a coupling transformer are the following:

- DC-AC inverter can work in lower voltage and the transformer can match the voltage level up to kV level
- Inverter AC side passive filter size can be decreased
- It can protect PE devices against grid side faults till certain level
- Losses of the device can be compromised with high-efficiency transformers

However, transformers are bulky and heavy units, and coupling transformer can increase the weight and size of final installation unit. In MV level and working with few MWs, the size and weight of coupling transformer itself (in 50 Hz) can be comparatively close or even higher than the rest of system. Isolated DAB converters with high-frequency transformers could also be an attractive solution, but as mentioned earlier, still DAB commercial solutions do not exist and are in research stage. In addition, it is worth mentioning that correct transformer energization principles should be considered in order to ensure safe and feasible operation conditions.

5.6.2 Grid Code Requirements

IEEE standard considering BESS systems for stationary applications is still in the drafting stage, but IEEE standard 1547-2018 [50] and ENTSO-E RfG requirements [54] can be used as a good reference for BESS system PE design from grid code requirements point of view. The standard defines general requirements for power quality (harmonics, reactive power, voltage levels and sag/swells) and DER response to power systems disturbances (such as faults, open phase conditions, voltages and frequency deviation and also islanding situations), i.e. FRT requirements. It can be used as reference to design PE unit, but its focus is mostly on grid side and talks about general requirements, and in most cases, it leaves the details to the transmission and/or distribution systems operators (TSOs and DSOs) [50].

5.6.3 BESS Integration Design

This section provides an overview on the design and development of power electronics converters for BESS integration to the MV distribution system. Control design and development of voltage source converter (VSC) and DC/DC bidirectional buck-boost converter are explored in detail.

5.6.3.1 BESS Integration Methodology

Li-ion BESSs are best suited for multiple energy storage applications in smart grids, which is evident from their characteristics explained earlier. Hence, they were modelled as flexible energy sources to meet short/medium time energy storage demands

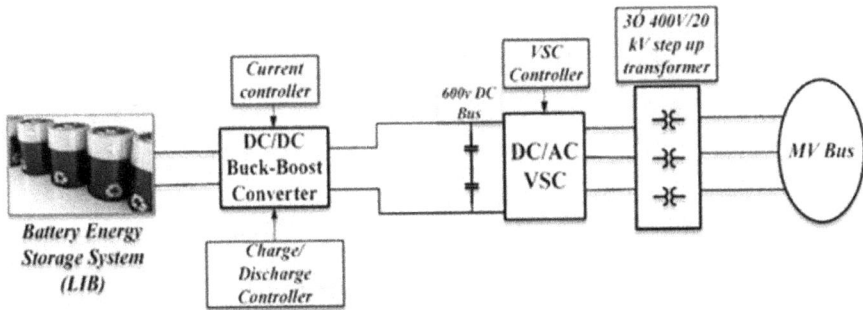

FIGURE 5.11   BESS grid integration methodology.

in smart grids. Figure 5.11 explains to complete BESS integration methodology in the MV distribution system. Li-ion BESSs are integrated into the MV bus by means of power electronics interfaces, i.e. through DC/DC bidirectional buck-boost converter whose low voltage side is connected to the BESS and the high voltage side is connected to the 600 V DC bus. DC/AC VSC converts 600 V_{DC} to 3-phase, 400 V_{RMS}. Coupling transformer converts VSC output to the MV bus voltage which is 21 kV in this case. Equivalent circuit model of Li-ion BESS [44] is utilized in the method. Power electronics converters controller designs are explained in the following section.

5.6.3.2   VSC Controller

VSC controller design is based on voltage-oriented control (VOC) technique [55] which is shown in Figure 5.12. Primary advantage of the VOC techniques lies in its and high static performance and fast transient response through its current control loop. Three-phase voltages and currents (i.e. I, abc and $Vabc$) are transformed into $dq0$ frames by means of Park's transformation [56] to I, dq and V, dq frames. Id, ref is provided by the PI- controller for V_{dc} management, and the Iq, ref is dictated by the reactive power demand for the MV grid application. Iq, ref and Id, ref control the active power (P) and reactive power (Q) outputs respectively. However, the controller's dynamic performance shall be affected by cross coupling between d- and q-axes components. Hence, V_d feed forward signal is provided to the d-component control loop. Similar feed forward signal shall be added to the q-component control loop; however, it is not needed in this application. VSC was designed to provide 1.5 times the nominal power to accommodate transient requirements from the grid.

5.6.3.3   Battery Charge and Discharge Controller

Bidirectional Buck-Boost converter acts as the battery charger (buck mode) while charging the batteries and the DC supply converter (boost mode) during discharging scenarios. In this case, it is developed as an average model system with single IGBT and its accompanying anti-parallel diode with a switching inductance [57]. Current control through the converter is provided by a simple PI-controller as shown in Figure 5.13. DC bus voltage is regulated and controlled by the VSC's d-component control loop, so the positive or negative sign carried by the PI- controller output defines the P-flow direction. Hence, an additional DC bus voltage control loop is

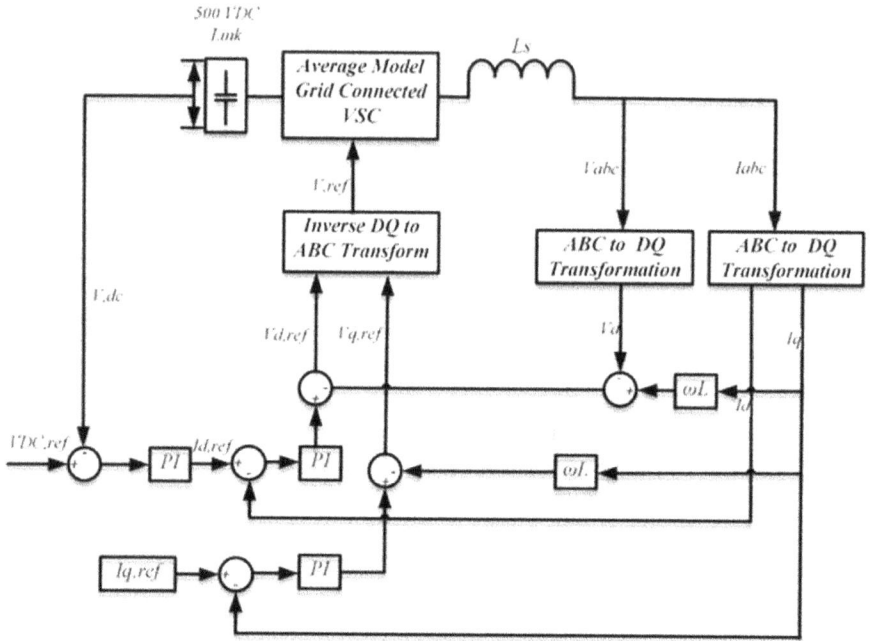

FIGURE 5.12   VSC controller.

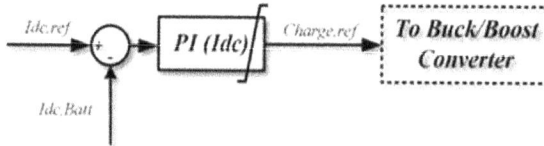

FIGURE 5.13   Battery charge discharge controller.

not necessary in this case in the battery discharging scenarios. During both control modes, i.e. charging or discharging modes, the PI-controller output delivers the duty cycle to the Buck or Boost converter.

To keep the Li-ion BESS in the safe operational threshold, BESS Discharge mode is set in the range between 20% and 90% of the battery SoC. While, charging the batteries, the maximum SoC is set at 90% which is provided by the constant current charging technique alone, thereby eliminating constant voltage charging sequence. Adding SoC of the BESS control loops, the battery charging and discharging currents are always maintained within the specifications provided by the manufacturers.

5.6.4   Case Study and Simulation Results

Integration of Li-ion BESSs to the MV distribution system is validated by means of simulation based on a case study. Grid side controllers provided the P_{REF} reference to the DC/DC converter stage which is associated with the battery operations. The nature of the designed use case was to validate the stability of VSC and the DC/DC-bidirectional buck boost converter during transient and steady-state system behaviour.

Table 5.2 presents the characteristics of the Li-ion BESS used in the study. It is sized for a nominal discharge (1 C) of 0.335 MW and a peak power discharge (3 C) of 1 MW. The DC/DC converter was rated for 3 C battery discharge, and the VSC is sized at 1.5 MVA. Battery's initial SoC is maintained at 50% which can accommodate both charge and discharge operations as commanded by the grid services.

Total simulation time span was considered at 80 s. The P_{REF} given to the BESS is shown in Table 5.3. It corresponds to the various C-rates of battery discharge and changes every 20 s during the simulation time period, thereby inducing transient instability conditions. In the final time period of simulation, BESS charges at a rate of 0.5 C. Such loads are not usually existent upon the BESSs; however, the simulation cases were developed considering extreme events or changes while supporting RESs such as PV/Wind power generation.

The performance of the Li-ion BESS integration through its PE converters is shown in Figure 5.14. Battery load current characteristics are shown in Figure 5.14a, where its magnitude changes as set by P_{REF}. Figure 5.14b shows Li-ion BESS voltage. Voltage fluctuations in the Li-ion BESS are evident. Such accurate voltage characteristics provide set-points for V_{dc} control in the DC bus. DC power output of the BESS is shown in Figure 5.14c, whose characteristics are defined based on the C-rates from Table 5.2. Such large step changes were chosen to observe the DC bus stability and the controller interactions during transient stability conditions. Li-ion BESS SoC changes are shown in Figure 5.14d. The DC bus voltage is presented in Figure 5.14e where it is constantly maintained at 600 V, despite changes in Li-ion BESS voltage

TABLE 5.2
Li-Ion BESS Specifications

Nominal voltage	312 V
Maximum voltage	354 V
Minimum voltage	236 V
Discharge energy (1 C)	335 kWh
Discharge current (1 C)	945 A

TABLE 5.3
Simulation Case Details

Simulation Time (s)	Active Power Ref. (kW)	Status
T1	670	Discharge
T2	330	Discharge
T3	168	Discharge
T4	−168	Charge

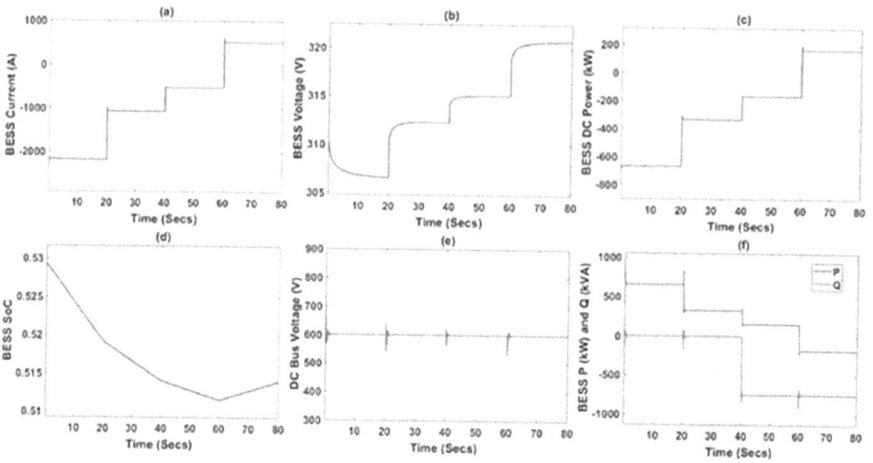

FIGURE 5.14 (a) Battery load current (b) Battery voltage (c) Battery DC power (d) Battery SoC (e) Battery DC bus voltage. (f) Battery *P* and *Q* characteristics.

and current dispatch characteristics. VSC is designed to provide both *P* & *Q* control capabilities, which is demonstrated after 40 s into simulation, where the converter starts to absorb reactive power as shown in Figure 5.14f. Therefore, overall design objectives of the Li-ion BESS integration to the MV distribution system are achieved by controlling active and reactive power flows in the power system, maintaining the safe operations of Li-ion batteries.

5.7 SUMMARY AND CONCLUSION

Integration of distributed generators, i.e. including the RES, has been ever-increasing in the LV and MV distribution systems. Higher penetration of DGs in the power grids comes with its own set of challenges, and multiple innovative ways are needed to mitigate those issues. ANM of the distribution system, by effective management and control of the available flexibilities, plays a crucial role in managing network challenges caused by higher DG penetration. ESSs form a key component in executing ANM schemes by providing both active and reactive power control–related flexibilities. Overall, this book chapter provides an overview on DG types, ANM control methodologies, ESS (fundamentals, classification and applications) and integration design and methodology for battery ESSs in the MV distribution grids.

REFERENCES

1. H. Khajeh, A. A. Foroud, and H. Firoozi, "Optimal participation of a wind power producer in a transmission-constrained electricity market," *2017 25th Iranian Conference on Electrical Engineering ICEE 2017*, no. ICEE20 17, pp. 1230–1235, 2017.
2. O. S. Nduka and B. C. Pal, "Quantitative evaluation of actual loss reduction benefits of a renewable heavy DG distribution network," *IEEE Transactions on Sustainable Energy*, vol. 9, no. 3, pp. 1384–1396, 2018.

3. B. Khan, G. Agnihotri, S. E. Mubeen, and G. Naidu, "A TCSC incorporated power flow model for embedded transmission usage and loss allocation," *AASRI Procedia*, vol. 7, pp. 45–50, 2014.

4. B. Khan, G. Agnihotri, G. Gupta, and P. Rathore, "A power flow tracing based method for transmission usage, loss & reliability margin allocation," *AASRI Procedia*, vol. 7, pp. 94–100, 2014.

5. B. Khan, G. Agnihotri, P. Rathore, A. Mishra, and G. Naidu, "A cooperative game theory approach for usage and reliability margin cost allocation under contingent restructured market," *International Review of Electrical Engineering*, vol 9, no. 4, pp. 854–862, 2014.

6. B. Khan and G. Agnihotri, "A comprehensive review of embedded transmission pricing methods based on power flow tracing techniques," *Chinese Journal of Engineering*, vol. 2013, Article ID 501587, 13 pages, 2013.

7. B. Khan, G. Agnihotri, and A. S. Mishra, "An approach for transmission loss and cost allocation by loss allocation index and co-operative game theory," *Journal of the Institution of Engineers (India): Series B*, vol. 97, pp. 41–46, 2016.

8. B. Khan and P. Singh, "Optimal power flow techniques under characterization of conventional and renewable energy sources: A comprehensive analysis," *Journal of Engineering*, vol. 2017, Article ID 9539506, 16 pages, 2017.

9. P. Singh and B. Khan, "Smart microgrid energy management using a novel artificial shark optimization," *Complexity*, vol. 2017, Article ID 2158926, 22 pages, 2017.

10. T. Molla, B. Khan, B. Moges, H. H. Alhelou, R. Zamani, and P. Siano, "Integrated optimization of smart home appliances with cost-effective energy management system," *CSEE Journal of Power and Energy Systems*, vol. 5, no. 2, pp. 249–258, June 2019.

11. Z. Tang, Y. Lin, M. Vosoogh, N. Parsa, A. Baziar, and B. Khan, "Securing microgrid optimal energy management using deep generative model," *IEEE Access*, vol. 9, pp. 63377–63387, 2021.

12. S. P. Bihari et al., "A comprehensive review of microgrid control mechanism and impact assessment for hybrid renewable energy integration," *IEEE Access*, vol. 9, pp. 88942–88958, 2021.

13. S. Montoya-Bueno, J. I. Muñoz-Hernández, and J. Contreras, "Uncertainty management of renewable distributed generation," *Journal of Cleaner Production*, vol. 138, pp. 103–118, 2016.

14. E. Lorenzo, *Solar Electricity: Engineering of Photovoltaic Systems*. Sevilla: Earthscan/James & James, 1994.

15. D. Çelik and M. E. Meral, "Current control based power management strategy for distributed power generation system," *Control Engineering Practice*, vol. 82, pp. 72–85, 2019.

16. S. Farajdadian and S. M. H. Hosseini, "Optimization of fuzzy-based MPPT controller via metaheuristic techniques for stand-alone PV systems," *International Journal of Hydrogen Energy*, vol. 44, no. 47, pp. 25457–25472, 2019.

17. R. AbdelHady, "Modeling and simulation of a micro grid-connected solar PV system," *Water Science*, vol. 31, no. 1, pp. 1–10, 2017.

18. M. Zaibi, G. Champenois, X. Roboam, J. Belhadj, and B. Sareni, "Smart power management of a hybrid photovoltaic/wind stand-alone system coupling battery storage and hydraulic network," *Mathematics and Computers in Simulation*, vol. 146, pp. 210–228, 2018.

19. F. Kalavani, B. Mohammadi-Ivatloo, and K. Zare, "Optimal stochastic scheduling of cryogenic energy storage with wind power in the presence of a demand response program," *Renewable Energy*, vol. 130, pp. 268–280, 2019.

20. D. Li, X. Xu, D. Yu, M. Dong, and H. Liu, "Rule based coordinated control of domestic combined micro-CHP and energy storage system for optimal daily cost," *Applied Science*, vol. 8, no. 1, p. 8, 2018.

21. W. Bai, M. R. Abedi, and K. Y. Lee, "Distributed generation system control strategies with PV and fuel cell in microgrid operation," *Control Engineering Practice*, vol. 53, pp. 184–193, 2016.
22. N. Mahmud and A. Zahedi, "Review of control strategies for voltage regulation of the smart distribution network with high penetration of renewable distributed generation," *Renewable and Sustainable Energy Reviews*, vol. 64, pp. 582–595, 2016.
23. R. A. Walling, R. Saint, R. C. Dugan, J. Burke, and L. A. Kojovic, "Summary of distributed resources impact on power delivery systems," *IEEE Transactions on Power Delivery*, vol. 23, no. 3, pp. 1636–1644, 2008.
24. S. K. Ibrahim, "Distribution system optimization with integrated distributed generation," in *Theses and Dissertations—Electrical and Computer Engineering, University of Kentucky*, p. 116, 2018.
25. S. Ruiz-Romero, A. Colmenar-Santos, F. Mur-Pérez, and Á. López-Rey, "Integration of distributed generation in the power distribution network: The need for smart grid control systems, communication and equipment for a smart city - Use cases," *Renewable and Sustainable Energy Reviews*, vol. 38, pp. 223–234, 2014.
26. S. N. Afifi, "Impact of hybrid distributed generation allocation on short circuit currents in distribution systems," PhD thesis, Brunel University, March, p. 214, 2017.
27. S. Abapour, S. Nojavan, and M. Abapour, "Multi-objective short-term scheduling of active distribution networks for benefit maximization of DisCos and DG owners considering demand response programs and energy storage system," *Journal of Modern Power Systems and Clean Energy*, vol. 6, no. 1, pp. 95–106, Jan. 2018.
28. H. Laaksonen, K. Sirviö, S. Aflecht, and P. Hovila, "Multi-objective active network management scheme studied in sundom smart grid with MV and LV connected DER units," in *25th International Conference on Electricity Distribution, CIRED*, Madrid, pp. 1–5, 2019.
29. I. Abdelmotteleb, T. Gomez, and J. P. Chaves-Avila, "Benefits of PV inverter volt-var control on distribution network operation," in *2017 IEEE Manchester PowerTech*, Powertech, 2017, pp. 15–20, 2017.
30. M. A. G. De Brito, L. Galotto, L. P. Sampaio, G. De Azevedo Melo, and C. A. Canesin, "Evaluation of the main MPPT techniques for photovoltaic applications," *IEEE Transactions on Industrial Electronics*, vol. 60, no. 3, pp. 1156–1167, 2013.
31. Y. Zhang, Y. Xu, H. Yang, and Z. Y. Dong, "Voltage regulation-oriented co-planning of distributed generation and battery storage in active distribution networks," *International Journal of Electrical Power & Energy Systems*, vol. 105, pp. 79–88, 2019.
32. H. Khajeh, H. Laaksonen, A. S. Gazafroud, and M. Shafie-Khah, "Towards flexibility trading at TSO-DSO-customer levels: A review," *Energies*, vol. 13, no. 1, pp. 1–19, 2019.
33. S. Dalhues et al., "Towards research and practice of flexibility in distribution systems: A review," *CSEE Journal of Power and Energy Systems*, vol. 5, no. 3, pp. 285–294, 2019.
34. H. Klinge Jacobsen and S. T. Schröder, "Curtailment of renewable generation: Economic optimality and incentives," *Energy Policy*, vol. 49, pp. 663–675, 2012.
35. M. A. Matos et al., "Control and management architectures," *Smart Grid Handbook*, pp. 1–24, 2016.
36. M. S. Guney and Y. Tepe, "Classification and assessment of energy storage systems," *Renewable & Sustainable Energy Reviews*, vol. 75, no. November 2016, pp. 1187–1197, 2017.
37. M. C. Argyrou, P. Christodoulides, and S. A. Kalogirou, "Energy storage for electricity generation and related processes: Technologies appraisal and grid scale applications," *Renewable & Sustainable Energy Reviews*, vol. 94, no. July, pp. 804–821, 2018.
38. M. Faisal, M. A. Hannan, P. J. Ker, A. Hussain, M. Bin Mansor, and F. Blaabjerg, "Review of energy storage system technologies in microgrid applications: Issues and challenges," *IEEE Access*, vol. 6, pp. 35143–35164, 2018.

39. M. G. Molina, "Energy storage and power electronics technologies: A strong combination to empower the transformation to the smart grid," *Proceedings of IEEE*, vol. 105, no. 11, pp. 2191–2219, 2017.

40. O. Palizban and K. Kauhaniemi, "Energy storage systems in modern grids—Matrix of technologies and applications," *Journal of Energy Storage*, vol. 6, pp. 248–259, 2016.

41. N. Günter and A. Marinopoulos, "Energy storage for grid services and applications: Classification, market review, metrics, and methodology for evaluation of deployment cases," *Journal of Energy Storage*, vol. 8, pp. 226–234, 2016.

42. European Commission, "Commission Staff Working Document; Energy storage – The role of electricity," Brussels, pp. 1–25, 2017.

43. M. Rahman, A. O. Oni, E. Gemechu, and A. Kumar, "Assessment of energy storage technologies : A review," *Energy Conversion and Management*, vol. 223, no. August, p. 113295, 2020.

44. C. Parthasarathy, H. Hafezi, and H. Laaksonen, "Lithium-ion BESS integration for smart grid applications - ECM modelling approach," in *2020 IEEE Power & Energy Society Innovative Smart* Grid Technologies Conference (ISGT), pp. 1–5, 2020.

45. G. Wang et al., "A review of power electronics for grid connection of utility-scale battery energy storage systems," *IEEE Transactions on Sustainable Energy*, vol. 7, no. 4, pp. 1778–1790, 2016.

46. A. Bessman, R. Soares, O. Wallmark, P. Svens, and G. Lindbergh, "Aging effects of AC harmonics on lithium-ion cells," *Journal of Energy Storage*, vol. 21, no. October 2018, pp. 741–749, 2019.

47. N. Mohan, T. M. Undeland, and W. P. Robbins, *Power Electronics: Converters, Applications, and Design*. Hoboken, NJ: John Wiley & Sons, 2003.

48. N. M. L. Tan, T. Abe, and H. Akagi, "Design and performance of a bidirectional isolated DC-DC converter for a battery energy storage system," *IEEE Transactions on Power Electronics*, vol. 27, no. 3, pp. 1237–1248, 2012.

49. J. Everts, F. Krismer, J. Van Den Keybus, J. Driesen, and J. W. Kolar, "Optimal zvs modulation of single-phase single-stage bidirectional dab ac-dc converters," *IEEE Transactions on Power Electronics*, vol. 29, no. 8, pp. 3954–3970, 2014.

50. "IEEE 1547-2018- IEEE standard for interconnection and interoperability of distributed energy resources with associated electric power systems interfaces." [Online]. Available: https://standards.ieee.org/standard/1547-2018.html. [Accessed: 10-April-2019].

51. M. A. Perez, S. Bernet, J. Rodriguez, S. Kouro, and R. Lizana, "Circuit topologies, modeling, control schemes, and applications of modular multilevel converters," *IEEE Transactions on Power Electronics*, vol. 30, no. 1, pp. 4–17, 2015.

52. L. Maharjan, S. Member, S. Inoue, and H. Akagi, "A transformerless energy storage system.pdf," in *IEEE Transactions on Industry Applications*, vol. 44, no. 5, pp. 1621–1630, September–October, 2008.

53. H. Akagi, "Classification, terminology, and application of the modular multilevel cascade converter (MMCC)," *IEEE Transactions on Power Electronics*, vol. 26, no. 11, pp. 3119–3130, 2011.

54. Entsoe, "Commission Regulation (EU) 2016/631- Establishing a network code on requirements for grid connection of generators," 14 April 2016. Available: https://www.entsoe.eu/network_codes/rfg/, [Accessed on: 11 May 2021].

55. F. Blaabjerg, R. Teodorescu, M. Liserre, and A. V. Timbus, "Overview of control and grid synchronization for distributed power generation systems," *IEEE Transactions on Industrial Electronics*, vol. 53, no. 5, pp. 1398–1409, 2006.

56. "Power electronics handbook | ScienceDirect." [Online]. Available: https://www.science-direct.com/book/9780128114070/power-electronics-handbook. [Accessed: 18-August-2020].

57. G. D'Antona, R. Faranda, H. Hafezi, and M. Bugliesi, "Experiment on bidirectional single phase converter applying model predictive current controller," *Energies*, vol. 9, no. 4, p. 233, 2016.

6 Internet of Things (IoT) in Renewable Energy Utilities towards Enhanced Energy Optimization

Ashok G. Matani

CONTENTS

DOI: 10.1201/9781003278030-6

6.1   INTRODUCTION

Companies in the business of renewable energy have been experiencing substantial global growth over the last couple of years. However, with tremendous scaling comes the challenge of sustaining profits and productivity. Managing these constantly expanding grids requires power supply companies to explore new ways and methods to optimize their capacities across remote locations. The best feasible alternative way companies can drive efficiency is by implementing the applications of Artificial Intelligence (AI) and Internet of Things (IoT) systems [1]. By implementing smart appliances and connected gadgets, companies can use high-tech sensors to amass massive amounts of real-time energy data and transmit it to the power grid – for advanced storage and analysis

- IoT sensors can enable real-time monitoring of power grids while providing decision makers with the opportunity to build data-driven optimization strategies
- They can also provide better transparency in the way the energy is being consumed. That allows people to understand their energy consumption habits and adjust them accordingly to optimize usage

6.2   THE BENEFITS OF IoT MANAGEMENT SYSTEMS

Energy management based on AI and IoT systems has a wide range of benefits for every part of the electric supply chain network, from power generation to the point when consumers pay their electricity bills [2].
   Some of the AI and IoT systems applications are:

- Reduced energy spending
- Minimizes carbon emission
- Fulfils governmental restrictions
- Integrates green energy
- Optimizes asset maintenance
- Automates processes
- Cuts operational expenses
- Gains visibility to energy use
- Identifies malfunctions in time and prevent them
- Effectively combat outages, accidents, blackouts
- Predicts consumption and spending and plan accordingly

6.3 CHALLENGES FOR IoT IN THE ENERGY SECTOR AND HOW TO OVERCOME THEM

Some of the disadvantages and drawbacks needed to take into account when considering the use of IoT technologies in the energy industry are as follows:

1. **Security** is a common threat to all IoT solutions. The APIs that connect devices into a unified network can be used as an entry point for targeted attacks.
2. **Connectivity**: Systems need to always be on, with a minimum delay for data processing and feedback.
3. **Integration** challenges may arise when there is a need to connect a new IoT network to the existing legacy systems that often rely on outdated technologies, thereby modernizing the current infrastructure.

6.4 SMART ENERGY SOLUTIONS IN VARIOUS AREAS

6.4.1 ENERGY SYSTEM MONITORING AND MAINTENANCE

AI and IoT systems applications in the energy industry are used to keep track of a number of system metrics, overall health, performance and efficiency thereby simplifying operations and maintenance. Be it a wind turbine or other critical equipment, it might be hard to identify a problem before the system goes down. Plus, checking for problems manually is an extremely wasteful and laborious process. General Electric is using AI and IoT systems application sensors to monitor the output and productivity of its equipment. By using data from the sensors, General Electric had implemented predictive maintenance models based on its Predix platform. As a result, the company can better forecast their maintenance needs, achieve better operating efficiency and minimize systems downtime. Figure 6.1 shows the applications of industrial IoT.

6.4.2 PROCESS AUTOMATION

AI and IoT systems applications facilitate to build completely autonomous energy plants. Smart sensors are used to monitor the system's performance in real time thereby automatically adjusting its efficiency using machine learning and AI. At the same time, AI and IoT systems applications-enabled drilling equipment can automatically tailor the drill depth and adjust to the external conditions using AI algorithms for optimal performance. As a result, the need to manually tamper with the equipment is minimized.

6.4.3 INCREASED EFFICIENCY

AI and IoT systems applications-enabled system developed by General Electric helps increase a coal power plant efficiency by up to 16% while reducing greenhouse gas emissions by 3%. This is achieved by optimizing fuel combustion and adjusting the process to the specifics of the fuel being burned, i.e. automatically regulating the oxygen flows in the boiler. This can make power plants more efficient and reduce waste.

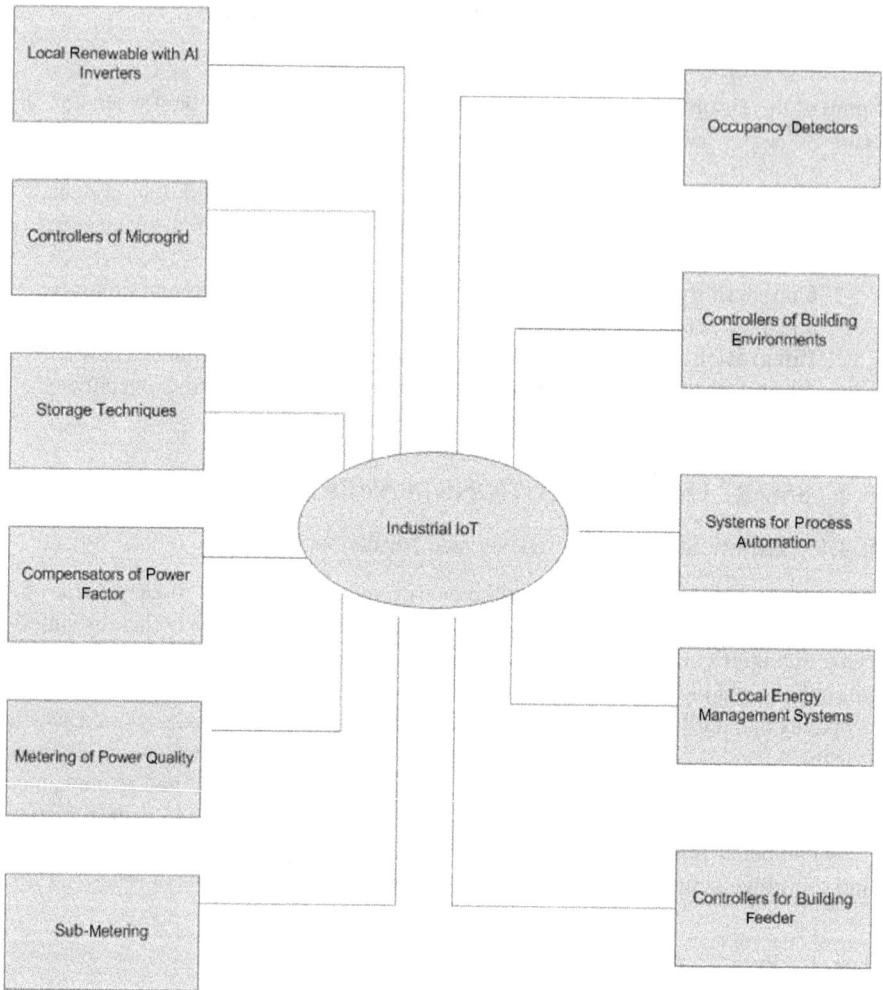

FIGURE 6.1   Applications of industrial IoT.

6.4.4   SAFETY AND DISASTER PREVENTION

AI and IoT systems are used in the energy industry to improve operational safety and prevent production accidents as well as eliminate their consequences. Pacific Gas and Electric Company (PG&E) has been using drones to explore methane leaks during the Northern California wildfires. The drones communicated data to the unified control system which, in turn, automatically isolated leaks until the damage was fixed by the crew on-site. The organization has been testing this approach to conduct routine infrastructure inspections as well. Safety drones are also used as a part of the hazard management system to lower employee risk on nuclear plants or at mining locations [3].

6.5 PRESENT STATUS OF EMERGING TRENDS IN ENERGY SECTOR IN INDIA

6.5.1 CREATING OPPORTUNITIES AND CHALLENGES FOR THE SECURITY OF THE GLOBAL ENERGY INFRASTRUCTURE

As renewable energy sources are now becoming part of the energy portfolio and are rapidly gaining market share, they are offering various advantages, such as energy mix diversification, with distributed generation growing rapidly worldwide, and its installed capacity is expected to double in the next few years.

Moreover, as and when the energy generation portfolio transitions and diversifies further, new challenges and opportunities are emerging, which need major changes in the electric utility business model and regulatory policies to ensure secure and reliable supply to the markets [4–13].

6.5.2 DIGITAL DISRUPTION CREATING THREATS AS WELL AS NEW OPPORTUNITIES

As the renewable energy technology is instrumental for realizing intelligent grids and interconnected assets, it has created new threats of possibility of cyberattacks. The increasing interconnectivity and proximity of energy systems are also creating conflicts and ripple effects on energy markets and prices. Emerging technologies, batteries and grid-embedded generation are also attracting the cyber security of grid systems more vulnerable. Global inexperience in handling of newly introduced large-scale cyberattacks, combined with the greater capabilities of state and non-state actors, has enhanced the opportunities of future wars, and attacks will have a larger cyber component [14,15].

6.5.3 NEW GLOBAL ENERGY SECURITY ORDER IS NEEDED FOR REBALANCING OF ENERGY SUPPLY AND DEMAND

Emerging price drops in oil prices have led to a remarkable shift in wealth from net oil exporters to oil importers. This has resulted in the development of unconventional sources of oil and gas, as well as the recent economic slowdown in emerging markets in India and China, resulting in price readjustments as compared to backdrop of a general shift in energy supply patterns. These geopolitical shifts and new distribution of powers and energy trade flows are creating emerging challenges and opportunities for energy security in the new energy architecture around the world [16].

6.6 IoT APPLICATIONS AREAS IN RENEWABLE ENERGY

6.6.1 AUTOMATION TO IMPROVE OVERALL PRODUCTION

Solar and wind energy are the most popular renewable energy sources due to abundance availability and reliability as compared to any other renewable energy sources. In 2019, Germany suffced a quarter of its energy demands from its windmill farms. The cost associated with energy production through these resources has also decreased significantly. From the year 1977 onwards, the cost of solar panels has reduced by 99%. Japan, Germany and China are the global leaders in using solar energy [17].

The assimilation of AI and IoT systems along with sensors in solar and wind energy systems application had increased their reliability. In order to maximize energy production, most of the solar panels use dual-axis trackers. These tracking systems calibrate the angle of solar panels and assist to receive the maximum solar radiation throughout the day [18].

AI and IoT systems can be used to remotely regulate and control these tracking systems to ensure maximum energy production efficiency. By using analytics solutions, the movement of the sun and solar radiation is tracked which is used to automatically adjust the angle of solar panels. Also AI and IoT systems in wind energy systems are used to monitor operating parameters affecting power generation.

6.6.2 SMART GRIDS FOR ELEVATED RENEWABLE IMPLEMENTATION

The growth of renewable energy is restricted due to less reliability of transmission and distribution systems. The traditional energy grids were built to support the one-way transmission of uniform energy from power plants and bill the customers once a month. Hence, nowadays these grids are not applied to support the varying electricity supply from renewable sources. AI and IoT systems have enabled the creation of smart grids supporting manual switching between renewable and long-established power plants to ensure an uninterrupted power supply. This switching in smart grids is supporting the varying nature of renewable energy and facilitates non-stop energy supply to the consumers.

6.6.3 IoT INCREASING THE ADOPTION OF RENEWABLE SYSTEMS

The development of smart grids through AI and IoT systems has escalated the growth of renewable energy sources. Because they offer benefits of power consumption monitoring and real-time alerting, this allows energy utilities to include renewable sources for energy distribution. Below are some of these benefits.

6.6.3.1 Contribution from End Consumers

Even the end consumers are now utilizing renewable energy sources to reduce their electricity bills and become self-dependent.

Many countries, like India, are providing solar subsidies to citizens to increase the adoption of renewable energy systems.

Countries are assisting the renewable energy users to develop solar stations on their rooftop and use them for personal electricity needs. Moreover, consumers can also discharge the excess electricity into the smart grids in exchange for money. This is helping countries to increase the overall adoption of renewable and create a greener environment for citizens to live in.

6.6.3.2   Balancing Supply and Demand

Smart grids allow energy utilities to provide consumers with a consistent power supply. The integration of AI and IoT systems in renewable energy assists the energy suppliers to accommodate electricity from renewable sources and suffice the end-consumer demands. The use of smart energy metres on a commercial level gives real-time consumption data to electricity suppliers. By using analytics and data processing solutions, they can also develop trends and patterns related to peak load conditions. Therefore, by using manual switching techniques, energy utilities have reduced the use of power plants during normal off-peak timings and run them when the electricity demand is extreme, resulting in synchronizing the demand and supply conditions in addition to reducing the emission limits of toxic substances in the environment.

6.6.3.3   Cost-Effectiveness

As per estimates, the global energy demands can be fulfilled by harnessing 1.2% of solar energy from the Sahara desert (around 110,400 km^2) thereby reducing losses linked with the transmission and distribution of electricity from such a remote location. Power losses in transmission lines can reach up to 10% for long distances thereby creating complications and challenges which are preventing the escalated growth of solar and renewable energy as a whole. Moreover, implementation of AI and IoT systems in solar energy will also reduce the cost of building and managing solar stations significantly. Real-time monitoring and predictive analytics features of AI and IoT systems are used to monitor parameters that can reduce the efficiency of the power station or result in unexpected breakdowns. Hence, utilities can cut some costs related to inspection and repairs and improve their efficiency [19].

6.6.4   REMOTE ASSET MONITORING AND MANAGEMENT IMPROVING RELIABILITY

Affixing AI and IoT systems sensors to generation, transmission and distribution equipment enables energy companies to monitor them remotely. These sensors measure parameters such as vibration, temperature and wear to optimize maintenance schedules. This preventative maintenance approach can significantly improve reliability by keeping equipment in optimal state and providing the opportunity to make repairs before it fails [20].

6.6.5   A MORE DISTRIBUTED GRID

The energy grid is becoming more distributed thanks to the rise of residential solar and other technologies. House owners and businesses can now generate their own electricity by placing solar panels on their rooftops or even building small wind turbines on their properties is increasingly distributed power system represents a major change for energy companies. In addition to managing a few large generators, they must also now manage a growing number of small generation resources located across the grid. Smart grid technology powered by IoT is helping to enable this distributed energy transformation. A smart grid uses IoT technology to detect changes in electricity supply and demand [21].

6.6.6 MORE INFORMED CUSTOMERS THEREBY IDENTIFYING WASTE AREAS

AI and IoT systems technology helps energy customers to be more informed about their energy usage. Internet-connected smart metres collect usage data and send it to both utilities and customers remotely. Because of smart metre technology, many energy companies are now sending customers detailed reports about their energy usage. Customers can also install smart devices in their homes or commercial buildings that measure the power consumed by each appliance and device. They can use this information to identify waste and especially power-hungry appliances to save on their energy bills. Other IoT devices, such as thermostats, can automatically optimize their operation to reduce energy use. Residential customers could potentially benefit the most from these technologies, as the U.S. residential sector representing 37% of energy usage. The commercial sectors using 35% and industrial sectors using 27% could benefit substantially as well [22].

6.6.7 IMPROVED GRID MANAGEMENT TO BUILD NEW INFRASTRUCTURE

AI and IoT systems technology enable the integration of more distributed resources into the grid thereby improving grid management. Placing sensors at substations and along distribution lines provides real-time power consumption data that energy companies are using to make decisions about voltage control, load switching, network configuration and more which are automated. Sensors located on the grid alert operators to outages, allowing them to turn off power to damaged lines to prevent electrocution, wildfires and other hazards. Smart switches can isolate problem areas automatically and reroute power to get the lights back on sooner [23].

Power usage data can also serve as the basis for load forecasting resulting in managing congestion along transmission and distribution lines and ensures that all of the connected generation plants meet requirements related to frequency and voltage control. This power consumption data can also help companies decide where to build new infrastructure and make infrastructure upgrades. The AI and IoT systems are transforming nearly every sector of our economy, including the one that powers – the energy sector. In the near future, the energy industry will be becoming smarter, more efficient, more distributed and more reliable, due to effective utilization of the AI and IoT systems [24].

6.7 MAHARASHTRA STATE USING DRONES TO INSPECT EHV POWER TRANSMISSION LINES AND TOWERS

Maharashtra State in India has become the first State to use drones for aerial surveillance and inspection of Extra High Voltage (EHV) power transmission lines and towers to reduce risk to staff, slash maintenance costs and minimize losses from outages. The Union Ministry of Home Affairs and the Director General of Civil Aviation have permitted the Maharashtra State Electricity Transmission Company Ltd. (MSETCL) to utilize drones to inspect faulty lines, reducing the risk posed to operating staff of MSETCL. Up till now, workers are using ladders and chairs to inspect transformers which are very risky in the hilly areas of the State. MSETCL is planning to provide a drone equipped with ultra HD cameras to capture high-resolution close-up photographs and videos of EHV lines and towers in each zone. This will allow better assessment of faults. In addition to zonal offices, the head office in Mumbai will be monitoring the drones to guide engineers and operators in case of any failure and guidance required. The proposed drones will assist in slashing maintenance costs and reducing various losses from outages. These drones have the potential to revolutionize the inspection of power lines and transmission towers. It will also allow aerial surveillance, which is more efficient than manually surveying power lines. Drones will also help detect defects at the incipient stage. The Maharashtra State Electricity Distribution Company Limited (MSEDCL) is planning for the electrification of remote and inaccessible tribal areas in Melghat and Gadchiroli of Maharashtra State which presently do not have electricity due to geographical hurdles due to forest clearances [25–27].

6.8 FLYING LONG-DISTANCE ROBOTS PLANNED FOR UTILIZATION BY POWER COMPANIES

Power utilities in Europe are planning to utilize long-distance drones to scour thousands of miles of grids for damage and leaks in an attempt to avoid network failures to avoid losses of billions of dollars per year. Italy's Snam, Europe's biggest gas utility, is trialling BVLOS drones because they fly beyond the visual line of sight of operators – in the Apennine hills around Genoa for scouting a 20 km stretch of pipeline. Snam and Électricité de France (EDF)'s network subsidiary Réseau de Transport d'Électricité (RTE) have tested prototypes of long-distance drones that fly at low altitudes over pipelines and power lines. France's RTE has also tested a long-distance drone, which flew about 50 km inspecting transmission lines and sent back data that allowed technicians to virtually model a section of the grid. The company has a budgeted plan for investing 4.8 million Euros ($5.6 million) on drone technology in the coming 2 years. At present, power companies largely use helicopters equipped with cameras to inspect their networks. They have also recently started occasionally using more basic drones that stay within sight of controllers and have a range of only about 500 m [24].

In-sight BVLOS drones costing nearly 20,000 Euros each and a fleet of dozens would be needed to monitor a network. According to Navigant Research, power grid companies are planning to spend over $13 billion a year on drones and robotics by 2026 globally from about $2 billion now. The energy sector loses about $170 billion every year due to network failures and forced shutdowns. Flying robots which travel dozens of kilometres without stopping are planned to be utilized by power companies. Power utilities in Europe are planning for long-distance drones to scour thousands of miles of grids for damage and leaks in an attempt to avoid network failures that cost them billions of dollars a year [28].

The popularity of renewable energy means it is being utilized in various sectors, as well as the need to monitor the myriad extra connections needed to link solar and wind parks to grids, which is motivating the energy utilities to adopt the advanced technology. These drones are 100 times faster than manual measurement, more accurate than helicopters and, with AI devices on board, could soon be able to fix problems [29,30].

6.9 CONCLUSIONS

Various applications of AI and IoT systems for saving energy are focused on opti-mizing the very source of power production. Stations, plants, solar fields and wind turbines also consume energy, require maintenance and a wide range of effort and resource-heavy works to keep them running. Using AI and IoT systems in this sector is the right way to maximize the performance. Resource management by implement-ing AI and IoT systems results in optimizing the performance of power grid includ-ing using sensors, data analytics, predictive maintenance and other practices.

REFERENCES

1. A. Ahmadi, L. Tiruta-Barna, E. Benetto, F. Capitanescu, and A. Marvuglia, "On the importance of integrating alternative renewable energy resources and their life cycle networks in the eco-design of conventional drinking water plants," *Journal of Cleaner Production*, vol. 135, no. 11, pp. 872–883, 2016.
2. A. D. Woldeyohannes, D. E. Woldemichael, and A. T. Baheta, "Sustainable renew-able energy resources utilization in rural areas," *Renewable and Sustainable Energy Reviews*, vol. 66, no. 12, pp. 1–9, 2016.
3. M. Aksoezen, M. Daniel, U. Hassler, and N. Kohler, "Building age as an indicator for energy consumption," *Energy and Buildings*, vol. 87, pp. 74–86, 2015.
4. B. Khan, G. Agnihotri, S. E. Mubeen, and G. Naidu, "A TCSC incorporated power flow model for embedded transmission usage and loss allocation," *AASRI Procedia*, vol. 7, pp. 45–50, 2014.
5. B. Khan, G. Agnihotri, G. Gupta, and P. Rathore, "A power flow tracing based method for transmission usage, loss & reliability margin allocation," *AASRI Procedia*, vol. 7, pp. 94–100, 2014.
6. B. Khan, G. Agnihotri, P. Rathore, A. Mishra, and G. Naidu, "A cooperative game theory approach for usage and reliability margin cost allocation under contingent restructured market," *International Review of Electrical Engineering*, vol. 9, no. 4, pp. 854–862, 2014.
7. B. Khan and G. Agnihotri, "A comprehensive review of embedded transmission pricing methods based on power flow tracing techniques," *Chinese Journal of Engineering*, vol. 2013, Article ID 501587, 13 pages, 2013.
8. B. Khan, G. Agnihotri, and A. S. Mishra, "An approach for transmission loss and cost allocation by loss allocation index and co-operative game theory," *Journal of the Institution of Engineers (India): Series B*, vol. 97, pp. 41–46, 2016.
9. B. Khan and P. Singh, "Optimal power flow techniques under characterization of conventional and renewable energy sources: A comprehensive analysis," *Journal of Engineering*, vol. 2017, Article ID 9539506, 16 pages, 2017.
10. P. Singh and B. Khan, "Smart microgrid energy management using a novel artificial shark optimization," *Complexity*, vol. 2017, Article ID 2158926, 22 pages, 2017.
11. T. Molla, B. Khan, B. Moges, H. H. Alhelou, R. Zamani, and P. Siano, "Integrated opti-mization of smart home appliances with cost-effective energy management system," *CSEE Journal of Power and Energy Systems*, vol. 5, no. 2, pp. 249–258, June 2019.
12. Z. Tang, Y. Lin, M. Vosoogh, N. Parsa, A. Baziar, and B. Khan, "Securing microgrid optimal energy management using deep generative model," *IEEE Access*, vol. 9, pp. 63377–63387, 2021.
13. S. P. Bihari et al., "A comprehensive review of microgrid control mechanism and impact assessment for hybrid renewable energy integration," *IEEE Access*, vol. 9, pp. 88942–88958, 2021.

14. S. Freitas, C. Catita, P. Redweik, and M.C. Brito, "Modelling solar potential in the urban environment: State-of-the-art review," *Renewable and Sustainable Energy Reviews*, vol. 41, pp. 915–931, 2015.

15. N. Fumo and M. R. Biswas, "Regression analysis for prediction of residential energy consumption," *Renewable and Sustainable Energy Reviews*, vol. 47, pp. 332–343, 2015.

16. M. A. Hassaan, "A GIS-based suitability analysis for siting a solid waste incineration power plant in an urban area case study: Alexandria Governorate, Egypt," *Journal of Geographic Information System*, vol. 7, no. 6, p. 643, 2015.

17. J. Pan, R. J. S. Paul, T. Vu, A. Saifullah, and M. Sha, "An Internet of Things framework for smart energy in buildings: Designs, prototype and experiments," *IEEE Internet of Things Journal*, vol. 2, no. 6, pp. 527–537, 2015.

18. M. K. Mattinen, J. Heljo, J. Vihola, A. Kurvinen, S. Lehtoranta, and A. Nissinen, "Modeling and visualization of residential sector energy consumption and greenhouse gas emissions," *Journal of Cleaner Production*, vol. 81, pp. 70–80, 2014.

19. W. Su, J. Wang, and J. Roh, "Stochastic energy scheduling in microgrids with intermittent renewable energy resources," *IEEE Transactions on Smart Grid*, vol. 5, no. 4, pp. 1876–1883, 2014.

20. P. Morano, M. Locurcio, and F. Tajani, "Energy production through roof-top wind turbines A GIS-based decision support model for planning investments in the City of Bari (Italy)," in *International Conference on Computational Science and Its Applications*, Springer, New York, NY, pp. 104–119, 2015.

21. S. Surender Reddy and P. R. Bijwe, "Day-ahead and real time optimal power flow considering renewable energy resources," *International Journal of Electrical Power & Energy Systems*, vol. 82, no. 11, pp. 400–408, 2016.

22. S. E. Collier, "The emerging enernet: Convergence of the smart grid with the internet of things," *IEEE Rural Electric Power Conference*, 2015, pp. 65–68.

23. Q. Sun, H. Li, Z. Ma, C. Wang, J. Campillo, Q. Zhang, F. Wallin, and J. Guo, "A comprehensive review of smart energy meters in intelligent energy networks," *IEEE Internet of Things Journal*, vol. 3, no. 4, pp. 464–479, 2015.

24. T. Adefarati and R. C. Bansal, "Integration of renewable distributed generators into the distribution system: A review," *IET Renewable Power Generation*, vol. 10, no. 7, pp. 873–884, 2016.

25. Various reports published in The Times of India Mumbai edition, 2020.

26. Various reports published in The Hindu Chennai edition, 2020.

27. Various reports of Government of Maharashtra, 2020.

28. Various reports of Government of India, 2020.

29. Various reports of Zee News T V channel, 2020.

30. Various reports of ABP News T V channel, 2020.

7 Congestion Management and Market Analysis in Deregulated Power System

Ashish Singh and Aashish Kumar Bohre

CONTENTS

DOI: 10.1201/9781003278030-7

7.1 INTRODUCTION

Power demand is rapidly increasing with fast industrialization and urbanization. Therefore, meeting the load demand is a huge challenge. Flexible AC Transmission Systems (FACTS) devices help with reactive power compensation and active power enhancement by maintaining the power quality of the system. FACTS devices are very costly, so their proper location in a power system is essential for maximum benefit.

With the deregulation of the power system, every player is coming into this sector, especially the generation sector. In a deregulated power system, the power exchange is already at its stability margin. Many power sellers are paying attention to the power quality of the grid, and thus, there are constant threats to the grid. Instability may result if the seller is not properly controlled [1–5].

FACTS devices are necessary for the proper power flow control for a reliable, stable, and efficient power system. With the optimal location of FACTS devices, there is an increase in real power exchange without expanding the generation capacity. In addition, FACTS devices increase the power transfer capability of the line, thus saving the construction cost of a new transmission line. FACTS devices reduce the stress on the system.

Congestion is one of the major setbacks of the transmission constraints, resulting in the thermal heating of the transmission line, thus increasing the losses. Congestion also increases the total cost of power transfer.

On power exchanges, two pricing methods are used to sell the power, i.e., the uniform pricing market clearing price (MCP) and the non-uniform pricing market clearing price (LMP). In the MCP, all the generators are paid the same price. At the node, the price is common; there is no provision for the congestion price. In LMP, all the nodes have different prices; this includes congestion pricing and the price of the losses with respect to the generator. Therefore, different generators have a different price for a fair transaction for the seller. The nodal price at a location is calculated by the extra change in power to that location considering the security constraints. The LMP varies at each node; buyers pay the price for the ISO based on the dispatched energy. ISO pays the sellers based on their LMP. The variations in the LMP are due to the congestion costs and losses. The LMP is the total of all the generation, congestion, and loss costs.

In a deregulated power system, system planning reliability is the most important parameter. Reliability is the inherent property of a system that describes its capability to accomplish its projected functions. The evaluation of reliability in the power systems tells us the wellness of the system's performance and its primary functioning for supplying power to its customers. Growing investment can reduce the likelihood of customers not receiving power for any reason during the planning or operational phases [6–10]. Over-investment will lead to a huge cost of operations, whereas under-investment will lead to lower reliability. Trading between these two aspects is a major challenge for planners, designers, and operators of power system.

7.2 LITERATURE REVIEW

In [11], the optimal location of TCSC is obtained for an IEEE-14 bus system by increasing the social benefit from market participants' system losses. Social benefit is defined as the summation of the generator's surplus, wholesale surplus, and customer surplus. In [12], they discussed the available transfer capability (ATC), which is expressed based on the linear sensitivity approach with the AC Power Transfer Distribution Factor (ACPTDF) method, and how TCSC is used to increase the ATC in the IEEE 30 bus system using PSO. ATC can be evaluated by the power transfer capacity of the line for commercial power uses. In [13], they discussed the TCSC placement in various cases in the base cases, and its results are compared with the base case for the minimization of the loss and generation cost in an IEEE 14 bus system using MiPower software using Netwon Raphson load flow methods. In [14], they discussed the relief of congestion in an N-1 contingency case for the 6-bus system by using TCSC and UPFC in Power World Simulator software. In addition, it is observed that the ATC limit is enhanced by using FACTS devices. In [15], contingency ranking is calculated by the performance index, which is found by the number of overloaded lines and the number of voltage violations on buses in an IEEE 14 bus system. TCSC is placed at the optimal location, and it results in better transmission efficiency and steady-state stability limits, a better voltage profile, and loss is reduced, also relieving the congestion.

With decentralization on the generation side, more renewable energy sources (RES) sellers are participating in the market. However, grid integration with the RES is the biggest challenge on the grid. In [16], they discussed the impact of the wind integration in a 14-bus system and its effect on TCSC by minimizing the total loss and optimal location of TCSC using the Generalized Algebraic Modeling System

(GAMS). In [17], he discussed the overview of the SSR. The series capacitor is the source of the subsynchronous resonance (SSR). An incident, such as a shaft break, has occurred due to the subsynchronous torsional interaction, which is when the negative damping of the series compensated transmission network exceeds the mechanical damping of the shafts at a certain subsynchronous torsional frequency. TCSC helps mitigate the SSR.

7.3 RESTRUCTURED POWER SYSTEM

7.3.1 Traditional Vertically Integrated Electric Industry

The government enacts specific laws and rules to govern the operations of a specific industry or company. Traditionally, the electricity sector has been under the control of central and state government and its operations are generation, transmission and distribution of the electrical energy. Such utilities have a vertically integrated electric utility where the energy flow is flowing from generation to consumer while the money flow is flowing from the generation side. Every sector gets there, accordingly. It is not easy to calculate the costs involved in generation, transmission, and consumption in these vertically integrated utilities. As a result, utilities typically charge consumers average tariffs based on their location and energy consumption.

Earlier, there was a monopoly on Indian electricity markets. The electrical companies were serving only a particular geographical area where they got massive returns. Due to the monopoly, there was no competition and no technical innovations in the electricity sector. The electricity industry was governed by a single entity from the generating, transmission, and distribution sectors, and other parties could not enter into the power market. Before independence, the electricity sectors were owned by the private sector in India, which supplied loads to urban areas only in confined areas. After independence, the government transferred the private electricity sector into the public sector in order to increase the electricity coverage area around India. To this end, state electricity boards (SEBs) were formed in states. The power sector faces too many challenges, including high energy and peak demand, poor power quality, and increased tariffs on industrial loads due to cross-subsidies. In addition, subsidies were provided by SEBs by pricing lower than the actual rate; thus, SEBs were a loss-making sector. The SEBs had to deal with a significant financial loss as a result of the disparity between the average tariffs and the average cost of power supply. The other factors contributing to the financial problems of the SEBs were over-staffing, significant T&D losses, too many consumers' unpaid bills, and political interference. From this continuous financial deficit arises the inability to further invest in the power sector. The Indian electricity sector's performance has deteriorated because of a major percentage of T&D losses, which consists of both technical and non-technical reasons. The technical losses are due to the lack of efficient design in the transmission and distribution lines. In contrast, the reasons for non-technical losses include thefts, lack of metering for all consumers, and interference points. This factor consistently degraded the performance and efficiency of the electricity sector and thus demanded a new regulatory framework for changes.

7.3.2   REASONS FOR DEREGULATION

7.3.2.1   Changes in Power Technology

Earlier, only larger generators were encouraged to produce power as they were efficient, and the cost of power production was low. Nowadays, with the change in technology, small generators are also contributing to the grid, and different sources of new generating plants are there. With environmental concerns, there is also constant encouragement in the renewable sectors.

7.3.2.2   Liberalization of Power Industries

With the liberalization of the Indian economy in 1991, it has proved to be an improved economic efficiency and market. Similarly, in the deregulated power sector, more private investment was attracted with the change in the Electricity Act. The generation plant reduces the considerable investment, and thus private sectors are allowed for investment in the power sector, proving that the Indian electricity market is now an efficient market and its generation capability has increased.

7.3.2.3   Problems with Monopoly Utility

Price Disparities: The rate of charge varies with different regions and different enterprises. A customer may have to pay more than in other regions, which leads to discrimination in the prices, but as customers have no choice, they are bound to purchase and pay the fixed rate.

7.3.2.4   Computerized Controls and Data Communications

One of the significant challenges in the unbundled structures was the control and monitoring of the whole systems. By using upgraded data communication technology, including the internet, the ability to monitor and disperse equipment costs is low.

7.3.3   DIFFERENT ENTITIES IN DEREGULATED MARKET

See Figure 7.1.

GENCO (Generating Company)
A GENCO is a company that owns one or more generators and sells electricity to a competitive market. After deregulation, large generating stations are allowed to produce electricity and sell the power.

TRANSCO (Transmission Company)
It transmits the power from the generating station to different load centres. They only transmit the power and get paid for energy transfer. They don't participate in the transaction between buyer and seller.

DISCO (Distribution Company)
The local power distribution company is the monopoly owner-operator for the local power delivery which sells power to the commercial and residential areas. In various places, the Disom combines the retail function, which allows them to buy the wholesale power either from the spot market or through direct contracts with the generating unit and sell it to the end-user.

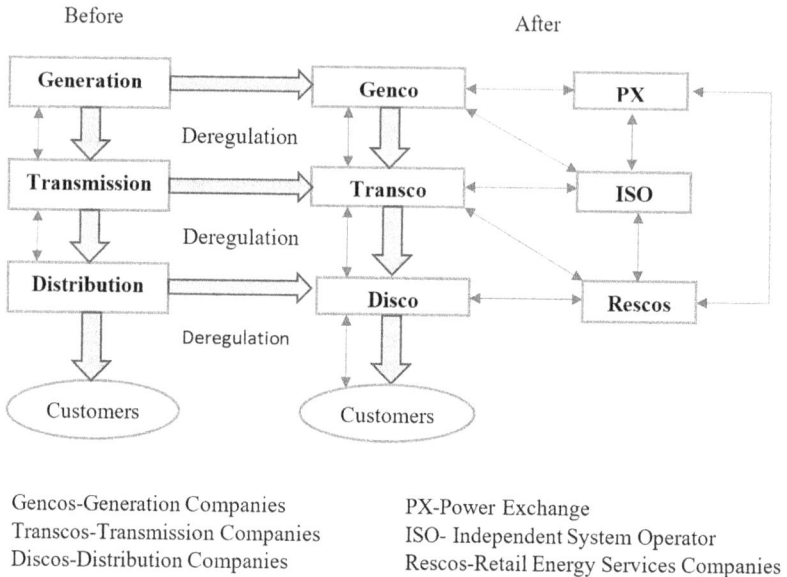

Gencos-Generation Companies   PX-Power Exchange
Transcos-Transmission Companies   ISO- Independent System Operator
Discos-Distribution Companies   Rescos-Retail Energy Services Companies

FIGURE 7.1   Regulated and deregulated power system.

RESCO (Retail Energy Service Company)

RESCO is an electric power retailer. Most of the retail departments are from conventional integrated utilities, and new companies are also allowed for better services. RESCO purchases a bulk amount of electricity from the generating unit and sells it directly to the buyers.

ISO (Independent System Operator)

The ISO is responsible for ensuring reliable and secure power delivery to consumers. The ISO doesn't take part in market trading and doesn't have any generating plants. ISO does not own the grid. An independent system operator (ISO) operates the power grid and spot market in an integrated way. An ISO is a single mandatory trading platform.

Customers

In a restructured power market, consumers have several options for purchasing electricity. Consumers may choose to purchase electricity through spot bidding or may directly buy from a genco or disco. With a deregulated power system, customers are also responsible for maintaining the power quality.

7.3.4 BENEFITS OF DEREGULATION

- Efficient Systems capacity
- Optimization of energy supplies take place.
- Electricity prices decrease.
- Consumers have choices.
- Efficiency is improved by to restructured power system in price signal.
- Energy flow takes place through surplus area to shortage area.

7.3.5 Structure of an Electricity Market in a Deregulated Environment

In the deregulated electricity market, there are many generation companies, thus making the power market more competitive and selling power accordingly [18]. No discrimination is allowed in a deregulated power system as all have open and fair access to the system. Presently, the generation is more competitive, but the transmission system monopoly is still there. This is because of the huge investment required in the transmission network. In addition, the ecological problem is still there. ISO operates the grid for open and fair access. It is the responsibility of the ISO to provide the network information to all market participants, such as available transfer capability (ATC), reliability margin, and total transfer capacity of the line, in real-time to ensure the safety of the grid.

In a deregulated power market, the electricity price fluctuations are not much higher due to the larger number of generation plants interconnected and long-term contracts. Buyers also have the option of directly purchasing power from the seller/ ISO through exchanges where trading is executed. In a deregulated market, two types of contracts are available: bilateral contracts and pool market contracts. Yet multilateral market contracts are also part of the proposal. Buyers and sellers trade power directly in a bilateral market under a negotiated price agreement. The trading platform agents request the ISO for the scheduled power to carry out the respective power trade, and if there are no threats to the system, the transaction is completed. Thus, financial transactions are carried out first, and then power trading. The seller has advanced knowledge of power trades for the respective schedule time. Thus, management is done accordingly, and financial transactions are carried out smoothly in advance. Therefore, the entire market economic model has an addition sum, thus making the market more competitive.

The average construction of a thermal power plant takes 4–5 years with a huge investment. The monopolist regime was investing in a limited value so as to protect the investments made. Gas power plants need less investment, and the amount of construction is reduced in the operating phase. It was possible for the generation expansion to continue with this smaller power production for the smaller enterprises. With rising environmental concerns, generation production is shifting toward renewable sources. Various policies, subsidies, and tariff-based schemes are provided to encourage renewable power plants.

With the restructured power systems, separation of generation and transmission has brought new investment and grid efficiency, and power quality has improved. Devices such as FACTS provide smooth control of the electrical transmission to the grid [19].

A supplier applies new technology and stronger incentives in order to minimize the cost and increase efficiency in an open market. In a competitive market, the bidding for power is done at a minimum cost. It is also more economical as it reduces the burden of the state-owned power sector, as the privatization of the market brings huge investments and also public funds can be diverted to other sectors.

7.3.5.1 The Electricity Act, 2003

Some of the main legislation in the Indian power sector are the Indian Electricity Act 1910, the Electricity (Supply) Act 1948, the Electricity Regulatory Commissions Act, 1998 and the Electricity Act 2003. The Electricity Act of 2003 changed the entire electricity market structure from a regulated to a deregulated market by forming separate generation, transmission, and distribution entities, with regulatory commissions regulating tariffs and granting licenses. State electricity boards (SEBs) were restructured. Thus, through this Act, more licences were given to the generation plants, open access and non-discriminatory implementation of the transmission lines and distribution sectors to change the electricity market. This Act provided a smooth way to attract more investment in this sector, removed the bureaucratic barriers to the development of electricity projects, and provided distribution reforms to remove its debts. The Indian electricity sector is more competitive through this Act because of the de-licensing of the thermal power plants, open access, eliminating the surcharges for the captive generation, and multi-year tariff.

7.3.5.2   The Electricity (Amendment) Act, June 2007

The government passed the Electricity (Amendment) Bill, 2007 [20] by amendments to the existing Electricity Act, 2003 in May 2007. The bill considered power theft as a punishable offence and called for strict action against it.

    The power sector is facing a T&D loss of nearly 30%, which is almost one-third of the power lost. Some of the losses are due to theft and technical reasons. A significant part of the electricity lost is due to theft and agricultural consumption that goes unaccounted for. Hence, giving priority to eliminating thefts of electricity will help to reduce distribution losses.

7.3.6   POWER TRADING AT EXCHANGE

7.3.6.1   The Real-Time Electricity Market (RTM)

The market segment features an auction session every 30 minutes and 48 auction sessions throughout the day through the Indian Power Exchange. The market provides flexibility to the distribution utilities to manage their power demand-supply fluctuations and meet their 24×7 power supply aspirations in the most flexible, efficient, dynamic, and cost-competitive way. For generators, the RTM market provides an opportunity to sell the requisitioned capacity, thereby enabling them to generate capacity efficiently.

7.3.6.2   India's Power Market Trading

Previously, there was a single buyer and seller, but the market shifted to a multi-buyer and seller. Nowadays, spot trading on the exchange is done with the aim of improving the liquidity and efficiency of trading [21]. Day-ahead market (DAM), in which trading takes place at 15-min intervals on the same day before the auction. Second is the term "ahead market," which involves trading ahead of 1 day, weekly, or intraday. This trading is done from 3 h ahead to 11 days ahead. In the Day Ahead Market, congestion is managed through market splitting and determining the Area Clearing Price (ACP) specific to a particular area. In power exchange, POC loss and state loss are considered during power transfer. POC loss is nearly 1.5% and state loss is nearly 4.85% of the power transfer. The majority of the buyers on the power exchange are textiles, manufacturers, metals, chemicals, autos, paper, and cement industries. While bidding on a power exchange, the spot price is purchased by the buyer, then

POC and state loss charges are added to the distribution loss if the voltage level is at or below 66 kV. Cross-subsidy surcharges are also levied for power purchased outside the discom. NLDC scheduling & operating charges and exchange are also levied. In addition, the charge varies between solar and non-solar.

7.3.7 PRESENT INDIAN ELECTRICITY MARKET

7.3.7.1 Renewable Sources
Renewable energy contributes 23.39% of the total generation. With the advancement of technology, the solar tariff can be achieved at as low as Rs 2.44/unit at Bhadla, Rajasthan. India has set the biggest renewable energy programme of 175 GW till 2022 [22].

7.3.7.2 Solar Energy
The National Institute of Solar Energy has estimated that India has a solar potential of nearly 748 GW considering 3% of the wasteland area that can be covered by solar PV modules. Solar energy is taking a central place in climate change through India's National Action Plan, with the National Solar Mission as a critical mission. In many parts of India, clear weather is present for nearly 300 days in a year; thus, the annual radiation varies from 1,600 to 2,200 kWh/m^2.

7.3.7.3 Off-Grid
Off-grid solar home and street lighting systems, solar power plants, solar pumps, and solar lanterns.

7.3.7.4 Grid Connected
A different step is taken to encourage solar power plants to be connected to the grid. India's ranking is fifth in the world in terms of solar power. Solar power capacity has increased more than 11 times in the past 5 years, to about 28.18 GW (March 2019) from 2.6 GW (March 2014). With the improvement in the technology, the market has increased, solar cell prices have dropped, and power generated through solar power in India is now more competitive and at grid parity.

7.3.7.5 Offshore Wind
India has a target of 5 GW of power from offshore wind installations by 2022 and 30 GW by 2030, which has been set for the confidence and boost of project developers in the Indian markets. The National Institute of Wind Energy (NIWE) estimated that total wind power potential is 302 GW at 100-m hub heights. An offshore wind turbine is big in size (in the range of 5–10 MW per turbine) as compared to the 2–3 MW of onshore wind turbines. While the cost per MW for offshore turbines is high because of the stronger structures and foundations needed in the marine environment, the desirable tariff will be achieved on account of the higher efficiency of these turbines after the development in that ecosystem.

7.3.7.6 Green Energy Corridors
The Green Energy Corridor Project is aimed at synchronizing power produced from renewable sources, such as solar and wind, with the conventional power stations

in the grid. For the evacuation of large-scale renewable energy, the Intra-State Transmission System (InSTS) project is set. In India, the bulk power transmission length has increased from 3,708 km in 1950 to about 265,000 km at present.

7.4 CONGESTION

Congestion is the condition that occurs when the transmission line reaches its upper limit. Thus, thermal heating is present and limits violate its security constraints. Contingency cases cause congestion, so power flow in the line is more than the ATC limits. When there is congestion, the national load dispatch centre issues a notice to the regional load dispatch centre, and a congestion charge is imposed for the extra power. Due to congestion, the cost of power transfer increases, and there is also a threat to the grid. If this power is more than its TTM limits as defined by the grid, this is a complete failure of that line. In an interconnected system, it may lead to a blackout. The ideal condition is when the power generated by the generator is consumed at the loads. When coordination between the transmission and the generation is not there, congestion occurs. Congestion leads to violations of security constraints, voltage limits, and reliability of the network. Thermal heating is present. The purpose of congestion management is to analyse the interconnected power system to identify the overloaded lines and problems arising from contingency.

Congestion management is important for relieving congestion. Congestion management is divided into two categories: the first is the preventive type, and the second is the corrective type. Thus, in preventive congestion management, day-ahead generation and demand-scheduled operations are managed considering the transmission congestion constraints (line limits), so that congestion is not avoided in the actual process. While in corrective congestion management, the day-ahead schedule is managed in a time block period of 15–30 min because of contingency. There are two ways of congestion management, i.e., the cost-free method and the non-cost-free method. The cost-free method consists of the use of FACTS devices, phase shifters, and the operation of transformer taps. Cost-free methods do not involve the generators and the distribution; here the cost involved is nominal. Non-technical methods are non-cost-free methods considering the security-constraints of generation dispatch, the network security factor method, congestion prices, and market-based methods, which include methods like Generator Rescheduling (GR), load shedding, Demand Responses (DR), Distributed Generation (DG), and nodal pricing [23].

7.5 CONTINGENCY

One of the important elements of the power systems during planning and operations is to strike a balance between the generation and the load side. An unpredictable condition or unwanted condition while in operation is known as a contingency. Contingency analysis is the process for calculating the impact of the occurrence or possibility of the contingency by detecting the operating reliability limits violation of the security constraints and thus making the operation of the power systems deliver the power reliability while also focusing on the planning for the overload condition. When the system is within the constraints, then it is the most economical condition for the power transfer.

The basic purpose is to operate the power system in a stable condition. Contingency occurs in the transmission system as a result of a generator outage, a transmission branch outage, an overloading condition, or any piece of equipment failure caused by internal or external causes such as a lightning strike, an object striking the transmission tower, or human error in relay setting. Any of the above disturbances may lead to a cascading effect resulting in tripping the interconnected systems and may result in a blackout. Contingency puts the whole system under stress [24].

7.6 THYRISTOR-CONTROLLED SERIES CAPACITOR (TCSC)

In 1986, Vithayathil proposed a method of "rapid adjustment of network impedance", which is a second-generation FACTS device which controls the line reactance by placing a variable reactance in series with the line. This variable reactance is an FC-TCR (Fixed Capacitor) combination with a mechanical switch capacitor placed in series [25,26].

The series compensation is preferred as an alternative to increase the power flow capability of the line in comparison with the shunt compensation, whose rating of the series compensation is very low [27–36]. TCSC reduces the sub-synchronous resonance, which is the main problem with series capacitors.

Transmission line admittance of the line where TCSC is connected is as follows:

$$G_{TCSC} + jB_{TCSC} = \frac{1}{R + j(X_{line} + X_{TCSC})} \qquad (7.1)$$

where

R and X_{line} are the resistance and reactance of the line without TCSC.

X_{TCSC} = reactance of the line with TCSC

The TCSC is connected in series with the transmission line conductor. TCSC plays a significant role in the control and operation of power systems, such as increasing the power flow transmission capacity, improving the stability of the system, reducing the loss of the system, and improving the voltage profile.

7.6.1 DEGREE OF SERIES COMPENSATION (K)

Degree of series compensation (K) is a ratio of effective reactance of TCSC $[X_{TCSC}(\alpha)]$ to the net reactance of the transmission line $[X_{TL}]$.

$$K = \frac{X_{TCSC}(\alpha)}{X_{TL}} \quad 0 < K < 1 \qquad (7.2)$$

To avoid series resonance, 100% compensation is not provided. With 50% series compensation, it almost doubles the steady-state transmitted power. Practically 70% of series compensation is suitable for line reactance compensation. Beyond 70% compensation, there is an increase in the risk of SSR problems. This SSR problem more often arises in cases of capacitive compensation than inductive compensation.

In practice, inductive reactance, X_L, is smaller than capacitive reactance, X_C. For the selection of the TCSC inductor value, X_L must be smaller than X_C to obtain both an effective inductive and capacitive reactance region in the TCSC device.

At the critical value of the firing angle, the resonance region is where TCSC should not be operated as it may cause excessively high voltage and current. While choosing the operating mode for a small value for the inductor, it will produce a high value of current harmonics in the thyristor branch. In contrast, the high value of the inductor doesn't give an appropriate amount of series compensation. The different modes of TCSC are as follows:

1. Blocking mode
2. Bypass mode
3. Capacitive boost mode (Capacitive mode)
4. Inductive boost mode (Inductive mode)

For blocked operating modes, triggering of the thyristor valves is not done, and thyristor is in the non-conducting states. For bypass mode, thyristors are at fully conducting states, and line current is flowing through the thyristor, and as TCSC has low inductive reactance. For variable control, thyristor is conducting in a controlled amount of inductive currents can passes by the capacitor, hence increases the overall reactance of the capacitive or inductive in a module.

Benefits of TCSC

- Enhance the power transmission capacity
- Improves system stability
- Reduces system loss
- Improves the voltage profile
- Damping of the power swing from local and inter-areas oscillation.
- Accurately regulating the power flows in the transmission lines
- Mitigating sub-synchronous resonances (SSR)
- Improves transient stability
- Active power oscillation damping
- Improving the post-contingency stability
- Dynamic control of power flow

7.7 MARKET ANALYSIS BY USING LMP

7.7.1 GENERATOR SENSITIVITY FACTOR (GSF)

The GSF helps to determine which generators will take part in the rescheduling in order to remove congestion. A positive value of the generator sensitivity factor indicates there is a decrease in the power flow in the congested line with the decrease in the generation, and on the other hand, the negative value of the generator sensitivity factor represents a decrease in the power flow in the congested line with the increase in the generation. The generators having a non-uniform and large magnitude of GSFG will take part in the rescheduling for the removal of congestion. The generator sensitivity factor (GSFG) is computed using the following procedure.

$$GSF_G = \frac{\Delta P_{ij}}{\Delta P_{Gg}} \tag{7.3}$$

$$P_{ij} = -V_i^2 G_{ij} + V_i V_j G_{ij} \cos(\theta_i - \theta_j) + V_i V_j B_{ij} \sin(\theta_i - \theta_j) \tag{7.4}$$

where θ_i and θ_j are bus voltage angle at bus i & j respectively. V_i and V_j are the bus voltages at i and j bus respectively. G_{ij} and B_{ij} are the conductance and susceptance of the line connected between i and j. Neglecting the PV coupling the GSF$_G$ Expression can be written as

$$GSF_g = \frac{dP_{ij}}{d\theta_i}\frac{d\theta_i}{dP_G} + \frac{dP_{ji}}{d\theta_j}\frac{d\theta_j}{dP_G} \tag{7.5}$$

where

$$\frac{dP_{ij}}{d\theta_i} = -V_i V_j G_{ij} \sin(\theta_i - \theta_j) + V_i V_j B_{ij} \cos(\theta_i - \theta_j) \tag{7.6}$$

$$\frac{dP_{ij}}{d\theta_j} = V_i V_j G_{ij} \sin(\theta_i - \theta_j) - V_i V_j B_{ij} \cos(\theta_i - \theta_j) \tag{7.7}$$

$$\frac{dP_{ij}}{d\theta_i} = -\frac{dP_{ij}}{d\theta_j} \tag{7.8}$$

GSF$_G$ denotes the amount of active power flowing in a transmission line connecting buses with ith and jth buses that would change due to active power injection by the generator. Based on its non-uniform and large-magnitude sensitivity value, system operators select the most sensitive generator for the power flow in the congested line. As a result, rescheduling the power output helps to alleviate congestion.

The slack bus is taken as a reference for obtaining the GSF. Thus, the slack bus generator sensitivity is always zero. Generators having a larger and non-uniform value of the sensitivity factor will take part in the congestion management. All the generators have non-uniform GSFG values, so all the generators will take part in congestion relief. The generators with positive GSFG values will decrease their generation, and the generators with negative GSF values will increase their generation.

7.7.2　Locational Marginal Pricing (LMP)

LMP is the incremental cost of a bus for serving minor changes in load while adhering to all security constraints. LMP is used for the power exchange and it manages the congestion. LMP on a bus is the shadow price of the power equation.

[26] Formalized paraphrase

The LMP consists of energy prices, marginal loss prices, and congestion costs.

In power exchange, LMP at the day-ahead market (DAM) is called "ex-ante LMP" as the calculation of LMP is done before the power transfer. In a real-time market, the LMP calculation is done after the event, so it is also symbolized as "post-LMP." In a wholesale energy market, the electricity price is constantly changing for various reasons, which include changing load demands, changes in generation offers and demand bids, an outage in the transmission system, and availability of the generators or generator outages. Most of the time, there are load variations [37].

$$\text{Generation cost} = a_i + b_i P_{Gi} + c_i P_{Gi}^2 \qquad (7.9)$$

where
P_{Gi} = output power of generator at ith bus
a_i, b_i, c_i = cost co-efficient

$$\beta_k = \frac{\text{Reduction in total cost}}{\text{Changes in contraint power flow}} \qquad (7.10)$$

$$\text{LMP}^{\text{Congestion}} = \sum \text{GSF}_{ik} \beta_k \qquad (7.11)$$

$$\text{LMP}^{\text{Losses}} = (\text{DF}_i - 1)\text{LMP}^{\text{references}} \qquad (7.12)$$

$$\text{LMP} = \text{Generation cost} + \text{congestion cost} + \text{losses cost} \qquad (7.13)$$

DF_i = delivery factor of ith bus w.r.t to reference bus
GSF_{ik} = GSF for ith bus on line k
k = congested transmissions line set
β_k = kth line constraint costs

7.8   SYSTEM INDEX

The optimal location of TCSC in an electrical power network is used to solve the congestion of the line by satisfying the thermal limits, voltage limits, and power transfer capability limits of the limit in a network. The optimal location is calculated based on the objective function by minimizing the loss, voltage deviation, and maximizing reliability in the system.

7.8.1   VOLTAGE DEVIATION INDEX

Considering the voltage constraints, an objective function is taken with reference to the base voltage. The voltage profile of a system must be within the limits.

$$V_{\text{Index}} = \frac{(V_{\text{TCSC}} - V_{\text{ref}})}{V_{\text{ref}}} \qquad (7.14)$$

V_{TCSC} = Voltage with TCSC
V_{ref} = Voltage for the base case

Minimization of the voltage deviation and voltage must be within the permissible limits, i.e., 5% deviation is allowed for the transmission network as given by the Indian grid code.

7.8.2 LOSSES

$$S_{\text{Index}} = \frac{S_{\text{TCSC}}}{S_{\text{outage}}} \tag{7.15}$$

S_{TCSC} = Total losses after TCSC is connected
S_{Base} = Total losses after contingency case

In a system losses should be minimum as possible.

7.8.3 RELIABILITY

Expected energy not supplied (EENS) is a fundamental index for the measure of services, reliability, and adequacy. EENS is the expected amount of energy not served to consumers by the system per year during that period, due to system capacity shortages or unexpected severe power outages, and is also known as "expected demand not supplied" [38].

The ENS for the customers is calculated as Eq. (7.16).

$$\text{ENS} = \alpha d \sum_{k=1}^{N} \lambda_k \left| I_{kp} \right| \times V_{\text{rated}} \tag{7.16}$$

where I_{kp} is the peak load current of a branch, λ_k is the rate of failure for k^{th} branch or line, and V_{rated} is the rated voltage of the system.

$$\alpha = \text{load factor}$$

$$d = \text{duration of repairing}$$

The system reliability is as below:

$$R = \left(1 - \frac{\text{ENS}}{\text{PD}}\right) \tag{7.17}$$

R = Reliability
PD = power demand.

Reliability of the system should be maximum.

7.8.4 WEIGHTING FUNCTION

A weighting function is formed considering the loss index, voltage deviation index, and reliability. This multi-objective function is considered to form a single objective function by giving weightage to each of the objective functions. The overall summation of the objective functions is equal to one.

$$W = 0.5 V_{\text{index}} + 0.3 S_{\text{index}} + 0.2 R^{-1} \tag{7.18}$$

This weighting function is used as a fitness function for optimal placement through PSO. The minimum fitness function and its corresponding value are considered for the best location of TCSC.

7.9   PARTICLE SWARM OPTIMIZATION (PSO)

In 1995, a metaheuristic algorithm was developed by James Kennedy and Russell Eberhart as the PSO algorithm for solving the complex mathematical problem based on the concept of swarm intelligence [39].

Studies were conducted on the social behaviours of animal groups such as fish and birds, and the availability of food search. In a certain area, food is available. Birds or fish find food all over the area without having the knowledge of food availability and its locations. Therefore, the birds and fish combined their self-experience with social interaction experience, as an effective way is to follow the bird nearest to the food. PSO uses this behaviour of birds and fishes to maximize the food in optimization problems. PSO uses the concept of social interaction and the movement and intelligence of the swarm in problem-solving. PSO is a population-based optimization technique that is used in a search space where particles are bird or fish groups. The particles (agents) that are formed by the swarm move in the search space to get the best solution. A fitness function is formed for the evaluation of the fitness of these particles, and particles have their own velocity that directs them to the optimum particles while also updating their own position accordingly. Each particle adjusts its place according to its own experience and that of the other particle, and updates its position accordingly in various dimensional spaces.

Every particle keeps the tracks and is always associated with the best solution obtained so far. This value is called a personal best (pbest). Moreover, the best value yet obtained by any of the particles is the global best (gbest). Each particle is accelerated with a weighted acceleration at every step toward its P_{best} and G_{best}.

$$v_i^{k+1} = w \times v_i + c_1 \times \text{rand}_1 \times \left(P_{\text{best}_i} - s_i^k\right) + c_2 \times \text{rand}_2 \times \left(G_{\text{best}_i} - s_i^k\right) \qquad (7.19)$$

where
   v_i^{k+1} = velocity of particles i at iteration
   w = weight function
   c_1 and c_2 = learning factor
   rand_1 and rand_2 = random number in between 0 and 1
   s_i^k = At iteration k, Current position of the particle i
   P_{best_i} = Local best of particle in the current iteration
   G_{best_i} = Global best found by the particles up to the current iteration

Large weighting inertia denotes larger global search ability whether the smaller weighting inertia represents the smaller local search.

Weight function is given by Eq. (7.20)

$$w = w_{\max} - \frac{\left(w_{\max} - w_{\min}\right)}{\text{it}_{\max}} \times \text{it}_{\text{present}} \qquad (7.20)$$

$w_{\max}$ = maximum weight
$w_{\min}$ = minimum weight

$it_{present}$ = current iteration number.
it_{max} = maximum iteration number

The update the new particles by Eq. (7.21)

$$s_i^{k+1} = s_i^k + v_i^{k+1}$$ (7.21)

7.10   RESULTS AND DISCUSSIONS

7.10.1   CASE 1: BASE CASE AND TCSC CASE

Load flow is done for the base case in a 14-bus system. The optimal location in the base case for the TCSC is at branch 17, i.e., in-between bus 9 and bus 14. The voltage profile has been improved for the base case and base case with TCSC. It is shown in Figure 7.2.

7.10.2   CASE 2: BASE CASE, $N-1$ CONTINGENCY CASE AND TCSC WITH $N-1$ CONTINGENCY CASES FOR 14 BUS SYSTEMS

In a 14-bus system, an outage at branch no. 9, i.e., in between bus 4 and bus 9, is created. The best results are obtained by placing TCSC at branch 15, which is located between buses 4 and 9. Voltage profile with $N-1$ contingency is deviated by a larger margin from the base case, and with TCSC, the voltage profile is improved by a greater margin (Figure 7.3).

7.10.2.1   Active Power and Reactive Power

Active power is flowing for the base case, and with an $N-1$ contingency at branch 9, the power flow is zero also. With TCSC, there is an active power enhancement in branch 15, as shown in Figure 7.4.

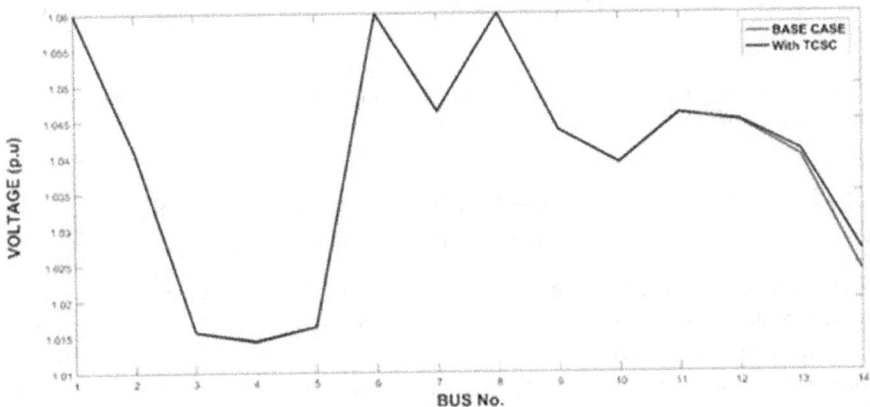

FIGURE 7.2   Voltage profile for the base case and with TCSC.

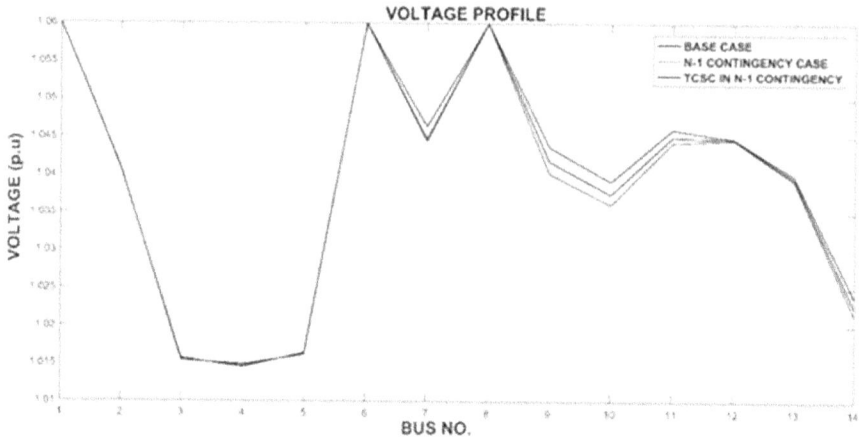

FIGURE 7.3   Voltage profile.

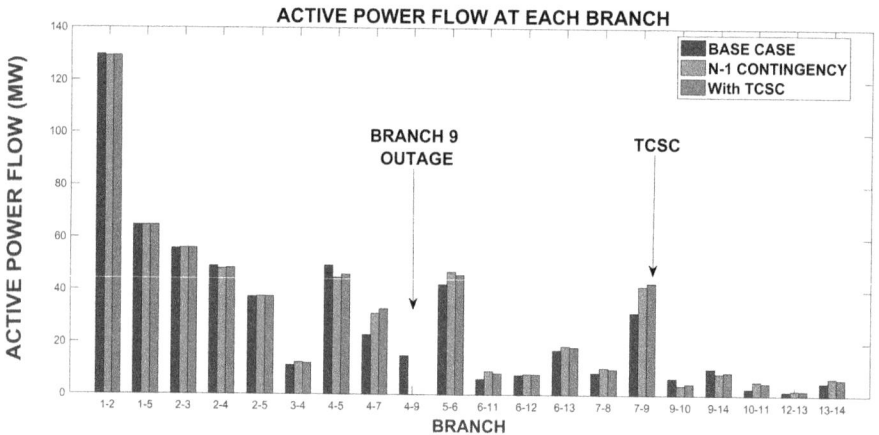

FIGURE 7.4   Active power profile for the base case, $N-1$ contingency case and with TCSC in contingency case.

With the contingency, the reactive power compensation is there at branch 15, as shown in Figure 7.5.

7.10.3   CASE 3: BASE CASE, $N-1$ CONTINGENCY CASE, AND TCSC WITH $N-1$ CONTINGENCY CASES FOR 30 BUS SYSTEMS

Here, load flow is solved for the base case in a 30-bus system. Then an outage is randomly created at branch 35, i.e., in-between bus 25 and bus 27. Thus, with the outage, the voltage profile is distorted by a larger margin, threatening the overall system and some buses, especially bus 26. A larger margin deviates the voltage, which means the bus is on the edge of collapse. To protect this system, TCSC is placed, whose location is obtained through PSO by forming a fitness function through a weighting

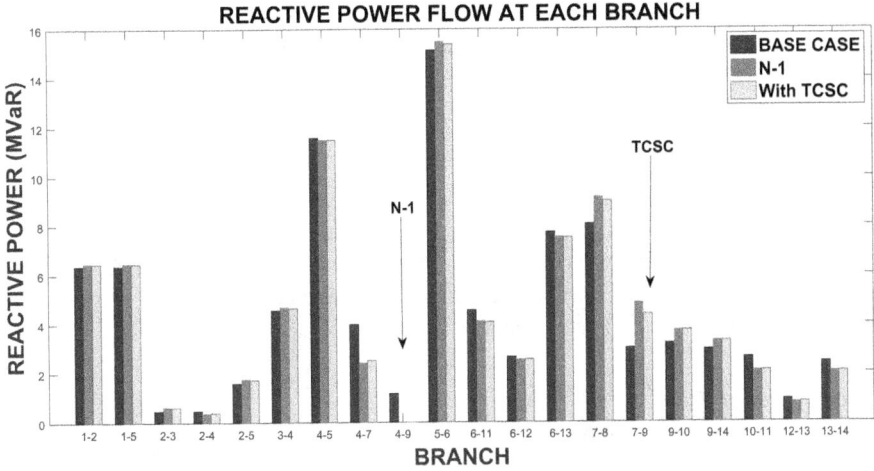

FIGURE 7.5 Reactive power profile for the base case, $N-1$ contingency case and with TCSC in contingency case.

function. The optimal location of the TCSC is obtained at branch 34, i.e., in-between bus 25 and bus 26. Thus, with TCSC, the voltage profile is improved, and the system is protected. The voltage profile for the 30 bus system considering all three cases is shown in Figure 7.6.

7.10.3.1 Losses
Due to congestion, the losses for the contingency case increase by a greater margin than the base case. As a result, thermal heating is more in the transmission line. With the use of TCSC, reactive power compensation is present in a system, and congestion

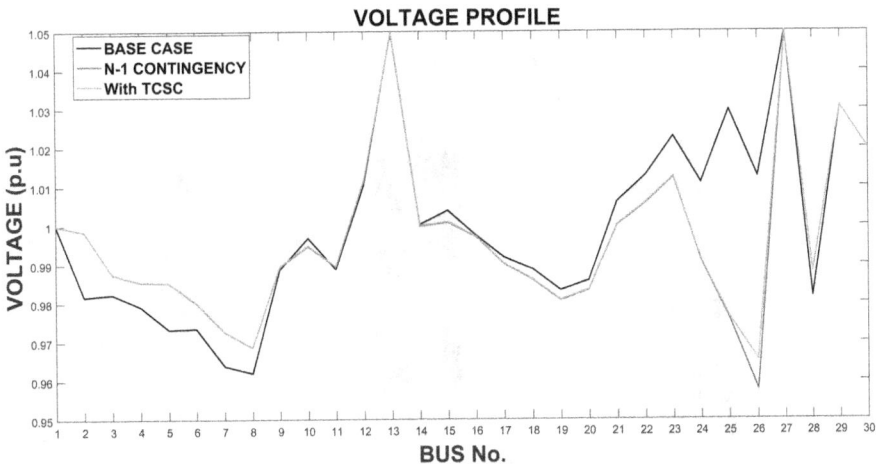

FIGURE 7.6 Active power profile for the base case, $N-1$ contingency case and with TCSC in contingency case.

is also removed; thus, thermal heating is reduced, so losses are reduced. Figure 7.7 presents the comparative analysis of losses for the base, $N-1$, and TCSC cases.

7.10.4   CASE 4: MARKET ANALYSIS

This is the market study for Case 3, i.e., the base case, $N-1$ contingency with an outage at branch 35 and TCSC at branch 34. A market analysis was done for the 30 bus systems. The 30 bus system has six generators. The LMP calculation decides price according to revenue earned and is calculated based on generation. Through the polynomial equation, the generation cost is calculated.

7.10.4.1   Base Case

Dispatch time duration: 1.00 h

Here, the generation cost is obtained through the polynomial equation. The generation cost at each bus differs, so the cost of generation is calculated accordingly. The price is obtained through the LMP calculation according to the total revenue collected by the different generators. The earning is the difference between revenue and generation costs, taking into account all congestion costs and marginal losses. This is the benefit of the LMP calculation, where the nodal pricing can be done easily. Market analysis for the base case is shown in Table 7.1.

7.10.4.2   N–1 Contingency Cases

With the contingency case, the LMP value will be changed from the base case to a larger margin, and the overall profit will decrease with the contingency case (Table 7.2).

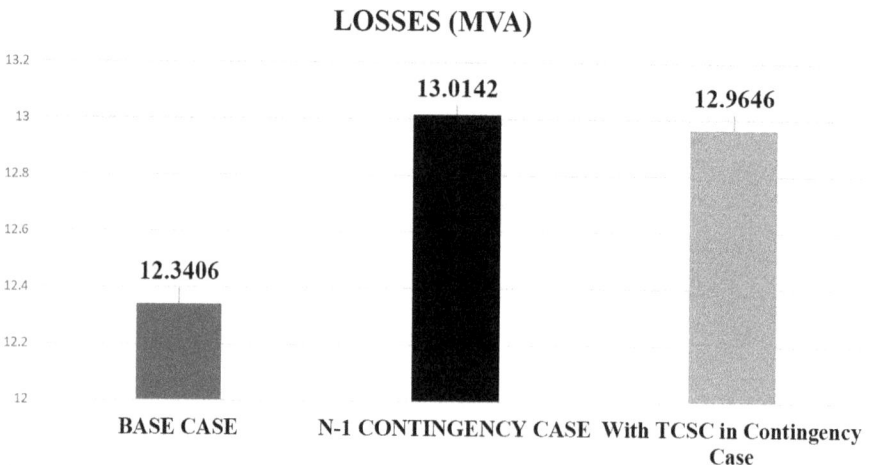

FIGURE 7.7   Losses for the base, $N-1$, TCSC cases.

TABLE 7.1
Market Details for Base Case

Gen	Bus	Pg (MW)	Price ($/MWh)	Revenue ($)	Generation Cost	Earnings ($)
1	1	43.45	3.600	156.43	124.67	31.76
2	2	48.00	3.614	173.47	124.32	49.15
3	22	20	4.015	80.30	45	35.30
4	27	44.89	4.076	182.96	162.70	20.26
5	23	19.84	4.050	80.35	69.36	10.99
6	13	16	3.874	61.98	54.40	7.58
Total		192.18		735.48	580.45	155.03

TABLE 7.2
Market Details for $N-1$ Contingency Case

T	Bus	Pg (MW)	Price ($/MWh)	Revenue ($)	Generation Cost	Earnings ($)
1	1	48	3.676	176.42	142.08	34.35
2	2	56.81	3.710	210.76	155.89	54.87
3	22	20	3.821	76.43	45	31.43
4	27	33	3.796	125.27	116.33	8.94
5	23	18	3.784	68.11	62.10	6.02
6	13	16	3.789	60.63	54.40	6.23
Total		191.81		717.63	575.79	141.83

7.10.4.3 TCSC

There is a significant increase in earnings with the TCSC. The increase in earnings is due to the change in the LMP. Thus, through LMP at each node, the prices are varied (Table 7.3).

7.11 CONCLUSION

In this chapter, detailed market analysis is performed for the contingency case and TCSC cases. Accordingly, the LMP and the revenue are calculated. It is observed that overall the revenue has decreased in the contingency case and with TCSC; the profit is higher as compared with the contingency case. Congestion is relieved with the help of TCSC, whose optimal location is found by PSO considering the weighting function in a 14-bus system and a 30-bus system. With TCSC, the losses are decreased, the voltage profile is improved, the active power is increased, and reactive power compensation is present. In Case 1, the voltage profile with TCSC in the base case is discussed. In Case 2, voltage profile, active power, and reactive power flow

TABLE 7.3

Market Details for TCSC Case

Gen	Bus	Pg (MW)	Price ($/MWh)	Revenue ($)	Generation Cost	Earnings ($)
1	1	48	3.676	176.42	142.08	34.35
					34.35	
2	2	56.81	3.710	210.76	155.89	54.87
3	22	20	4.125	82.50	45	37.50
4	27	33	3.796	125.27	116.33	8.94
5	23	18	3.784	68.11	62.10	6.02
6	13	16	3.789	60.63	54.40	6.23
Total		**191.81**		**723.69**	**575.79**	**147.90**

are discussed in the contingency case and with TCSC in the contingency case, and thus an improvement in the voltage profile, active power improvement, and reactive power compensation is obtained for a 14-bus system. In Case 3, the base case, along with the contingency case and TCSC, is discussed, and voltage profile improvement is obtained. Losses are increased with contingency and decreased with TCSC. In case 4, a detailed market analysis is obtained considering Case 3 for the contingency and TCSC location.

REFERENCES

1. B. Khan, G. Agnihotri, S. E. Mubeen, and G. Naidu, "A TCSC incorporated power flow model for embedded transmission usage and loss allocation," *AASRI Procedia*, vol. 7, pp. 45–50, 2014.
2. B. Khan, G. Agnihotri, G. Gupta, and P. Rathore, "A power flow tracing based method for transmission usage, loss & reliability margin allocation," *AASRI Procedia*, vol. 7, pp. 94–100, 2014.
3. B. Khan, G. Agnihotri, P. Rathore, A. Mishra, and G. Naidu, "A cooperative game theory approach for usage and reliability margin cost allocation under contingent restructured market," *International Review of Electrical Engineering*, vol. 9, no. 4, pp. 854–862, 2014.
4. B. Khan and G. Agnihotri, "A comprehensive review of embedded transmission pricing methods based on power flow tracing techniques," *Chinese Journal of Engineering*, vol. 2013, Article ID 501587, 13 pages, 2013.
5. B. Khan, G. Agnihotri, and A.S. Mishra, "An approach for transmission loss and cost allocation by loss allocation index and co-operative game theory," *Journal of The Institution of Engineers (India): Series B*, vol. 97, pp. 41–46, 2016.
6. B. Khan and G. Agnihotri, "A novel transmission loss allocation method based on transmission usage," *2012 IEEE Fifth Power India Conference*, 2012, pp. 1–3.
7. P. Rathore, G. Agnihotri, B. Khan, and G. Naidu, "Transmission usage and cost allocation using shapley value and tracing method: A comparison," *Electrical and Electronics Engineering: An International Journal (ELELIJ)*, vol. 3, pp. 11–29, 2014.

8. B. Khan and G. Agnihotri, "An approach for transmission usage & loss allocation by graph theory," *WSEAS Transactions on Power Systems*, vol. 9, pp. 44–53, 2014.

9. S. Khare, B. Khan, and G. Agnihotri, "A shapley value approach for transmission usage cost allocation under contingent restructured market," *2015 International Conference on Futuristic Trends on Computational Analysis and Knowledge Management (ABLAZE)*, Noida, 2015, pp. 170–173.

10. B. Khan, G. Agnihotri, and G. Gupta, "A multipurpose matrices methodology for transmission usage, loss and reliability margin allocation in restructured environment," *Electrical & Computer Engineering: An International Journal*, vol. 2, no 3, p. 11, September 2013.

11. S. Sachan and C. P. Gupta, "Influence of optimally placed TCSC on social welfare in deregulated market," *2015 IEEE UP Section Conference on Electrical Computer and Electronics (UPCON)*. IEEE, 2015.

12. M. M. Karthiga, S. C. Raja, and P. Venkatesh. "Enhancement of available transfer capability using TCSC devices in deregulated power market," *2017 Innovations in Power and Advanced Computing Technologies (i-PACT)*. IEEE, 2017.

13. J. Singh and Y. P. Verma, "Power flow management for grid stability using TCSC device," *2018 IEEE 8th Power India International Conference (PIICON)*. IEEE, 2018.

14. S. Thote and M. Jape. "Computation of ATC under N-1 contingency and congestion removal using series and shunt compensation," *2017 IEEE International Conference on Electrical, Instrumentation and Communication Engineering (ICEICE)*. IEEE, 2017.

15. P. Choudekar, S. K. Sinha, and A. Siddiqui. "Transmission line efficiency improvement and congestion management under critical contingency condition by optimal placement of TCSC," *2016 7th India International Conference on Power Electronics (IICPE)*. IEEE, 2016.

16. S. Pati and R. Dahiya. "Impact of wind integration on placement of TCSC," *2016 International Conference on Emerging Trends in Electrical Electronics & Sustainable Energy Systems (ICETEESES)*. IEEE, 2016.

17. K. Clark, "Overview of subsynchronous resonance related phenomena," *PES T&D 2012*. IEEE, 2012.

18. K. Bhattacharya, M. H. J. Bollen, and J. E. Daalder. *Operation of Restructured Power Systems*. Springer Science & Business Media, 2012. Springer, Boston, MA, United States.

19. Central Electricity Regulatory Commission, http://www.cercind.gov.in/Act-with-amendment.pdf.

20. http://www.mserc.gov.in/acts/no5_electricity_act_2007.pdf.

21. India Energy Exchange March 2020 report, https://www.iexindia.com/Uploads/Presentation/26_03_2020IEX-Electricity-Presentation-2020.pdf.

22. https://mnre.gov.in/.

23. K. R. S. Reddy, N. P. Padhy, and R. N. Patel, "Congestion management in deregulated power system using FACTS devices," *2006 IEEE Power India Conference*. IEEE, 2006.

24. V. J. Mishra and M. D. Khardenvis. "Contingency analysis of power system," *2012 IEEE Students' Conference on Electrical, Electronics and Computer Science*. IEEE, 2012.

25. N. G. Hingorani and L. Gyuyi. *Understanding Facts: Concepts and Technology of Flexible AC Transmission Systems*. Wiley, 1999. Springer, Boston, MA, United States.

26. D. Shanthini, "Analysis of locational marginal pricing based DCOPF," *Journal of Electrical System*, vol. 8, no. 3, pp. 2013–2016, 2015.

27. O. P. Mahela, B. Khan, H. H. Alhelou, and P. Siano, "Power quality assessment and event detection in distribution network with wind energy penetration using stockwell transform and fuzzy clustering," *IEEE Transactions on Industrial Informatics*, vol. 16, no. 11, pp. 6922–6932, November 2020.

28. O. P. Mahela, B. Khan, H. Haes Alhelou, and S. Tanwar, "Assessment of power quality in the utility grid integrated with wind energy generation," *IET Power Electronics*, vol. 13, no. 13, pp. 2917–2925, 14 October 2020.

29. O. P. Mahela et al., "Recognition of power quality issues associated with grid integrated solar photovoltaic plant in experimental framework," *IEEE Systems Journal*. doi:10.1109/JSYST.2020.3027203.

30. O. P. Mahela, A. G. Shaik, B. Khan, R. Mahla, and H. H. Alhelou, "Recognition of complex power quality disturbances using S-transform based ruled decision tree," *IEEE Access*, vol. 8, pp. 173530–173547, 2020.

31. O. P. Mahela, B. Khan, H. H. Alhelou, S. Tanwar, and S. Padmanaban, "Harmonic mitigation and power quality improvement in utility grid with solar energy penetration using distribution static compensator," *IET Power Electron*, vol. 14, pp. 912–922, 2021.

32. R. K. Pachauri et al., "Impact of partial shading on various PV array configurations and different modeling approaches: A comprehensive review," *IEEE Access*, vol. 8, pp. 181375–181403, 2020.

33. R. K. Pachauri, O. P. Mahela, B. Khan, A. Kumar, S. Agarwal, H. H. Alhelou, and J. Bai, "Development of arduino assisted data acquisition system for solar photovoltaic array characterization under partial shading conditions," *Computers & Electrical Engineering*, vol. 92, p. 107175, 2021.

34. Pachauri, R.K., et al., "Shade dispersion methodologies for performance improvement of classical total cross-tied photovoltaic array configuration under partial shading conditions," *IET Renewable Power Generation*, vol. 15, pp. 1796–1811, 2021.

35. T. F. Agajie, B. Khan, H. H. Alhelou, and O. P. Mahela, "Optimal expansion planning of distribution system using grid-based multi-objective harmony search algorithm," *Computers & Electrical Engineering*, vol. 87, p. 106823, 2020.

36. B. Khan, H. H. Alhelou, and F. Mebrahtu, "A holistic analysis of distribution system reliability assessment methods with conventional and renewable energy sources," *AIMS Energy*, vol. 7, no. 4, pp. 413–429, 2019.

37. Y. Fu and Z. Li, "Different models and properties on LMP calculations," *2006 IEEE Power Engineering Society General Meeting*. IEEE, 2006.

38. A. K. Bohre, G. Agnihotri, M. Dubey, and S. Kalambe, "Impacts of the load models on optimal planning of distributed generation in distribution system," *Advances in Artificial Intelligence*, vol. 2015, pp. 1–10, 2015.

39. A. K. Bohre, G. Agnihotri, M. Dubey, and S. Kalambe, "Assessment of intricate DG planning with practical load models by using PSO," *Electrical & Computer Engineering: An International Journal (ECIJ)*, vol. 4, no. 2, pp. 15–22, June 2015.

8 Grid Synchronization of Photovoltaic System with Harmonics Mitigation Techniques for Power Quality Improvement

Vijayakumar Gali, Anand Sharma,
Sunil Kumar Gupta, Manoj Gupta,
and Madisa V. G. Varaprasad

CONTENTS

8.1 INTRODUCTION

Electricity demand is constantly increasing day by day due to increase of population across the world, industrialization, increased usage of machinery in the fields such as agriculture, etc. The conventional energy sources including coal, oil, etc. are depleting, creating environmental pollution which leads to global warming [1,2]. Various countries are investing so much money on finding new technologies from non-conventional energy sources like wind, solar, biogas and geothermal efficiently. Among the afore-mentioned non-conventional energy sources, solar is the best source of energy which is available throughout the year especially in countries like India, with no pollution, less maintenance, etc., and 35 MW power can be generated from $1\,km^2$ of area. Due its vast features, many countries encourage people by giving subsidies to install solar panels on their rooftops, office buildings, educational institutions, etc. [3,4]. These installed solar power plants are connected to the local grid and feed the generated power. Technological advancement and introduction of new electronic devices into the electrical power system network inject more even harmonics into the power grid [5,6]. The even harmonics create electromagnetic interference (EMI) problems on electronic gadgets and create noise on telecommunication network which leads to overheat of the devices and damage the equipment [7–16]. There has been so much of research to eliminate odd harmonics from the power system network. However, less research has been reported in the litera-ture on the effect of even harmonics and elimination techniques [17].

The solar photovoltaic (PV) system is associated to the grid through voltage source inverters (VSI). These inverters work to feed the active power generated by PV array during day time and also work as a shunt active power filter (APF) during night-time when the solar power generation is off. The shunt APF works to eliminate differ-ent power quality (PQ) like current harmonics, excessive reactive power, low power factor, etc. [18]. The shunt APF will work with control techniques. The reference current generation techniques are single-phase p-q theory, Fryze power theory, unit power factor technique, etc. However, the performance of shunt APF is extremely influenced by external disturbances in grid voltages [19,20]. If the aforementioned reference current generation technique fails to give accurate reference current, then the shunt APF is not able to mitigate power quality problems. In this chapter a modi-fied composite observer filter is proposed for filtering the even order harmonics and enhancing the grid synchronizing capability of PV system during adverse grid and load conditions. This modified composite observer filter works with instantaneous reactive power theory (IRPT) for reference current generation. The generated refer-ence source currents are compared with actual grid currents and generate the error signals. These error signals go to the Pulse Width Modulation (PWM) generator and the PWM generator produces controlling pulses for VSI.

The construction of this chapter is as follows. Section 8.2 depicts the architecture of grid tied PV system. The effect of even harmonics on the power system network components is analysed in Section 8.3. Section 8.4 describes the proposed modified composite observer-based IRPT. A complete simulation study has been carried out using MATLAB® Simulink® software, which are depicted in Section 8.5. A complete summary of chapter is included in Section 8.6.

8.2 ARCHITECTURE OF GRID TIED PV SYSTEM

The architecture of grid tied PV system is shown in Figure 8.1. The PV array is con-nected to the grid through VSI system. The power system network is having different types of loads like electric vehicle charging stations, diesel generator system, etc. The VSI can acts a shunt APF to mitigate the even harmonics incorporate proposed modified composite observed based IRPT.

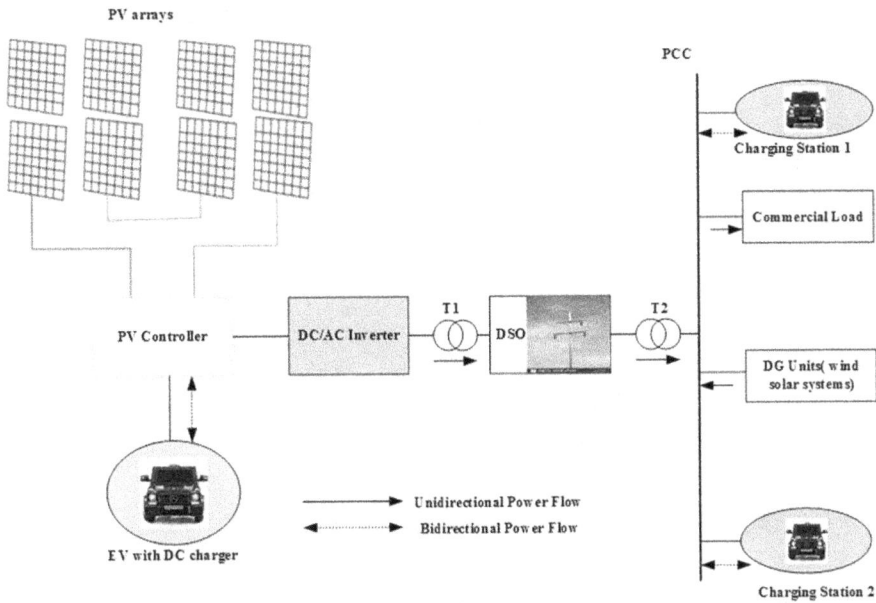

FIGURE 8.1   Grid-connected PV system.

8.3   ANALYSIS OF EVEN ORDER HARMONICS

Even harmonic generators are high-frequency switching converters, AC arc-furnaces, health care equipment, telecommunication equipment, etc. These types of equipment create dominant even harmonics in the electric distribution system. The complex waveform of distorted grid current is having fundamental waveform and dominant even harmonic component waveform. In case, if the fundamental grid supply current waveform is $I_g = I_{gmax} (2 \, \pi f t)$, then the even harmonic component waveform can be represented as follows:

$$I_{g2} = I_{g2max} (2 \times 2\pi f t) = I_{g2max} (4\pi f t) = I_{g2max} (2\pi t)$$

$$I_{g4} = I_{g4max} (4 \times 2\pi f t) = I_{g4max} (8\pi f t) = I_{g4max} (4\pi t)$$

(8.1)

The harmonics are generally classified by their names, for example second harmonic components as 100 Hz when the fundamental is 50 Hz signal, and harmonic sequence defines the phase rotation. Two sets of sequence components are available: one is positive sequence and the other is negative sequence. The 4th, 7th and 10th components are positive sequence components which rotate in the same direction. These positive harmonic sequence components are responsible for overheating of conductors and semiconductor devices, create EMI problems, etc. On the other hand, negative harmonic sequence components (2nd, 5th and 8th) are responsible for weakening of the flux in the motor field which degrades the efficiency of the motors, etc.

8.4 PROPOSED MODIFIED COMPOSITE OBSERVER-BASED IRPT

8.4.1 Modified Composite Observer Filter

The power system network is undergoing different disturbances due to interconnection of different linear and non-linear loads, conventional and more of non-conventional energy integration. The modified composite observer filter is designed to extract the fundamental frequency component from the grid system voltages with better solar PV power integration. This modified composite observer filter has N-dimensional vector with a set of grid voltages as shown in Eq. (8.2):

$$\psi(t) = \left[\psi_0(t), \psi_1(t), \psi_2(t), \psi_3(t), \ldots, \psi_p(t) \ldots, \psi_N(t)\right] \tag{8.2}$$

$$\psi(t) = \sum_{k=0}^{k=N} \psi_p(t) \tag{8.3}$$

The output is derived from the set of state space model which can be represented as follows:

$$\dot{y}(t) = H \times y(t)$$

$$\psi(t) = H'(t) \times y(t) \tag{8.4}$$

where H is an N-dimensional matrix. It can be written as:

$$H = \begin{bmatrix} H_0 & 0 & 0 & - & 0 & - & 0 \\ 0 & H_1 & 0 & - & 0 & - & 0 \\ 0 & 0 & H_2 & - & 0 & - & 0 \\ - & - & - & - & - & - & - \\ 0 & 0 & 0 & - & H_k & - & 0 \\ - & - & - & - & - & - & - \\ 0 & 0 & 0 & - & 0 & - & H_N \end{bmatrix}; H_0 = 0 \tag{8.5}$$

The matrix H is written as:

$$H_k = \begin{bmatrix} 0 & j.\psi \\ -j.\psi & 0 \end{bmatrix} \tag{8.6}$$

$$[K]^t = \begin{bmatrix} 1 1 0 1 0 - - - 1 0 \end{bmatrix} \tag{8.7}$$

The kth subsystem block can be represented as follows:

$$\dot{y}_k(t) = H_k \times y_k$$

$$\psi_k(t) = H_k^t \times y_k \tag{8.8}$$

$H_p^t = \begin{bmatrix} 1 & 0 \end{bmatrix}$ and the state space vector $y_k(t)$ is represented as follows:

$$y_p(t) = \begin{bmatrix} y_{p1}(t) \\ y_{p2}(t) \end{bmatrix} \tag{8.9}$$

The modified composite observer filter is modelled for set of N blocks with $N+1$ variables and its $N+1$ individual output variables are as follows:

$$\dot{\psi}(t) = \sum_{k=0}^{k=N} \dot{\psi}_p(t) \tag{8.10}$$

The error can be calculated as:

$$e(t) = \left[\psi(t) - \dot{\psi}(t) \right] \tag{8.11}$$

The modified composite observer filter is modelled as follows:

$$\hat{\dot{y}} = H_k \dot{y}_k + C_k e(t)$$

$$\dot{\psi}(t) H_k^t \dot{y}_k;\ k = 0,1,2,3,\ldots\ldots N \tag{8.12}$$

The filter can be defined briefly as follows:

$$\hat{\dot{y}} = H \times \dot{y} + D \times e(t)$$

$$\dot{\psi}(t) H_k^t \times \dot{y};\ k = 0,1,2,3,\ldots\ldots N \tag{8.13}$$

D is the gain of composite filter. Composite filter error ($\dot{E}$) can be defined as follows:

$$\dot{e}(t) = \left[H - DH^t \right] \times e(t) \tag{8.14}$$

The block diagram of modified composite observer filter is shown in Figure 8.2. This filter is having various frequency tuning components to tune fundamental frequency signals. Hence, the mason's gain formula is used to obtain open loop gain of composite observer filter as follows:

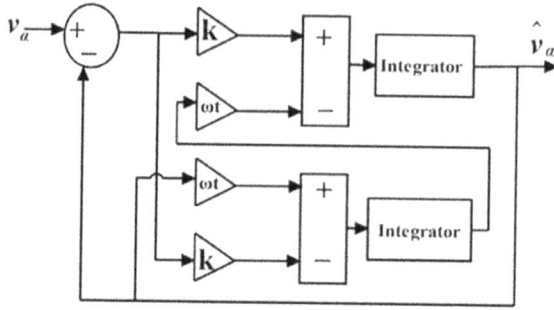

FIGURE 8.2 Modified composite observer filter.

$$\frac{\hat{u}_\alpha}{u_\alpha} = \frac{\dfrac{k}{s} + \dfrac{k\omega}{s^4}}{1 - \left\{\dfrac{-k}{s} - \dfrac{\omega^2}{S^6} - \dfrac{k\omega}{S^4}\right\}} \tag{8.15}$$

$$\frac{\hat{u}_\alpha(s)}{u_\alpha(s)} = \frac{\dfrac{k}{s} + \dfrac{k\omega}{s^4}}{1 - \left\{\dfrac{-k}{s} - \dfrac{\omega^2}{S^6} - \dfrac{k\omega}{S^4}\right\}} = \left[\frac{G(s)}{1 + G(s)}\right] \tag{8.16}$$

$$G(s) = \left[\frac{kS^5 + k\omega S^2}{S^6 + \omega^2}\right] \tag{8.17}$$

Figure 8.3a and b shows the Bode and Nyquist plots. It is observed from these plots that the phase margin influences the performance. Hence the PV grid-connected system will experience overshoot and ringing. A MATLAB SISO tool is used for calculating phase margin, which is tuned to 70° for the gain of 36.

8.4.2 INSTANTANEOUS REACTIVE POWER THEORY

A modified control scheme is implemented to mitigate PQ problems with shunt APF. There are many control techniques that are available in the literature to generate reference currents. The generalized *p-q* control technique [21] is a simple and effective solution for generation of reference currents. In *p-q* theory, the a-b-c coordinates are converted to two-phase system coordinates using parks transformation as follows:

$$\begin{bmatrix} u_\alpha \\ u_\beta \end{bmatrix} = \sqrt{\frac{2}{3}} \begin{bmatrix} 1 & \dfrac{-1}{2} & \dfrac{-1}{2} \\ 0 & \dfrac{\sqrt{3}}{2} & \dfrac{-\sqrt{3}}{2} \end{bmatrix} \begin{bmatrix} u_{sa} \\ u_{sb} \\ u_{sc} \end{bmatrix} \tag{8.18}$$

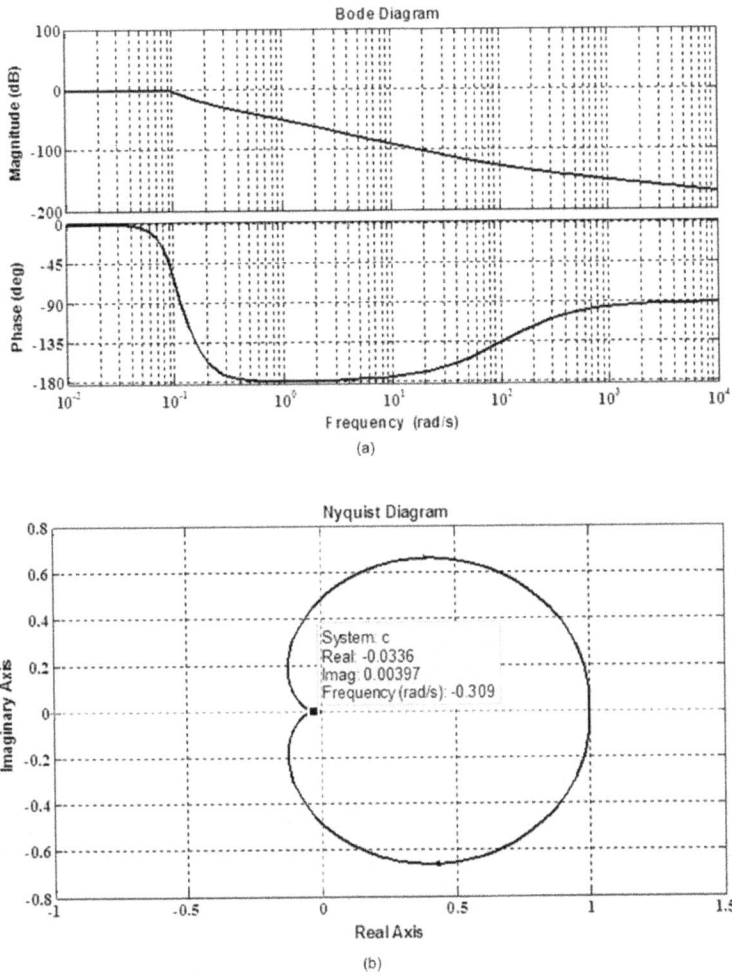

FIGURE 8.3 (a) Bode plot for the predictive tuned filter and (b) Nyquist plot.

The load currents in orthogonal components are i_α and i_β, as follows:

$$\begin{bmatrix} i_\alpha \\ i_\beta \end{bmatrix} = \sqrt{\frac{2}{3}} \begin{bmatrix} 1 & \dfrac{-1}{2} & \dfrac{-1}{2} \\ 0 & \dfrac{\sqrt{3}}{2} & \dfrac{-\sqrt{3}}{2} \end{bmatrix} \begin{bmatrix} i_{La} \\ i_{Lb} \\ i_{Lc} \end{bmatrix} \tag{8.19}$$

The instant actual AC power is calculated as:

$$P_{ac} = u_\alpha i_\alpha + u_\beta i_\beta \tag{8.20}$$

The low-pass filter used for extracting AC and DC powers is as follows:

$$P_{dc(loss)} = \left[U_{dc,ref} - U_{dc} \right] \left[k_p - \frac{K_i}{s} \right] \tag{8.21}$$

The total power is shown as follows:

$$P = \bar{P}_{ac} + P_{dc(loss)} \tag{8.22}$$

$$\begin{bmatrix} i_{s\alpha} \\ i_{s\beta} \end{bmatrix} = \frac{1}{u_\alpha^2 + u_\beta^2} \begin{bmatrix} u_\alpha & u_\beta \\ u_\beta & -u_\alpha \end{bmatrix} \begin{bmatrix} p \\ 0 \end{bmatrix} \tag{8.23}$$

From above Eq. (8.23), the $i_{s\alpha}$ and $i_{s\beta}$ currents are used for calculating the reference currents as:

$$\begin{bmatrix} i_{sa}^* \\ i_{sb}^* \\ i_{sc}^* \end{bmatrix} = \sqrt{\frac{2}{3}} \begin{bmatrix} 1 & 0 \\ \frac{-1}{2} & \frac{\sqrt{3}}{2} \\ \frac{1}{2} & \frac{-\sqrt{3}}{2} \end{bmatrix} \begin{bmatrix} i_{sa} \\ i_{s\beta} \end{bmatrix} \tag{8.24}$$

The actual and generated reference currents are compared and error will be generated. This error will go through the hysteresis controller for switching pulse generation for power electronic switches of Shunt active power filter (SAPF).

8.5   SIMULATION PERFORMANCE AND RESULTS DISCUSSION

The proposed modified composite observer filter based IRPT is developed for grid-connected PV system using MATLAB/Simulink software. The grid is connected with various types of loads where these loads dominantly inject even harmonic components into the grid. The proposed system is tested in three different conditions: firstly, it is tested under non-linear load with change in solar insolation; secondly, half-wave rectifier loads with distorted grid supply; and thirdly, non-linear load with distorted and unbalanced load conditions.

8.5.1   PV INVERTER MULTIFUNCTIONAL OPERATION

Performance of PV inverter, acting in double role, while feeding the active power to the grid during the PV power generation takes place and when the PV generation is off, the inverter acts as a shunt APF to mitigate various PQ issues in this section. The performance of proposed modified composite filter in multifunction operation is shown in Figure 8.4. When the PV system generation is ON from $t=1$ to 1.5 s,

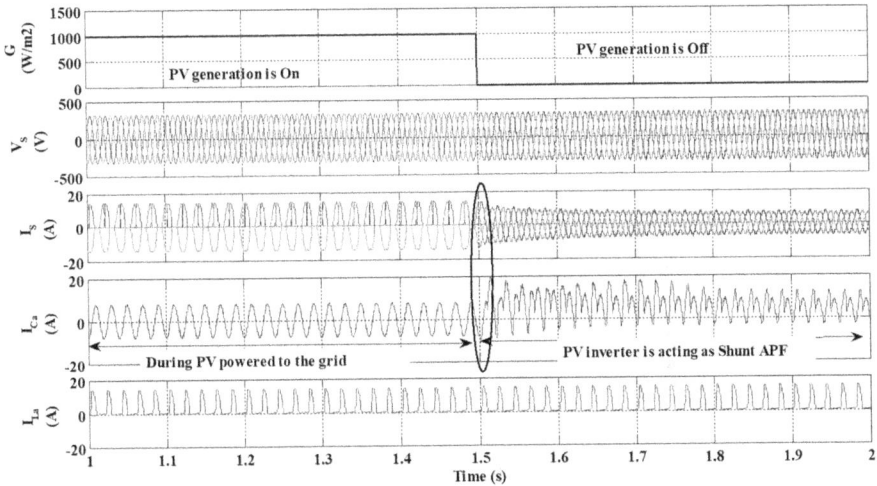

FIGURE 8.4 Performance of multifunctional operation.

the PV inverter successfully feeds the generated PV power to the grid. Also, when the PV power generation is off from $t = 1.5$ to 2 s, the same PV inverter changes its operation from feeding the active power to the grid to shunt APF by eliminating the even harmonic components from the grid system. The modified composite observer filter frequency components are tuned to eliminate even order harmonic from the power system network. The harmonic fast Fourier transform (FFT) analysis is made to observe the severity of the problem. The phase-a grid supply total harmonic distortion (THD) is 71.48 as shown in Figure 8.5a. The proposed modified composite observer filter successfully extracts the dominant even order harmonics for generation of reference currents, IRPT generates the reference source currents and the generated reference source currents are compared with actual source currents. The difference between actual and generated reference source currents feed to the PWM block for generation of controlling signal for PV inverter. The THD of grid phase-a supply after mitigation of harmonic is shown in Figure 8.5b. With this performance parameter, it can be observed that the proposed controller is robust in multifunctional operation, ease of operation and effective in mitigating even order harmonics.

8.5.2 Non-linear Load with Adverse Condition of Grid Supply

The modified composite observer filter based IRPT control strategy is tested in adverse conditions of grid supply. In case the grid network experiences unbalanced supply voltages when it goes under any failure, the integration and multifunction capability of PV system should not be affected. The performance of system parameters are shown in Figure 8.6.

The modified composite filter based IRPT extracts the positive sequence components from the unbalanced supply voltages and these extracted signals are used in IRPT for source reference currents. The source currents remain sinusoidal after

Fundamental (50Hz) = 16.08 (Peak), THD= 71.48%

(a)

Fundamental (50Hz) = 15.56 (Peak) , THD= 4.13%

(b)

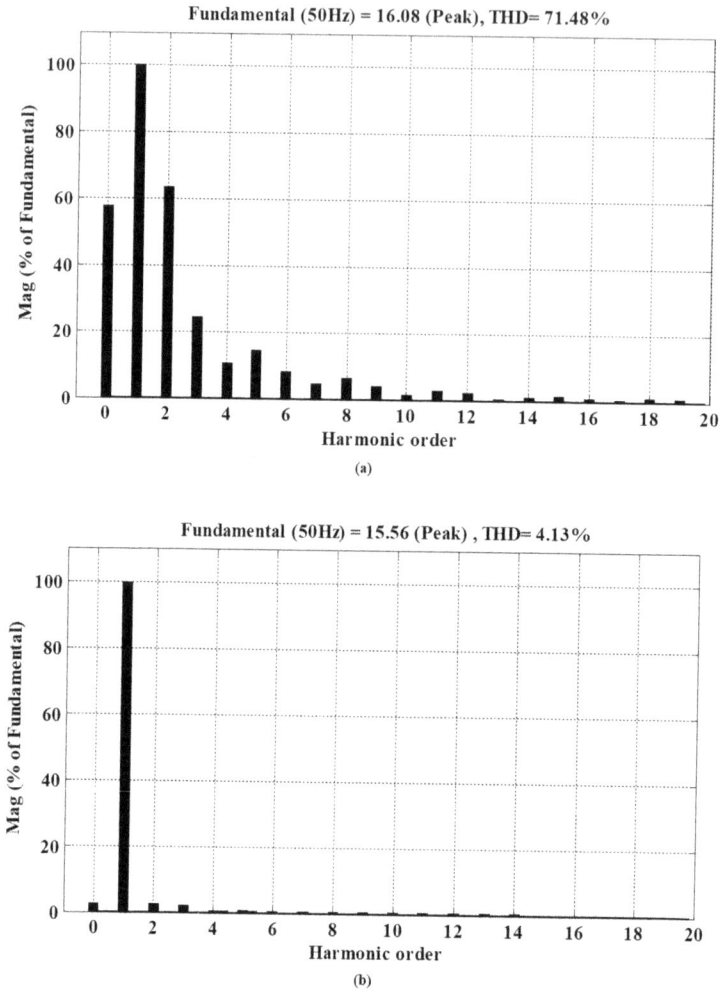

FIGURE 8.5   (a) THD of Grid's Phase 'a' supply before mitigation and (b) THD of grid's phase-a supply after mitigation of harmonic.

compensation. The THD spectral diagrams of three-phase source currents are shown in Figure 8.7a–c. The MATALB simulation results show that the proposed modified composite observer-based IRPT is effective to operate in multifunctional adverse conditions of grid voltages, mitigating the even order harmonics.

8.5.3   NON-LINEAR LOAD WITH ADVERSE CONDITION OF GRID SUPPLY AND NON-LINEAR LOAD

Grid system network undergoes failure of line due to line to ground fault condition, etc. The superiority of proposed modified composite filter is tested with adverse conditions of grid supply and non-linear load. The modified composite filters filter

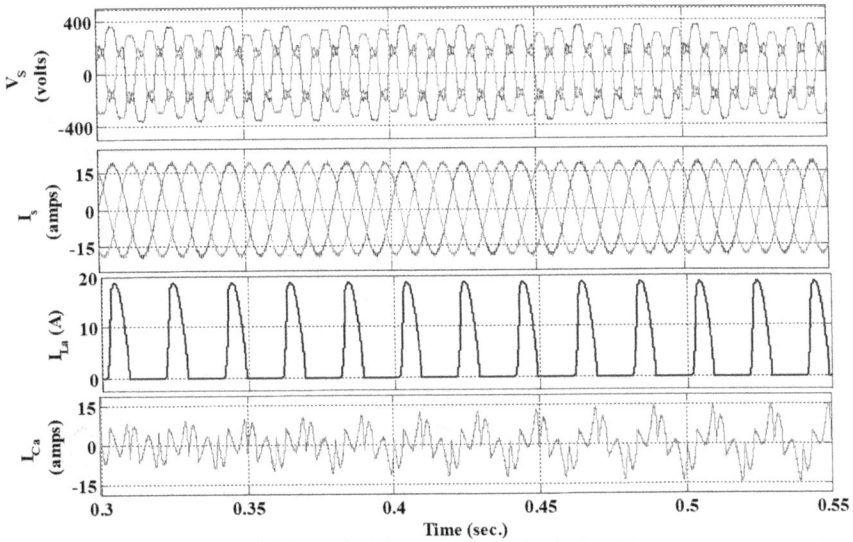

FIGURE 8.6 Performance parameters of modified composite filter based IRPT during grid unbalanced voltages.

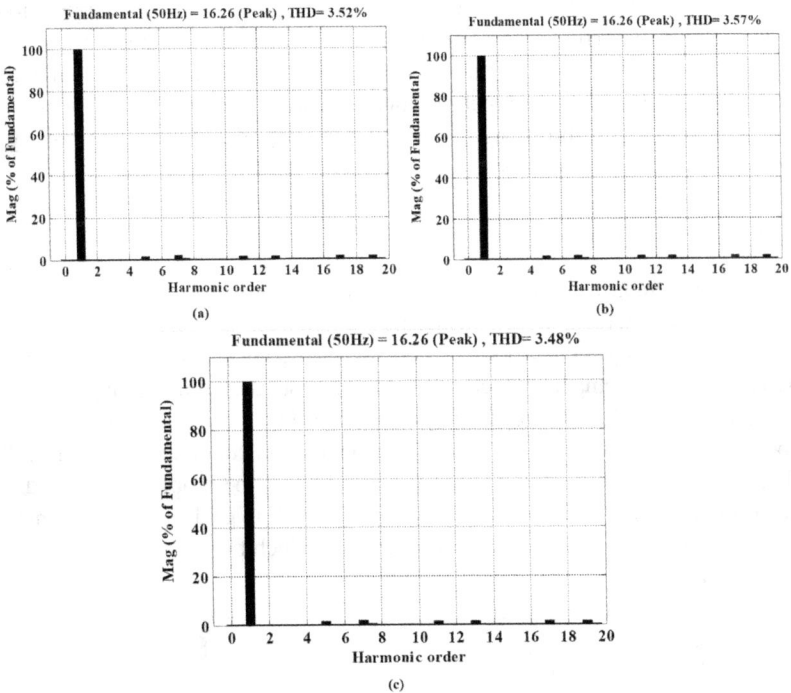

FIGURE 8.7 Harmonic spectral analysis (a) source current-a, (b) source current-b and (c) source current-c.

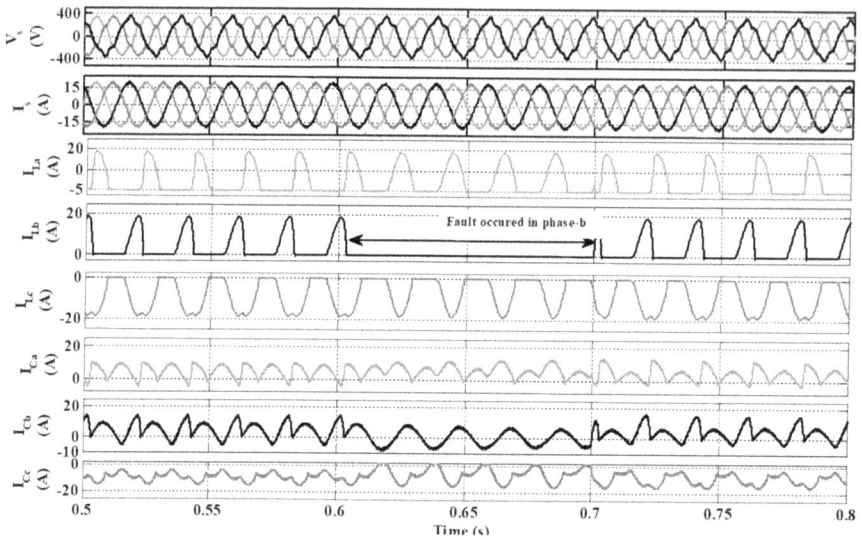

FIGURE 8.8 Performance PV system under unbalanced load and distorted grid supply voltages.

excerpts of the fundamental components of grid voltages and supply currents to generate the reference currents. The simulation performance parameters are shown in Figure 8.8. The grid supply currents remain sinusoidal and balanced which makes it a healthy system. It is observed that the control strategy is effective in synchronizing the PV with grid and eliminating the even order harmonics from the power system network under adverse conditions of load and grid supply.

8.6 CONCLUSION

In this chapter, a modified composite observer filter based IRPT is proposed for multifunctional PV system. Due to advancement in technology, rapid use of electronic equipment leads to inject more of even order harmonics into the grid power system network. The modified composite filter is tuned to filter out even order harmonic from the power system network. The proposed system is developed using MATLAB software. The performance of proposed system is tested in various conditions of supply voltage and load. It is observed from the simulation results that the proposed modified composite observer filter is effective in synchronizing the PV to grid and in multifunctional operation also compensating even order harmonics which further reduces the effect of EMI problems.

REFERENCES

1. R. Inglesi-Lotz, "The impact of renewable energy consumption to economic growth: A panel data application," *Energy Economics*, vol. 53, pp. 58–63, 2016.
2. I. Dincer, "Renewable energy and sustainable development: A crucial review," *Renewable and Sustainable Energy Reviews*, vol. 4, no. 2, pp. 157–175, 2000.

3. Y. H. Yoon and J. M. Kim, "Photovoltaic system application performance in extreme environments like desert conditions," *Journal of International Council on Electrical Engineering*, vol. 6, no. 1, pp. 214–223, 2016.

4. S. K. Yadav and U. Bajpai, "Energy, economic and environmental performance of a solar rooftop photovoltaic system in India," *International Journal of Sustainable Energy*, vol. 39, no. 1, pp. 51–66, 2020.

5. J. Selvaraj and N. A. Rahim, "Multilevel inverter for grid-connected PV system employing digital PI controller," *IEEE Transactions on Industrial Electronics*, vol. 56, no. 1, pp. 149–158, 2008.

6. M. Sharma, A. Achra, V. Gali, and M. Gupta, "Design and performance analysis of interleaved inverter topology for photovoltaic applications," in *Proceedings of IEEE International Conference on Power Electronics & IoT Applications in Renewable Energy and its Control (PARC)*, pp. 180–185, 2020.

7. O. P. Mahela, B. Khan, H. H. Alhelou, and P. Siano, "Power quality assessment and event detection in distribution network with wind energy penetration using stockwell transform and fuzzy clustering," *IEEE Transactions on Industrial Informatics*, vol. 16, no. 11, pp. 6922–6932, November 2020.

8. O. P. Mahela, B. Khan, H. Haes Alhelou, and S. Tanwar, "Assessment of power quality in the utility grid integrated with wind energy generation," *IET Power Electronics*, vol. 13, no. 13, pp. 2917–2925, 14 October 2020.

9. O. P. Mahela et al., "Recognition of power quality issues associated with grid integrated solar photovoltaic plant in experimental framework," *IEEE Systems Journal*. doi:10.1109/JSYST.2020.3027203.

10. O. P. Mahela, A. G. Shaik, B. Khan, R. Mahla, and H. H. Alhelou, "Recognition of complex power quality disturbances using S-transform based ruled decision tree," *IEEE Access*, vol. 8, pp. 173530–173547, 2020.

11. O. P. Mahela, B. Khan, H. H. Alhelou, S. Tanwar, and S. Padmanaban, "Harmonic mitigation and power quality improvement in utility grid with solar energy penetration using distribution static compensator," *IET Power Electron*, vol. 14, pp. 912–922, 2021.

12. R. K. Pachauri et al., "Impact of partial shading on various PV array configurations and different modeling approaches: A comprehensive review," *IEEE Access*, vol. 8, pp. 181375–181403, 2020.

13. R. K. Pachauri, O. P. Mahela, B. Khan, A. Kumar, S. Agarwal, H. H. Alhelou, and J. Bai, "Development of arduino assisted data acquisition system for solar photovoltaic array characterization under partial shading conditions," *Computers & Electrical Engineering*, vol. 92, p. 107175, 2021.

14. Pachauri, R.K., et al., "Shade dispersion methodologies for performance improvement of classical total cross-tied photovoltaic array configuration under partial shading conditions," *IET Renewable Power Generation*, vol. 15, pp. 1796–1811, 2021.

15. B. Khan, G. Agnihotri, P. Rathore, A. Mishra, and G. Naidu, "A cooperative game theory approach for usage and reliability margin cost allocation under contingent restructured market," *International Review of Electrical Engineering*, vol. 9, no. 4, pp. 854–862, 2014.

16. B. Khan, G. Agnihotri, and A. S. Mishra, "An approach for transmission loss and cost allocation by loss allocation index and co-operative game theory," *Journal of the Institution of Engineers (India): Series B*, vol. 97, pp. 41–46, 2016.

17. F. Wu, D. Sun, L. Zhang, and J. Duan, "Influence of plugging DC offset estimation integrator in single-phase EPLL and alternative scheme to eliminate effect of input DC offset and harmonics," *IEEE Transactions on Industrial Electronics*, vol. 62, no. 8, pp. 4823–4831, 2015.

18. V. Gali, N. Gupta, and R. A. Gupta, "Experimental investigations on multitudinal sliding mode controller-based interleaved shunt APF to mitigate shoot-through and PQ problems under distorted supply voltage conditions," *International Transactions on Electrical Energy Systems*, vol. 29, no. 1, p. e2701, 2019.

19. S. K. Chauhan, M. C. Shah, R. R. Tiwari, and P. N. Tekwani, "Analysis, design and digital implementation of a shunt active power filter with different schemes of reference current generation," *IET Power Electronics*, vol. 7, no. 3, pp. 627–639, 2013.

20. H.D. Taghirad, S. F. Atashzar, and M. Shahbazi, "Robust solution to three-dimensional pose estimation using composite extended Kalman observer and Kalman filter," *IET Computer Vision*, vol. 6, no. 2, pp. 140–152, 2012.

21. C. Xie, X. Zhao, M. Savaghebi, L. Meng, J. M. Guerrero, and J. C. Vasquez, "Multirate fractional-order repetitive control of shunt active power filter suitable for microgrid applications," *IEEE Journal of Emerging and Selected Topics in Power Electronics*, vol. 5, no. 2, pp. 809–819, 2016.

9 A Comprehensive Formal Reliability Study of Advanced Metering Infrastructure on Smart Grid

Abhishek Singh, Karm Veer Arya, Manish Gaur, and Vineet Kansal

CONTENTS

9.1 INTRODUCTION

The power sector in India is awakening to the need for smart grid. The blackouts in the north and east grids in the past raised an alarm reminding us that power sector reforms in the country have not been pursued with the sincerity and intensity which was required in the last Five-Year Plan. Advanced Metering Infrastructure (AMI) as part of "smart grid" innovativeness provides an answer for a few issues identified with power distribution. The two major problems in the Indian context are first uncontrolled high percentage (>35%) of aggregate losses and second mismatch in electricity availability and demand [4]. In the smart network framework, the remote correspondence system can assume a significant job in detecting/estimating the status (e.g., energy use, voltage oscillation, and force hardware harm) from various gadgets such as substations, smart meters, and sensors. The remote correspondence framework would be utilized to pass control signals to the various segments of the framework.

Traditional grid reliability analysis is a mature area, but smart grid reliability analysis still has more scope for development and testing [2]. The biggest challenges are the high computational requirements, due to the two-way communication involved in smart grid [22–24]. The reliability investigation of the communicating system framework helps demand-side management (DSM). There are various methods available for testing the reliability of smart network security frameworks including design compare, numerically and life cycle analysis. The numerical techniques, however, depend basically on certain iterative methodologies that produce outcomes. Exactness of the outcomes defines a life time of the smart network. The design compare is a more reliable method for testing smart network. Formal strategies are fit for defeating the previously mentioned imperative of incorrectness and have been broadly used to guarantee accuracy in applications, equipment, and physical structures [1]. To the best of our understanding, no previous research is available on the reliability study of advanced metering systems for smart grids. To obtain this, reliability study of AMI probabilistic model checker generally utilized proper method for analysing Markovian models of AMI component. To guarantee the exact outcomes in the field of smart grid, quality examination helps DSM [26–28]. To identify the effectiveness of this concept, Probabilistic Symbolic Model Checker (PRISM) probabilistic model checker is used to enhance the correctness of smart grid. The model checker can be used for modelling on both sides, i.e., at the customer side and the utility side of smart grid. Dependability analysis is another method for smart grid analysis that includes reliable quality estimations. AMI analysis of smart grid can be utilized to reduce the technical and commercial loss on the smart grid [5]. In a nutshell, the formal reliable study planning for smart grid system provides optimization which guarantees financially cost safe activity.

The remaining part of this chapter is structured as follows. Initially in the next section, a survey of the related work in the field of reliability and formal verification of AMI on smart grid is presented. After that, the major functionality of AMI and various issues are discussed in Section 9.3. The factors affecting the reliability of AMI on the basis of availability of component is presented in Section 9.4. In Section 9.5, we present technical assessment and possible solution to AMI challenges which we obtain in our study. Finally, the chapter is summarized in Section 9.6.

9.2 RELATED WORK IN FORMAL VERIFICATION OF SMART GRID AND PRISM MODEL CHECKER

Formal strategies have been generally used to investigate complex system, essentially because of the desperate need for precise examination in this domain. Formal verification of system can be performed either by model checking [30–36] or by mathematical proofs [45–52] which guarantee system properties. In [15], the authors considered the basic structure and execution of a smart grid substation communication and organization and assessed its reliability using the Reliability Block Diagrams (RBDs) and Fault Tree (FT) methods. The capacities of the diagnostic reliable quality investigation strategy were expanded in [16] by joining essential strategy with RBD and Markov displaying methods, for the determination examination of matrices. Distributed Probabilistic-Control Hybrid Automata is used with labelled transition system to investigate smart grid communication analysis framework utilizing RBDs [6]. The authors in [7] proposed a conventional FT-based dependability investigation technique for mechanical parameters to improve the bottlenecks of variable frequency devices which are power-automated components generally used to build effective framework. Coloured Petri nets have likewise been utilized to display dynamic RBDs [17] to depict the dynamic static quality conduct of frameworks.

Model checking [3,9] is a popularly used formal method strategy which fundamentally checks responsive framework. This framework displays a conduct based on the schedules and their conditions. The model checker incorporates the constrained state model of the system that must be a disjoint set of the normal structure properties generally conveyed in a transient method of reasoning. The model checker normally encourages if the properties hold for the given structure and gives an error in an event of a falling property. The essential confirmation standard behind model checking is to construct an accurate state-based model of the given structure and exhaustively check the given property for each state of this model. The assessment is obtained from model checking, which explains model checking is one of the most famous and widely used valid confirmation technique. PRISM is a probabilistic model checker for the analysis of irregular frameworks [43]. The arbitrary frameworks are displayed utilizing the state-based PRISM language, and their properties are indicated as probabilistic properties of the states. A model is distinguished by indicating the nature of the investigation, i.e., Continuous Time Markov Chain, Discrete Time Markov Chain, Markov Decision Procedure, Stochastic Multi-player Games, and so on [25].

Model checking is applicable to system structures those must be bounded or static state machines. Another critical hindrance of the model checking approach is state space explosion. The state space of a structure can be huge, or inconsistent. Thus, it ends up being computationally hard to research the entire state space with restricted resources of time and memory. This issue is commonly settled by working with theoretical, numerical, and semantics models, which guarantee correctness of the examination.

9.3 ADVANCED METERING INFRASTRUCTURE (AMI) FUNCTIONALITY AND ISSUES IN ENERGY DISTRIBUTION

AMI (Advanced Metering System) is the collective term for describing the entire infrastructure from the smart meter to the two-way communication network for controlling centre equipment and all applications that allow the collection and transmission of information about energy usage in near real time. AMI allows two-way consumer communications, which are the backbone of the smart grid. It is required that meters

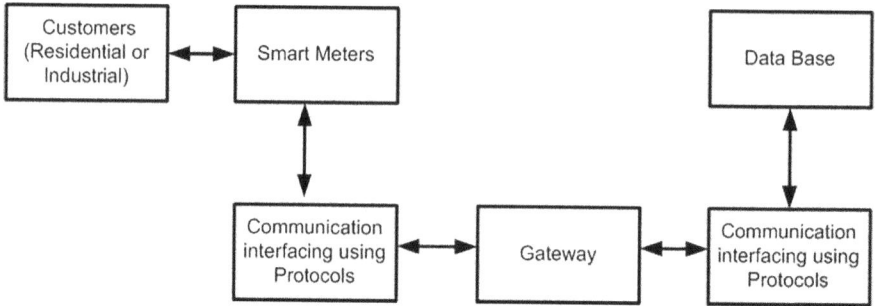

FIGURE 9.1 AMI block diagram (conventional).

supporting a common communication protocol are able to connect with the Head-End System (HES). To achieve this functionality, a robust system of communication devices along with two-way communication network is required [44] (Figure 9.1).

Based on the study [29] of existing literature, an opinion of experts at the national level was also sought on the main concerned areas of the distribution sector in India. It is expected that AMI system must broadly meet the following requirements.

9.3.1 AMI Functionalities

- **Energy Audits and Accounting** primarily keeps the record of monitoring and locating energy leakages at distribution transformers and low-tension and high-tension feeder distribution points in a timely manner. It also records the consumer energy consumption data of every 15 or 30 min intervals to forecast energy demands for the future.
- **Demand Response or DSM** to achieve a flatter demand curve using time of day multi-tariff metering and load connect/disconnect at consumer and low-tension feeder level for direct load control. Immediate and temporary disconnection of consumers on overloading would mean that only undisciplined consumers face power outages and not the entire population connected to a high-tension feeder.
- **Quick communication of abnormality alarms** from meters to the HES to locate and prevent theft and quick resolution of fault.
- **Efficient billing cycle** supporting postpaid and prepaid payment models.
- **Integration of Green Energy** (distributed sources of renewable energy) into the grid [29].
- The metering infrastructure in India would require following three elementary components:
- **Device Language Messaging Service (DLMS)** meters with fast communication port (9.6 kbps is the minimum and default baud rate specified in Indian Standard). Meters for Indian requirements are clearly classified and defined in the recently published specification of Indian collaborative semantic map matching (COSEM), "Data Exchange for Electricity Meter Reading, Tariff and Load Control, Companion Specification" (Doc: IS 15959:2011).

This standard is proposed for use as an ally to IEC 62056 arrangement of norms and gives inclusion to organized displaying of metering functionalities accessible at correspondence interfaces, with systems for distinguishing proof of these information objects (Object identification system (OBIS) codes) lastly for nearby or far off trade of these information messages by moving over different layers of correspondence channels [42].

- **Modems support two-way communication network** to connect the meters through HES. This is the supreme challenging portion of the AMI deployment. Acquiring data from different meters and reaching every 4h data from each meter to HES with more than 99% reliability in 24h is the most difficult task. The network would be secure, self-healing, self-discovering, robust, and reliable. The important and urgent commands to meters from the HES must also reach within a few seconds. To include existing non-DLMS meters under the AMI umbrella, the API to communicate with specific meters would reside in the modems [39,40].
- **HES and the Meter Data Management (MDM)** Software is used for data collection and making available in cloud [38], scheduling, meter configuration, and setting threshold values are important functions which a HES performs. On one side, the head-end interfaces with different types of meters via a collector or directly with meters equipped with a modem. On the other side, the HES interfaces with the MDM System (MDMS) on the standard Common Information Model defined under IEC 61968/70 [41].

9.3.2 Issues in Energy Distribution

It is observed that the main concern is: the occurrence of more than 30% of aggregate technical and commercial losses at the national level and the availability of power for all types of consumers. Inability to manage the peak loads also features among the three major issues. These three major issues with the energy distribution sector in India which AMI can address are listed below:

- **First Issue:** *Aggregate Technical and Commercial Losses*
  AMI can not only identify tampering of meters, it can be used to raise on-event alarms to detect tampers to make the prosecution process much more effective with the help of a load balancing algorithm [8]. AMI business processes can be utilized to deterministically and probabilistically determine potential tampers in the system which is otherwise not detected by individual meter data. Network mapping and data aggregation can be used to detect network hooking. A robust IT infrastructure can be used to close the information loop to determine unmetered or illegal connections.
- **Second Issue:** *Availability of Power*
  AMI can be used to enable efficient DSM – eventually leading to flatter demand curves and better capacity utilization. The demand flatness can also be enabled by incentivizing schemes like time-of-use tariffs and demand response programmes. Modern distributed demand management techniques riding on intelligent information and communications technology

(ICT) processes can resolve issues with distributed generation and storage enhancing availability.

- **Third Issue:** *Wastage of Energy*

    AMI can enable energy conservation in a big way. Research shows that nearly 15% energy reduction has been obtained just by providing measured information in money terms to consumers. Today most people don't conserve because they don't know their energy costs are going. Smart systems can not only measure but also intelligently defer loads to optimize costs against comfort.

9.4 RELIABILITY CHALLENGES AND ANALYSIS IN AMI

Reliability is an assessment of how reliable a communication system can perform data transfers according to specific requirements [54–63]. The reliability of a network can be increased by designing a hybrid network which has Power Line Carrier (PLC) as well as Radio Frequency (RF) communication interoperable devices communicating in the same seamless network. A hybrid system of RF and PLC communication may succeed in India where only PLC cannot be tried with success because of poor state of distribution network, susceptibility to noise, and very low data rate offered by PLC [18–21]. RF alone does not seem the best fit because its performance in relation to data exchange and traffic load within the buildings is suspected [53]. Reliability analysis is a technique to represent the error-free data movement in the corresponding framework as indicated by the particular necessities [11–14].

Smart grid applications like conveyed mechanization and anticipation of profoundly dependable information can bear the power outages in data movement. The major challenges of reliability analysis are as follows.

9.4.1 RELIABILITY ANALYSIS OF AMI

AMI is a framework that screen, gather, and dissect power creation and utilization. The principal part of an AMI is the system supporting two-way correspondence among sensors (e.g., smart meter) and the AMI elements.

9.4.2 COST AND AVAILABILITY STUDY

There will be cost identified with disappointment and inaccessibility of the interchange's framework. For instance, if meter information cannot be moved to open utility, an additional expense occurs in problematic force conveyance. Likewise, if the sensor and actuator arrangement cannot work appropriately, serious harm can happen to the force lattice.

9.4.3 REDUNDANCY AND NETWORK DESIGN

On the basis of reliability and availability analysis, network design allows us to improve the cost of power distribution and minimize the cost of power failure.

9.4.4 RELIABILITY UNDER SAFETY ATTACKS

Safety is another basic question in the smart meter framework. Denial of Service attack can prompt the damage of asset accessible for moving information and message.

9.4.5 RISK

Failure causes damage of a network. In this way, it is critical to recognize, survey, and organize the wellsprings of disappointment and their effect on framework execution.

Determining the reliability quality of the smart grid network is troublesome in light of the fact that it is comprised of thousands of parts, and despite the fact that the disappointment method of every part might be known, they are not autonomous of one another. Along these lines, the models required to lead dependability investigation are hard to create and are frequently unpredictable enough to make the strategies for investigation very lumbering. Consider the availability execution measure from the hypothesis of unwavering quality investigation [10]. The accessibility is estimated in consistent condition of the working system on the particular timespan of failure and fix part, system, and framework.

The availability of each component in a system can be calculated by following expression:

$$\text{Availability}(A) = \frac{\text{Uptime of component}}{\text{Uptime of component} + \text{Downtime of component}}$$

Uptime is also identified as the mean time between failures, and the downtime is identified as the mean time between repairs. Failure rate (F) can be calculated from availability as:

$$F = (1 - A)$$

Here, availability is the probability that the smart meter can direct power request to the MDMS. Figure 9.2 displays a dependency diagram which is helpful to evaluate the availability of smart grid AMI. A Dependency Diagram defines the contribution of each variable in the computation of device availability. The components may be linked in different series-parallel combinations depending on the system structure.

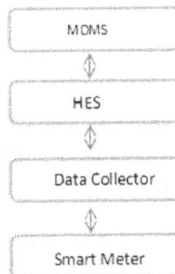

FIGURE 9.2 Dependency diagram for Advanced Metering Infrastructure.

9.4.6 Availability of AMI

The availability of an AMI is calculated by combining the availability of the different components of the system and can be computed as:

$$A_{AMI} = \left[\text{Availability of MDMS} \times \text{Availability of HES} \times \text{Availability of Data Collector}\right.$$
$$\left. \times \text{Availability of Smart Meter}\right]$$

Another method to analyse the reliability is the FT analysis by simplifying the dependency diagram of the AMI. When the association between a HES gateway and a smart meter is inaccessible and the MDMS cannot gather the genuine power request from a smart meter, then the evaluated request will be utilized.

9.5 TECHNOLOGY ASSESSMENT AND RELIABILITY ENHANCEMENT OF AMI NETWORK

RF and PLC media are used to implement the two-way communication network. RF network can use the free bands in India as allowed by Govt. of India in its Gazette. The various technical assessments obtained on the basis of the above section are as follows.

9.5.1 PLC Communication (PLCC)

Previous technologies such as FSK (Frequency Shift Keying), S-FSK (Spread Spectrum Frequency Shift Keying) in power line communication suffered from issues of low data rate (data rate < 2,400 bps) and susceptibility to narrowband noise. Latest third-generation PLC technology uses multicarrier modulation technique called narrowband Orthogonal Frequency Division Multiplexing. This overcomes the problem of interference by narrowband noise, and data rates >10 kbps are also made possible.

9.5.2 Upcoming Technologies and Supporting Protocols in PLC

PRIME is an open and royalty-free standard for advanced narrowband PLC communication. The three building blocks of PRIME are physical layer (PHY), Media Access Control (MAC), and Convergence layers. The PHY layer of PRIME is based on Orthogonal Frequency Division Multiplexing modulation technique achieving data rates up to 130 kbps. The MAC layer is optimized for low-voltage PLC networking.

The reliability of a network can be enhanced by designing hybrid network communication. Figure 9.3 describes network topology having grouped meters sharing a common modem. Expandability for Home Area Network implementation is possible with a ZigBee compliant hub which communicates with Neighbourhood Area Network on one end and with other home devices on the other end.

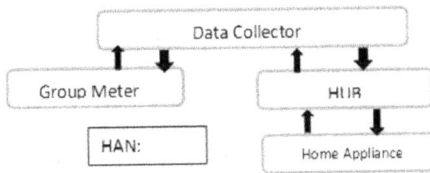

FIGURE 9.3 Group meter network topology.

9.6 CONCLUSION

In this chapter, formal verification and reliability analysis of AMI is presented. The important factors to enhance power generation and reduce the price of power supply in smart grid system have been discussed in detail. Formal verification of smart grid system plays an important role to guarantee system safety and properties. Reliability analysis of AMI is used to identify the availability of data communication network of AMI, and consequently identify the power demand in case of data communication network failures. The probabilistic model checking tools are helpful to analyse the system's critical safety properties and related issues.

REFERENCES

1. A. Khurram, H. Ali, A. Tariq, and O. Hasan. Formal Reliability Analysis of Protective Relays in Power Distribution Systems. In: C. Pecheur, M. Dierkes (eds) *Formal Methods for Industrial Critical Systems. FMICS 2013*. Lecture Notes in Computer Science, vol. 8187. Springer, Berlin, Heidelberg, 2013.
2. P. M. Anderson and S. K. Agarwal. An improved model for protective system reliability. *Reliability*, 41(3): 422–426, 1992.
3. C. Baier and J. Katoen. *Principles of Model Checking*. MIT Press, 2008.
4. L. Chen, K. Zhang, Y. Xia, and G. Hu. Scheme Design and Real Time Performance Analysis of Information Communication Network used in Substation Area Backup Protection. In: *Proceedings of Power Engineering and Automation Conference*, pp. 1–4, 2012.
5. N. Tyagi, A. Rana, and V. Kansal. Resourceful System-level Optimized Placement of Virtual Machines for Cloud Computing. In: *Proceedings of 2018 4th IEEE International Conference on Applied and Theoretical Computing and Communication Technology (iCATccT)*, pp. 99–102, 2018.
6. A. Platzer, J. Martins, and J. Leite. Statistical Model Checking for Distributed Probabilistic-Control Hybrid Automata with Smart Grid Applications. In: Shengchao Qin and Zongyan Qiu (eds) *Formal Methods and Software Engineering, Lecture Notes in Computer Science*, vol. 6991, Springer, Berlin, Heidelberg, pp. 131–146, 2011.
7. G. Norman, J. Rutten, M. M. Kwiatkowska, and D. Parker. *Mathematical Techniques for Analyzing Concurrent and Probabilistic Systems. CRM Monograph Series*, vol. 23. American Mathematical Society, New York, 2004.
8. N. Tyagi, A. Rana, and V. Kansal. Weighted Optimization Load Balancing Algorithm in Virtualization. *Journal of Engineering and Applied Science*, 14(22): 8386–8402, 2019.
9. M. Kwiatkowska, G. Norman, and D. Parker. PRISM: Probabilistic Symbolic Model Checker. In: Field T., Harrison P.G., Bradley J., Harder U. (eds) *Computer Performance Evaluation: Modelling Techniques and Tools. Lecture Notes in Computer Science*, vol. 2324. Springer, Berlin, Heidelberg, pp. 200–204, 2002.

10. D. Niyato, P. Wang, and E. Hossain. Reliability Analysis and Redundancy Design of Smart Grid Wireless Communications System for Demand Side Management. *IEEE Wireless Communications*, 19(3): 38–46, 2012.
11. K. Poulsen. Tracking the Blackout Bug. SecurityFocus2004. Available at URL: http://www.securityfocus.com/news/8412.
12. R. Robidoux, H. Xu, L. Xing, and M. Zhou. Automated Modeling of Dynamic Reliability Block Diagrams using Colored Petri Nets. *IEEE Transaction on Systems, Man and Cybernetics, Part A: Systems and Humans*, 40(2): 337–351, 2010.
13. J. Pan R. Yu, Y. Chen and R. W. Vesel. Generic Reliability Evaluation Method for Industrial Grids with Variable Frequency Drives. *IEEE Wireless Communications*, 5(4B): 83–88, 2013.
14. J. Wafler and P. E. Heegaard. A Combined Structural and Dynamic Modelling Approach for Dependability Analysis in Smart Grid. In: *Proceedings of ACM Symposium on Applied Computing*, SAC '13, pp. 660–665, 2013.
15. E. Yüksel, H. Zhu, H. R. Nielson, H. Huang and F. Nielson. Modelling and Analysis of Smart Grid: A Stochastic Model Checking Case Study. In: *Sixth International Symposium on Theoretical Aspects of Software Engineering*, Beijing, pp. 25–32, 2012.
16. R. Zeng, Y. Jiang, C. Lin, and X. Shen. Dependability Analysis of Control Center Networks in Smart Grid using Stochastic Petri Nets. *IEEE Transaction on Parallel and Distributed Systems*, 23(9): 1721–1730, 2012.
17. G. Zhanjun, C. Qing, and L. Zhaofei. Fault Diagnosis Method for Smart Substation. In: *Proceedings of 2011 International Conference on Advanced Power System Automation and Protection*, Beijing, pp. 427–430, 2011.
18. S. Dharmaraja, V. Jindal, and U. Varshney. Reliability and Survivability Analysis for UMTS Networks: An Analytical Approach. *IEEE Transaction on Network and Service Management*, 5(3): 132–42, 2008.
19. G. Egeland and P. Engelstad. The Availability and Reliability of Wireless Multi-Hop Networks with Stochastic Link Failures. *IEEE Journal of Selected Area in Communication*, 27(7): 1132–46, 2009.
20. O. M. Al-Kofahi and A. E. Kamal. Survivability Strategies in Multihop Wireless Networks. *IEEE Wireless Communication*, 17(5): 71–80, 2010.
21. A. G. Bruce. Reliability Analysis of Electric Utility SCADA Systems. *IEEE Transaction on Power Systems*, 13(3): 844–49, 1998.
22. Z. Xie, G. Manimaran, V. Vittal, A. G. Phadke, and V. Centeno. An Information Architecture for Future Power Systems and Its Reliability Analysis. *IEEE Transaction on Power Systems*, 17(3): 857–863, 2002.
23. L. Saker and S. E. Elayoubi. Sleep Mode Implementation Issues in Green Base Stations. In: *Proceedings of the 21st Annual IEEE International Symposium on Personal, Indoor and Mobile Radio Communications (PIMRC 2010)*, Istanbul, Turkey, pp. 1681–1686, 2010.
24. S. Sengupta, S. Das, M. Nasir, A. V. Vasilakos, and W. Pedrycz. An evolutionary Multiobjective Sleep-scheduling Scheme for Differentiated Coverage in Wireless Sensor Networks. *IEEE Transactions on Systems, Man, and Cybernetics – Part C: Applications and Reviews*, 42(6): 1093–1102, 2012.
25. J. W. Stahlhut, T. J. Browne, G. T. Heydt, and V. Vittal. Latency Viewed as A Stochastic Process and its Impact on Wide Area Power System Control Signals. *IEEE Transactions on Power Systems*, 23(1): 84–91, 2008.
26. R. Billinton, M. Fotuhi-Firuzabad, and L. Bertling. Bibliography on the Application of Probability Methods in Power System Reliability Evaluation. *IEEE Transactions on Power System*, 16(4): 595–602, 2001.
27. A. G. Phadke and R. M. de Moraes. The Wide World of Wide-Area Measurement. *IEEE Power Energy Magazine*, 6(5): 52–65, 2008.

28. X. Xie, Y. Xin, J. Xiao, J. Wu, and Y. Han. WAMS Applications in Chinese Power System. *IEEE Power Energy Magazine*, 4(1): 54–63, 2006.

29. A. Singh and P. Pandey. Energy Efficient Resource Management Techniques in Cloud Environment for Web Based Community by Machine Learning: A Survey. *International Journal of Web Based Communities*, 15(3): 1, 2019.

30. A. Hartmanns and H. Hermanns. A Modest Approach to Checking Probabilistic Timed Automata. In: *Proceedings of QEST'09*, pp. 187–196, 2009.

31. J. Heath, M. Kwiatkowska, G. Norman, D., Parker and O. Tymchyshyn. Probabilistic Model Checking of Complex Biological Pathways. *Theoretical Computer Science*, 391(3): 239–257, 2008.

32. T. Herault, R. Lassaigne, F. Magniette, and S. Peyronnet. Approximate Probabilistic Model Checking. In: Steffen B. and Levi G. (eds) *Verification, Model Checking, and Abstract Interpretation. VMCAI 2004*. Lecture Notes in Computer Science, vol. 2937. Springer, Berlin, Heidelberg, pp. 307–329, 2004.

33. J. Katoen, I. S. Zapreev, E. M. Hahn, H. Hermanns, and D. N. Jansen. The Ins and Outs of the Probabilistic Model Checker MRMC. In: *2009 Sixth International Conference on the Quantitative Evaluation of Systems*, Budapest, pp. 167–176, 2009.

34. M. Kattenbelt, M. Kwiatkowska G. Norman, and D. Parker. Abstraction Refinement for Probabilistic Software. In: Jones N.D., Müller-Olm M. (eds) *Verification, Model Checking, and Abstract Interpretation. VMCAI 2009*. Lecture Notes in Computer Science, vol. 5403. Springer, Berlin, Heidelberg, pp. 1–15, 2008.

35. A. Mohsenian-Rad, V. W. S. Wong, J. Jatskevich, R. Schober, and A. Leon-Garcia. Autonomous Demand-Side Management based on Game-Theoretic Energy Consumption Scheduling for the Future Smart Grid. *IEEE Transaction on Smart Grid*, 1(3): 320–31, 2010.

36. A. Singh, P. Pandey, and B. Gupta. Energy Management System for Analysis and Reporting in the Advanced Metering. In: A. Luhach, D. Singh, P. A. Hsiung, K. Hawari, P. Lingras, and P. Singh (eds) *Advanced Informatics for Computing Research. ICAICR 2018*. Communications in Computer and Information Science, vol. 955. Springer, Singapore, pp. 44–53, 2018.

37. N. Tyagi, A. Rana, and V. Kansal. Load Distribution Challenges with Virtual Computing. In: V. Solanki, M. Hoang, Z. Lu, and P. Pattnaik (eds) *Intelligent Computing in Engineering. Advances in Intelligent Systems and Computing*, vol. 1125, Springer, Singapore, pp. 51–56, 2020.

38. W. Wang, Y. Xu, and M. Khanna. A Survey on the Communication Architectures in Smart Grid. *Computer Networks*, 55: 3604–3629, 2011.

39. C. Wu, H. Ho, S. Lee, and S. Chang. Energy-Aware Link Scheduling Algorithms to Maximize Life Cycle of Relay Nodes on IEEE 802.16e Mesh Networks. In: *Proceedings of 26th IEEE International Conference on Advanced Information Networking and Applications*, pp. 51–58, 2012.

40. E. Zio and G. Sansavini. Vulnerability of Smart Grids with Variable Generation and Consumption: A System of Systems Perspective. *IEEE Transaction on Systems, Man, and Cybernetics: Systems*, 43(3):477–487, 2013.

41. D. Becker and T. L. Saxton. CIM Standard for Model Exchange between Planning and Operations. In: *2008 IEEE Power and Energy Society General Meeting - Conversion and Delivery of Electrical Energy in the 21st Century*, Pittsburgh, PA, 2008, pp. 1–5, doi:10.1109/PES.2008.4596090.

42. The PRISM model checker website. Available: http://www.prismmodelchecker.org/.

43. OPNET Technologies, Inc. Opnet simulator. Available: http://www.opnet.com.

44. C. A. R. Hoare. *Communicating Sequential Processes*. Hoboken, NJ: Prentice-Hall, 1985.

45. I. Lanese and D. Sangiorgi. An Operational Semantics for a Calculus for Wireless Systems. *Theoretical Computer Science*, 411(19): 1928–1948, 2010.

46. S. Liu, X. Wu, Q. Li, H. Zhu, and Q. Wang. Formal Approaches to Wireless Sensor Networks. In: *Proceedings of IEEE 5th International Conference on Secure Software Integration and Reliability*, pp. 11–18, 2011.
47. M. Merro. An Observational Theory for Mobile Ad Hoc Networks (Full Version). *Information & Computation*, 207(2): 194–208, 2009.
48. N. Mezzetti and D. Sangiorgi. Towards A Calculus for Wireless Systems. *Electronic Notes in Theoretical Computer Science*, 158: 331–353, 2006.
49. R. Milner. *A Calculus of Communicating Systems. Lecture Notes in Computer Science*, vol. 92, Springer, Berlin Heidelberg, 1980.
50. S. Nanz and C. Hankin. A Framework for Security Analysis of Mobile Wireless Networks. *Theoretical Computer Science*, 367(1–2): 203–227, 2006.
51. H. R. Nielson, F. Nielson, and R. Vigo. A Calculus for Quality. In: Păsăreanu C.S., Salaün G. (eds) *FACS 2012, Lecture Notes in Computer Science*, vol. 7684, Springer, Berlin Heidelberg, 2012.
52. N. Tyagi, A. Rana, and V. Kansal. Creating Elasticity with Enhanced Weighted Optimization Load Balancing Algorithm in Cloud Computing. In: *Proceedings of IEEE Amity International Conference on Artificial Intelligence (AICAI2019)*, pp. 600–604, 2019.
53. "Data Exchange for Electricity Meter Reading, Tariff and Load Control." Available: https://www.cescoorissa.com/tenders/ps93/Guide_lines.pdf.
54. T. F. Agajie, B. Khan, H. H. Alhelou, and O. P. Mahela. Optimal Expansion Planning of Distribution System Using Grid-Based Multi-Objective Harmony Search Algorithm. *Computers & Electrical Engineering*, 87: 106823, 2020.
55. B. Khan, H. H. Alhelou, and F. Mebrahtu. A Holistic Analysis of Distribution System Reliability Assessment Methods with Conventional and Renewable Energy Sources. *AIMS Energy*, 7(4): 413–429, 2019.
56. D. Anteneh and B. Khan. Reliability Enhancement of Distribution Substation by Using Network Reconfiguration a Case Study at Debre Berhan Distribution Substation. *International Journal of Economy, Energy and Environment*, 4(2): 33–40, 2019.
57. B. Khan, G. Agnihotri, G. Gupta, and P. Rathore. A Power Flow Tracing based Method for Transmission Usage, Loss & Reliability Margin Allocation. *AASRI Procedia*, 7: 94–100, 2014.
58. B. Khan, G. Agnihotri, P. Rathore, A. Mishra, and G. Naidu. A Cooperative Game Theory Approach for Usage and Reliability Margin Cost Allocation under Contingent Restructured Market. *International Review of Electrical Engineering*, 9(4): 854–862, 2014.
59. B. Khan and G. Agnihotri. A Comprehensive Review of Embedded Transmission Pricing Methods Based on Power Flow Tracing Techniques. *Chinese Journal of Engineering*, 2013, Article ID 501587: 13, 2013.
60. T. F. Agajie, B. Khan, J. M. Guerrero, and O. P. Mahela. Reliability Enhancement and Voltage Profile Improvement of Distribution Network using Optimal Capacity Allocation and Placement of Distributed Energy Resources. *Computers & Electrical Engineering*, 93: 107295, 2021.
61. S. Khare, B. Khan, and G. Agnihotri. A Shapley Value Approach for Transmission Usage Cost Allocation under Contingent Restructured Market. In: *2015 International Conference on Futuristic Trends on Computational Analysis and Knowledge Management (ABLAZE)*, Noida, pp. 170–173, 2015.
62. B. Khan, G. Agnihotri, S. E. Mubeen, and G. Naidu. A TCSC Incorporated Power Flow Model for Embedded Transmission Usage and Loss Allocation. *AASRI Procedia*, 7: 45–50, 2014.
63. B. Khan and G. Agnihotri. A Novel Transmission Loss Allocation Method Based on Transmission Usage. In: *2012 IEEE Fifth Power India Conference*, pp. 1–3, 2012.

10 An Optimized Approach for Restructuring of Transmission System to Mitigate Renewable Energy Constraints

*Sanju Verma, Om Prakash Mahela,
Sunil Agarwal, Akhil Ranjan Garg,
and Rahul Garg*

CONTENTS

DOI: 10.1201/9781003278030-10

10.1 INTRODUCTION

The integration of renewable energy (RE) into the network of power systems in the western parts of Rajasthan, India, is continuously increasing. Solar energy is integrated into the grid in bulk in the Bhadla region, whereas the maximum wind power generation is integrated into the grid in the Jaisalmer and Barmer regions. The transmission network developed by the Rajasthan Rajya Vidyut Prasaran Nigam Limited (RVPN) and the Power Grid Corporation of India (PGCIL) together is enough to evacuate solar power from the Bhadla region [1–4]. However, in the Barmer region, the transmission system developed by the RVPN has constraints on the evacuation of RE power. Hence, to evacuate RE power in bulk from the Barmer region to the load centres, additional transmission systems are required, which can be assessed by restructuring the existing transmission system to incorporate additional transmission elements. Many different models are available to perform restructuring of power system networks, specifically in India, which helps to adjoin load requirements. This will help in meeting the electricity requirements of society and the market in Rajasthan by evacuating the RE from the generating wind power plants to the load centres [5]. This also helps to meet the increased load demand and evacuation of increased generation, either conventional or RE. The methods and approaches used for the purpose of planning as well as operations, which were well established during the era of the past decades, are changing. The need has also arisen that India should recognize and meet these forthcoming challenges [6–15]. To make the market competitive, various methods have been adopted for the purpose of restructuring the industry of power systems. Different types of rectifies have been taken in the past, looking at the setup of the organizations, financial condition of the utilities, control structure adopted and coordination system [16]. For various reasons, the restructuring of power system networks has been carried out in different countries around the world. The Romanian government has taken a step towards the restructuring of the network of state utilities and assets of the state-owned power generation system, which had been considered in 2010. This has been achieved with the help of the creation of two companies at the national level. These companies help all the power generation assets be owned by the state governments or by the two companies owned by the state and related to coal-mining. This is also sped up by the three distribution companies owned by the state in regional electricity. The Romanian Ministry of Economy, Trade and Business Environment has taken a decision on the restructuring of the majority of assets of power generation in the country [17]. The authors analysed the welfare-related complications associated with the implications of reforms to the power sector in a condition where there are both public-sector undertakings and utilities in the private sector [18]. This study is related to the restructuring of the state power sector in West Bengal, India. A 15-year period has been considered for the review, which is associated with the pre-restructuring period and post-restructuring period. Analysis of the panel data related to four categories of customer shows the negative impact associated with the disintegration on welfare-related activities due to the increased cost of the transmission and increased rates of the tariff, which results in a decreased number of power consumers and power producers. The authors of [19] investigated the process used to implement

various governmental schemes from a network perspective in Haoqiao, a poor village in northern China. The authors mainly focused on interactions between consumers and their impacts on the restructuring process in rural areas. A design and
implementation of a system protection scheme for the Kawai-Kalisindh-Chhabra
Thermal Complex in Rajasthan is reported [20]. Restructuring of the power network
affects the system's stability, power flows, power quality, protection, etc., which
have been investigated in the literature [21–41].

Transmission constraints for RE evacuation can be mitigated by restructuring of the existing transmission system and the addition of new transmission elements, which can be achieved by different options (proposals). The optimal proposal
that can evacuate maximum power and provide stability to the grid is essentially
required. This is assessed by a detailed load flow study on the transmission system
of Rajasthan by considering different options. Hence, this research work is targeted
at restructuring the existing transmission network in the Barmer region of Rajasthan
to evacuate the RE power from the generating stations to the load centres with minimum cost and minimum losses in the network. This will give a techno-economically
feasible solution. Different options have been analysed and compared to the existing
power system network in Rajasthan in terms of losses to find the optimized solution
for evacuating RE power to load centres in the Barmer region of Rajasthan. The
study was performed using the MiPower software. The entire transmission network
of Rajasthan, considering the interstate connections, is considered and designed
using MiPower software. Different possible alternatives are analysed with different
scenarios of RE generation because in western Rajasthan, the RE penetration on the
utility grid is high. A short-circuit study of various proposals has been compared
with the base case for investigating the impact of restructuring on the protection
schemes. An optimized solution is advised to evacuate the RE power.

10.2 STUDY OBJECTIVE AND PROBLEM FORMULATION

The integration of wind power into the transmission system in the Barmer region of
Rajasthan is continuously increasing, and the transmission system is fully loaded.
The RE power is integrated into the 400kV grid substation (GSS) at Barmer.
Furthermore, the RE power is integrated into the 400kV GSS Akal and the 400kV
GSS Jaisalmer-II, which are linked to the 400kV GSS Barmer through 400kV transmission lines. Power flows from these two 400kV GSSs to the 400kV GSS Barmer
during the peak generation of RE power. Hence, there is an urgent requirement to
provide an alternative path for the evacuation of power from the 400kV GSS Barmer
to the load centers. A snapshot of the power map of the Barmer region is shown in
Figure 10.1 [42]. It is observed that in the region around 220kV GSS Dhorimanna,
there is a heavy load centre and existing transmission elements in the region are fully
loaded. Hence, the evacuation of the RE power from the 400kV GSS Barmer to the
Dhorimanna region through alternative transmission lines would help to mitigate the
evacuation constraints. Therefore, the main objective of this study is to investigate
the proposal for the creation of the new 220kV GSS in the Dhorimmanna region
with suitable 220kV interconnections so that load requirements of the region may be
met by the use of RE power from the 400kV GSS Barmer. This can be achieved by

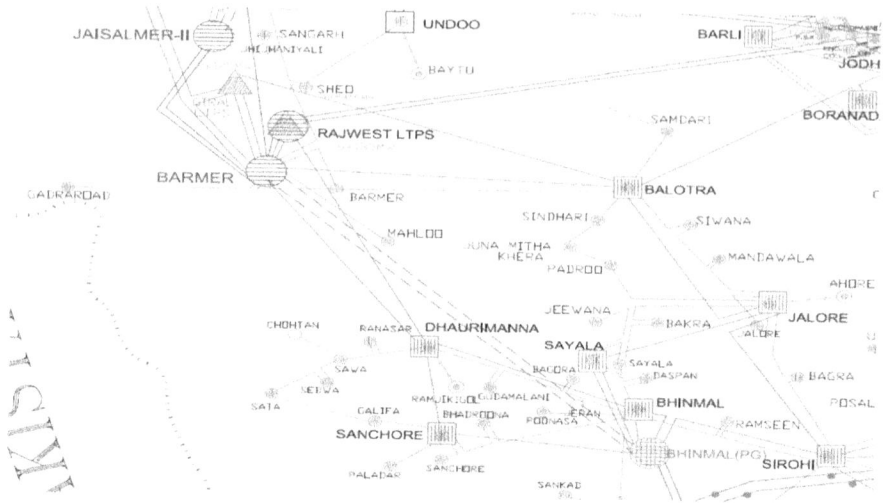

FIGURE 10.1 Power map of the Barmer region of the Rajasthan state of India.

investigating the new proposals through load flow studies and testing these options through short-circuit studies.

10.3 BASE TRANSMISSION SYSTEM

The existing network of the transmission system with voltage levels of 132, 220, 400, and 765 kV is modelled for the proposed study using the MiPower software. Further, the conventional generators installed on the power plants, including thermal, nuclear, and hydel, are considered in the modelling. Further, the RE power integrated into the transmission network is also considered for the study. All the transmission lines and grid substations (GSS) rated at voltage levels, such as 765, 400, 220, and 132 kV, are considered. This system contains the power system network developed by both central transmission utility (CTU) and the state transmission utility (STU). In Rajasthan, the RVPN performs the responsibility of the STU for the development of the transmission network. On behalf of CTU, Power Grid Corporation of India (PGCIL) is in charge for developing India's high voltage level network.

10.4 PROPOSED OPTIMIZED RESTRUCTURING METHOD

To perform the optimized approach for restructuring of the transmission system in the Barmer region of Rajasthan, different proposals have been framed looking at the minimum lengths of the lines to minimize the cost, right of way (RoW) problems in constructing the proposed transmission lines and reliability of power supply even during the contingency conditions. Here, the contingency condition refers to the outage of one or more lines. As per grid code guidelines of the central electricity regulatory commission (CERC), the transmission utility should comply with the $N-1$ contingency (network sufficient to meet the load demand and evacuate power with outage of one line) network as far as demand-side load management is considered. However, for the generators, the transmission utility should comply with the $N-1-1$ and $N-2$ contingency conditions. Hence, the possible proposals have taken care of the $N-1$ contingency. Line-in-line-out (LILO) lines simply indicate that the circuit of the existing line is broken at a point and that these new points are taken to the new location. The following are the possible proposals being considered for the study:

10.4.1 OPTION 1

The following transmission system is considered in option 1:-

- 1×160 MVA, 220/132 kV Power Transformer and $1 \times /25$ MVA, 132/33 kV Power Transformer at 220 kV GSS Sawa (Proposed).
- 100 km 220 kV D/C line from 400 kV GSS Barmer to 220 kV GSS Sawa (Proposed).
- 5 km LILO of existing 132 kV S/C Sawa (132 kV GSS)-Choutan line at 220 kV GSS Sawa (Proposed).
- 5 km LILO of existing 132 kV S/C Sawa (132 kV GSS)-Ranasar line at 220 kV GSS Sawa (Proposed).

10.4.2 OPTION 2

Following transmission system is considered in the option 2:-

- 1×160 MVA, 220/132 kV Power Transformer and $1 \times 20/25$ MVA, 132/33 kV Power Transformer at 220 kV GSS Sawa (Proposed).
- 50 km LILO of 220 KV S/C Dhorimanna-Sanchore line at 220 KV GSS Sawa.
- 5 km LILO of existing 132 kV S/C Sawa (132 kV GSS)-Choutan line at 220 kV GSS Sawa (Proposed).
- 5 km LILO of existing 132 kV S/C Sawa (132 kV GSS)-Ranasar line at 220 kV GSS Sawa (Proposed).

10.4.3 OPTION 3

The following transmission system is considered in option 3:-

- 1×160 MVA, 220/132 kV Power Transformer and $1 \times 20/25$ MVA, 132/33 kV Power Transformer at 220 kV GSS Sawa (Proposed).
- 100 km 220 kV D/C line from 400 kV GSS Barmer to 220 kV GSS Sawa (Proposed).
- 5 km LILO of existing 132 kV S/C Sawa (132 kV GSS)-Choutan line at 220 kV GSS Sawa (Proposed).
- 5 km LILO of existing 132 kV S/C Sawa (132 kV GSS)-Ranasar line at 220 kV GSS Sawa (Proposed).
- 50 km LILO of 220 kV S/C Dhorimanna-Sanchore line at 220 kV GSS Sawa (Proposed).

TABLE 10.1
Load Reflected at ISTS Buses

S. No.	Name of Bus	Voltage Level (kV)	Load (MW)
1	400 kV GSS Zerda	400	600
2	400 kV GSS RAPP-D	400	900
3	765 kV GSS Chhitorgarh	765	270
4	765 kV GSS Phagi	765	750
5	765 kV GSS Bikaner	765 kV	2,000

10.5 INVESTIGATION OF FEASIBILITY OF THE PROPOSALS THROUGH LOAD FLOW STUDIES

The technical feasibility of the possible proposals has been investigated with the help of a study of load flow. Studies of load flow are carried out for a total system load of 14,430 MW in Rajasthan, corresponding to the financial year (FY) 2021–2022. RE has a high penetration in the utility system in western Rajasthan, and the level of RE penetration is steadily increasing. Hence, the impact of the high penetration level of RE needs to be investigated to test the strength of the proposed transmission system. Therefore, a load flow study has been carried out in the scenario of high generation wind and solar power (75%) has been considered. While performing the study, the following loads representing the power transfer on the interstate tie lines have been switched ON: In conditions of high generation of wind and solar power, Rajasthan exports power to other states. Hence, these loads will absorb the power that is being exported to other states (Table 10.1).

10.6 SIMULATION RESULTS OF LOAD FLOW STUDY

Results of the load flow study for the restructuring of the Rajasthan transmission system in the Barmer region of the western part of Rajasthan are included and discussed in this section. Load flow results in the generation of high RE power (75% of total installed capacity) are considered and presented in the chapter. Load flow results related to the base case without considering the proposed network are also introduced in the section. The load flow study has been performed using the fast decoupled load flow method. The study was carried out for the base case and various proposals. A load flow study for the base case and all feasible proposals has been carried out in the three different operating scenarios of renewable power generation to test the strength of the proposed system with respect to different RE penetration levels. Studies of load flow are carried out by considering a total system load of 14,430 MW corresponding to FY 2021–2022. The projected load is decided by the central electricity regulatory commission for a particular state. For Rajasthan, the projected load corresponding to the years 2021–2022 is equal to 14,430 MW [43].

10.6.1   Load Flow Results for Base Transmission Network

Results of the load flow study [44–48] were carried out using the MiPower software without considering the proposed 220 kV GSS at Sawa, Barmer district, and considering all the transmission systems of the Rajasthan, including the transmission system of CTU, existing in the territory of Rajasthan, were obtained. Further, generators of the state and central sectors operating in the territory of Rajasthan are also considered while carrying out the load flow study. Results of power flow for the base case with high wind and solar power generation (75%) are obtained using the load flow study and shown in Figure 10.2. The power flow associated with the important EHV transformers installed at different GSS in the region is tabulated in Table 10.2. It is inferred that loading on all the 400/220 kV transformers at Barmer is low due to the non-availability of a sufficient downstream network. Further, loading on the 400/220 kV transformer at 400 kV GSS Bhinmal (PG) is very high. Loadings on the 220/132 kV transformers at the 220 kV GSS Barmer and Dhorimanna are high. Hence, additional downstream transmission systems will help to utilize the full capacity of the 400/220 kV transformer installed on the 400 kV GSS Barmer and, simultaneously, it will also help to reduce loading on the 220/132 kV transformers installed on the 400 kV GSS Barmer and 220 kV GSS Dhorimanna.

The flow of power on the EHV transmission lines in the Barmer region in the presence of 75% RE power is tabulated in Table 10.3. Here, the positive (+) sign indicates the export of power from the bus, and the negative sign (−) indicates the import of power from the bus. The flow of reactive power is not considered because this power will not play any role in the planning of the power system network. For power system planning, active power flow is an important key factor.

From Table 10.3, it is observed that loading on the 220 kV Barmer-Dhorimanna line is above the rated capacity of the line during the high RE scenario. Furthermore, loading on the 132 kV Dhorimanna-Ranasar line is continuously high. Hence, the creation of a 220 kV GSS in the region of Sawa, Barmer will help to reduce loading on these highly loaded lines and simultaneously evacuate the RE power from the 400 kV GSS Barmer to the load centres.

Total losses of the Rajasthan transmission system for the base case during the operating scenarios of high RE (75%) are tabulated in Table 10.4. This is because system losses during the operating scenario of high RE power generation have become equal to 711.796 MW. This is due to the export of the power to other states on the inter-state tie system (ISTS) lines. The study of the base case presented in this section will help to carry out the study of possible proposed cases to select a feasible proposal to evacuate RE power from the Barmer region to the load centres.

10.6.2   Load Flow Results for Transmission Network Considered under Option 1

Results of the load flow study carried out using the MiPower software by considering the transmission with the creation of a proposed 220 kV GSS at Sawa, Barmer district, and associated transmission line are obtained. The existing transmission

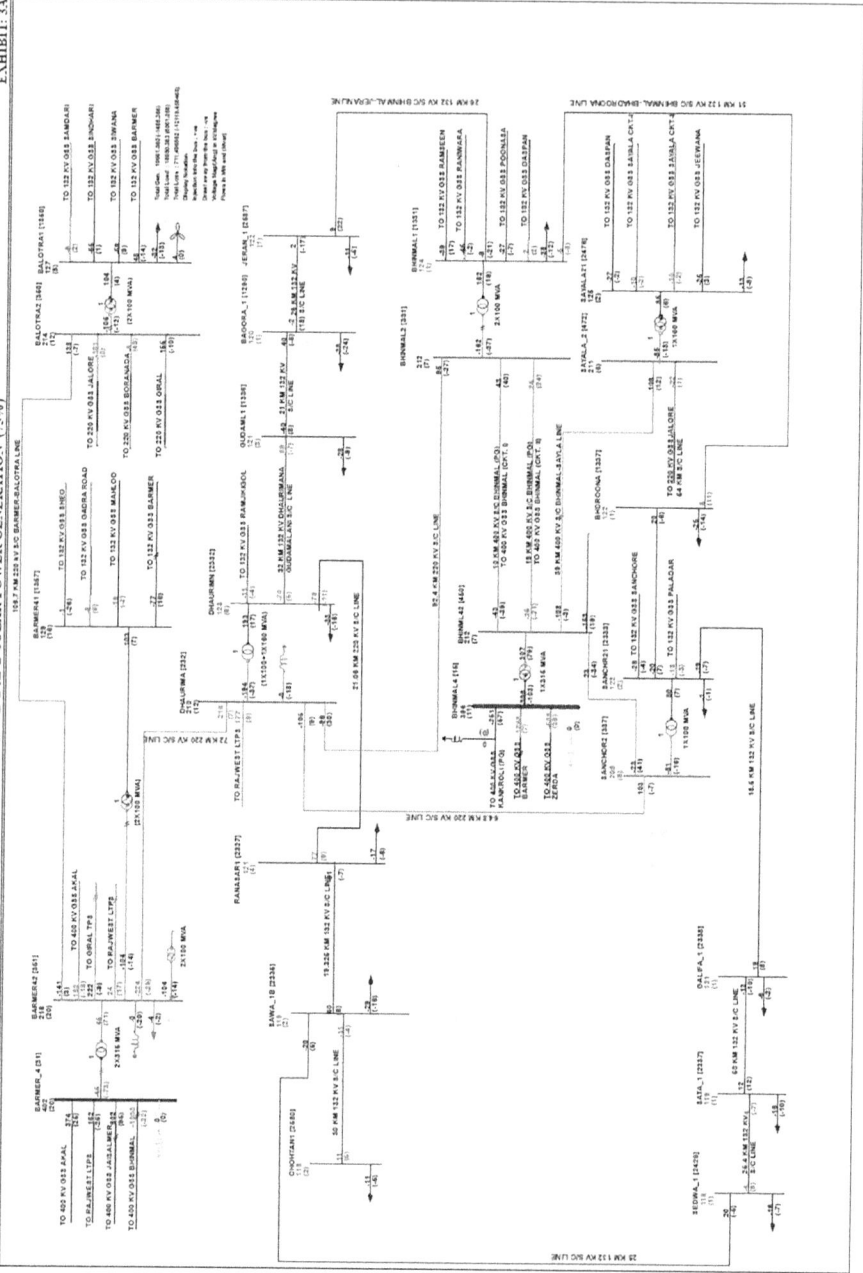

FIGURE 10.2 Power flow plots for base case.

TABLE 10.2
Power Flow on the EHV Transformers

		Power Flow (MW)			
S.No.	EHV Transformer	Base Case	Option 1	Option 2	Option 3
1	630 MVA, 400/220 kV Transformer at 400 kV GSS Barmer	45	69	40	67
2	315 MVA, 400/220 kV Transformer at 400 kV GSS Bhinmal (PG)	308	284	323	289
3	200 MVA, 220/132 kV Transformer at 400 kV GSS Barmer	105	101	104	101
4	260 MVA, 220/132 kV Transformer at Dhaurimna GSS	194	133	164	153
5	200 MVA 220/132 kV transformer at 220 kV GSS Bhinmal	162	152	157	153
6	160 MVA, 220/132 kV transformer at 220 kV GSS Sayala	85	84	84	83
7	100 MVA, 220 kV transformer at 220 kV GSS Sanchor	81	66	61	62
8	160 MVA, 220/132 kV at 220 kV GS Sawa (Proposed)	-	108	68	91

TABLE 10.3
Power Flow on the Transmission Lines

		Power Flow (MW)			
S. No.	Transmission Line	Base Case	Option 1	Option 2	Option 3
1	400 kV Barmer-Akal line	−374	−375	−375	−375
2	400 kV Barmer-Rajwest line	−152	−151	−156	−153
3	400 kV Barmer-Jaisalmer line	−602	−605	−603	−605
4	400 kV Barmer-Bhinmal line	1,083	1,062	1,094	1,066
5	220 kV Barmer-Balotra line	141	135	143	136
6	220 kV Barmer-Akal line	−183	−188	−183	−188
7	220 kV Barmer-Giral LTPS line	−222	−231	−220	−230
8	220 kV Barmer-Rajwest line	−24	−54	−26	−68
9	220 kV Barmer-Dhaurimanna line	224	192	217	170
10	132 kV Barmer-Sheo line	1	1	1	1
11	132 kV Barmer-Gadra road line	9	9	9	9
12	132 kV Barmer-Mahloo line	16	16	16	16
13	132 kV Barner (400 kV GSS)-Barmer (132 kV GSS) line	77	75	77	75
14	220 kV Balotra-Jalore line	181	173	184	174
15	220 kV Balotra-Boranada line	8	7	7	7
16	220 kV Balotra-Giral line	−155	−151	−156	−151
17	132 kV Balotra-Samdari line	9	9	9	9
18	132 kV Balotra-Sindhari line	55	54	55	54
19	132 kV Balotra-Siwana line	59	57	60	58
20	132 kV Balotra-Barmer line	−48	−46	−48	−46
21	132 kV Chohtan-Sawa line	−11	-	-	-
22	132 kV Sawa-Sedwa line	20	46	40	45

(Continued)

TABLE 10.3 (*Continued*)
Power Flow on the Transmission Lines

S. No.	Transmission Line	Base Case	Option 1	Option 2	Option 3
		Power Flow (MW)			
23	132 kV Sawa-Ranasar line	−60	-	-	-
24	132 kV Ranasar-Dhaurimann line	−77	−6	−39	−22
25	220 kV Dhaurimanna-Rajwest LTPS line	−177	−155	−172	−139
26	220 kV Dhaurimanna-Sanchore line	96	104	-	-
27	220 kV Dhaurimanna-Bhinmal line	96	105	109	123
28	132 kV Dhauriman-Ramji ki Gol line	11	11	11	11
29	132 kV Dhauriman Gudamalani line	70	83	80	86
30	132 kV Gudamalani – Bagora line	40	52	49	55
31	132 kV Bagora-Jeran line	2	13	10	16
32	132 kV Jeran-Bhinmal line	−9	3	−1	5
33	400 kV Bhinmal-Kankroli line	251	253	248	253
34	400 kV Bhinmal-Barmer line	−1,068	−1,047	−1,078	−1,051
35	400 kV Bhinmal-Zerda line	508	510	507	510
36	220 kV Bhinmal (RVPN) - Bhinmal (PG ckt-1) line	−43	−31	−32	−21
37	220 kV Bhinmal (RVPN)-Bhinmal (PG ckt-2) line	−25	−18	−18	−12
38	220 kV Bhinmal-Sayala line	108	112	103	111
39	220 kV Bhinmal-Sanchore line	−23	−37	19	−14
40	132 kV Bhinmal-Ramseen line	39	46	41	46
41	132 kV Bhinmal-Raniwara line	45	45	45	45
42	132 kV Bhinmal-Poonasa line	27	27	27	27
43	132 kV Bhinmal-Daspan line	−2	3	0	3
44	220 kV Sanchore- Bhdroona line	20	30	19	26
45	132 kV Sanchore (220 kV GSS)-Sanchore (132 kV GSS) line	28	27	27	27
46	132 kV Sanchore-Paladar line	13	13	13	13
47	132 kV Sanchore-Galifa line	19	−6	−1	−6
48	220 kV Sayala-Jalore line	22	28	19	27
49	132 kV Sayala-Daspan line	27	23	26	23
50	132 kV Sayala-Sayala ckt-1	10	12	10	12
51	132 kV Sayala-Sayala ckt-2	10	12	10	11
52	132 kV Sayala-Jeewana line	25	24	24	24
53	132 kV Galifa-Sata line	13	−12	−7	−12
54	132 kV Sata-Sedwa line	−4	−28	−23	−28
55	220 kV Sawa – Barmer (Proposed Line)	-	−108	-	−137
56	132 kV Sawa-Chohtan (Proposed Line)	-	11	11	11
57	132 kV Sawa-Sawa (Proposed Line) Ckt-I	-	46	42	46
58	132 kV Sawa-Sawa (Proposed Line) Ckt-II	-	28	26	28
59	132 kV Sawa-Ranasar (Proposed Line)	-	11	−23	−6
60	220 kV Sawa-Dhaurimanna (Proposed Line)	-	-	−109	−30
61	220 kV Sawab-220 kV Sanchore (Proposed Line)	-	-	42	77

TABLE 10.4

Total Transmission Losses with Base Case

S. No.	Transformer details	Base Case	Proposal 1	Proposal 2	Proposal 3
1	Total System losses (MW)	711.496	703.685	708.862	703.488
2	Loss saving compared to Base case (MW)	-	7.811	2.634	8.008

system of Rajasthan, including the transmission system of CTU existing in the territory of Rajasthan, is considered in addition to the transmission system of option 1. Further, generators of the state and central sector generators operating in the territory of Rajasthan are also considered while carrying out the load flow study for proposed option 1. Plots of power flow are obtained as the output of the load flow study corresponding to proposal 1 with high wind and solar power generation (75%) and are shown in Figure 10.3.

Power flow associated with the important EHV transformers installed at different GSS in the region during the study of proposed option 1 is tabulated in Table 10.2. It is inferred that loading on all the 400/220 kV transformers at Barmer has increased due to the availability of additional downstream networks. Further, loading on the 400/220 kV transformer at 400 kV GSS Bhinmal (PG) has been reduced significantly. Loading on the 220/132 kV transformers at the 220 kV GSS Barmer and Dhorimanna has also been reduced. Hence, additional downstream transmission systems resulted in the utilization of the full capacity of the 400/220 kV transformer installed on the 400 kV GSS Barmer, and simultaneously it will also help to reduce loading on the 220/132 kV transformers installed on the 400 kV GSS Barmer and 220 kV GSS Dhorimanna.

The flow power on the EHV transmission lines in the Barmer region for the proposed option 1 is tabulated in Table 10.3. From Table 10.3, it is observed that the loading on the 220 kV Barmer-Dhorimanna line has been reduced now that it is within the permissible limits, which is slightly higher than the rated capacity of the line during the high RE scenario. However, the high RE scenario persists for 2–3 h, which is considered a short duration, and lines can be overloaded by up to 30% for a short duration. Further, loading on the 132 kV Dhorimanna-Ranasar line has been reduced and it is within the permissible limits for all the scenarios of RE power generation. Hence, the creation of a 220 kV GSS in the region of Sawa, Barmer has resulted in reducing the loading on the highly loaded lines and will simultaneously evacuate the RE power from the 400 kV GSS Barmer to the load centres.

Total losses of the Rajasthan transmission system during all the operating scenarios of high RE power are tabulated in Table 10.4. It is assumed that the system loss has reached 703.685 MW. This is due to the export of the power to other states on the inter-state tie system (ISTS) lines. Further, the loss has been reduced compared to the base case.

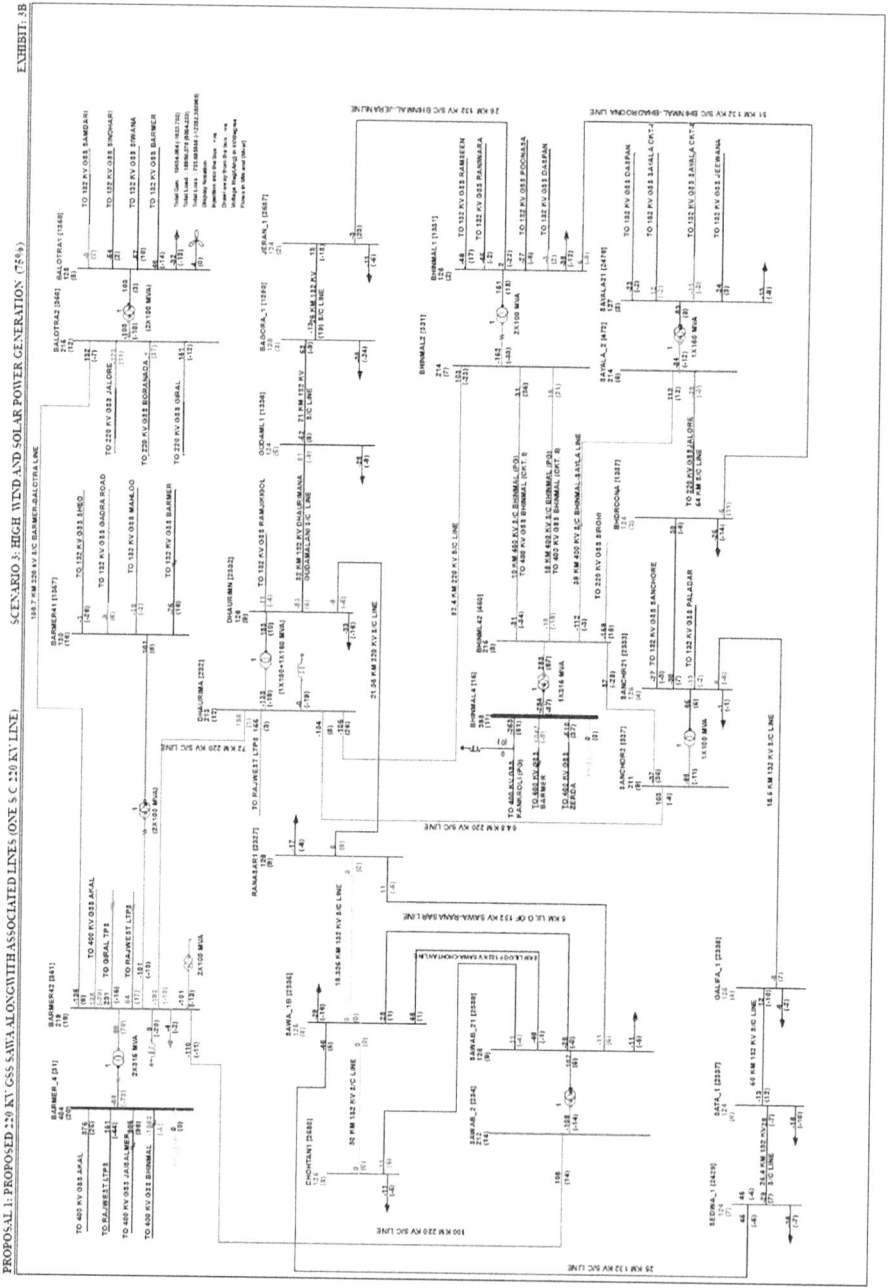

FIGURE 10.3   Power flow plots for option 1.

10.6.3 LOAD FLOW RESULTS FOR TRANSMISSION NETWORK CONSIDERED UNDER PROPOSAL 2

The results of the load flow study were carried out using the MiPower software by considering the transmission system with the creation of the proposed 220 kV GSS at Sawa and the associated transmission line. The existing transmission system of the Rajasthan, including the transmission system of CTU existing in the territory of Rajasthan, is considered in addition to the transmission system of the proposed option 2. Further, the generators of the state and central sector generators operating in the territory of Rajasthan are also considered while carrying out the load flow study for proposed option 2. Plots of power flow are obtained as the output of the load flow study corresponding to the proposal 1 with high wind and solar power generation (75%) and are shown in Figure 10.4.

Power flow associated with the important EHV transformers installed at different GSS in the region during the study of proposed option 2 is tabulated in Table 10.2. It is inferred that loading on all the 400/220 kV transformers has not changed significantly. Further, loading on the 400/220 kV transformer at 400 kV GSS Bhinmal (PG) has increased. Loading on the 220/132 kV transformers at the 220 kV GSS Barmer has not changed significantly. However, loading on the 220/132 kV transformer at Dhorimanna is reduced. Hence, additional transmission systems have given relief in loading on the 220 kV GSS Dhaurimanna, and significant relief has not been observed on the 400 kV GSS Barmer and 400 kV GSS Bhinmal.

The flow power on the EHV transmission lines in the Barmer region for the proposed option 2 is tabulated in Table 10.3. The flow of reactive power is not considered because this power will not play any role in the planning of the power system network. For power system planning, active power flow is an important key factor. From Table 10.3, it is observed that loading on the 220 kV Barmer-Dhorimanna line is reduced marginally. However, during the high RE scenario, the loading on this line is high and above the permissible limits. Loading on the 132 kV Dhorimanna-Ranasar line has been reduced to within the permissible limits for all the scenarios of RE power generation. Hence, the creation of a 220 kV GSS in the region of Sawa, Barmer has resulted in reducing loading on the transmission elements associated with the 220 kV GSS Dhorimanna.

Total losses of the Rajasthan transmission system during all the operating scenarios of high RE are for option 2 are tabulated in Table 10.4. It is inferred that system loss has become equal to 708.862 MW. This is due to the export of the power to other states on the inter-state tie system (ISTS) lines. Furthermore, the loss has been reduced in comparison to the base case in all scenarios of RE power generation.

10.6.4 LOAD FLOW RESULTS FOR TRANSMISSION NETWORK CONSIDERED UNDER PROPOSAL 3

The results of the load flow study were carried out using the MiPower software by considering the transmission with the creation of a proposed 220 kV GSS at Sawa and the associated transmission line. The existing transmission system of the Rajasthan, including the transmission system of CTU existing in the territory of Rajasthan, is

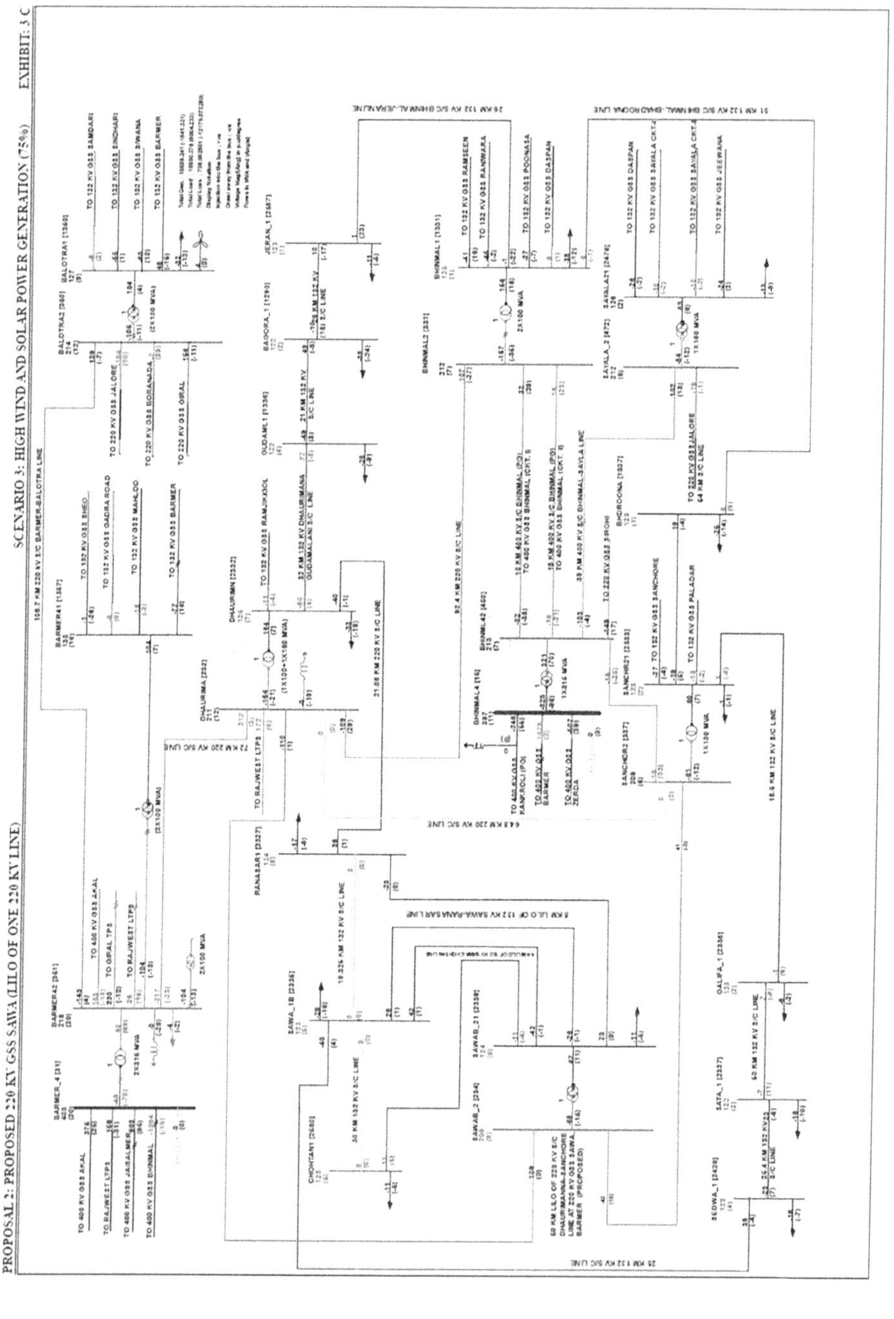

FIGURE 10.4 Power flow plots for option 2.

considered in addition to the transmission system of Option 3. Further, the generators of the state and central sector generators operating in the territory of Rajasthan are also considered while carrying out the load flow study for proposed option 3. Plots of power flow are obtained as the output of the load flow study corresponding to option 3 with high wind and solar power generation (75%) and are shown in Figure 10.5.

Power flow associated with the important EHV transformers installed at different GSS in the region during the study of proposed option 3 is tabulated in Table 10.2. It is inferred that loading on all the 400/220 kV transformers at Barmer has increased due to the availability of additional downstream networks. Further, loading on the 400/220 kV transformer at 400 kV GSS Bhinmal (PG) has been reduced significantly. Loading on the 220/132 kV transformers at the 220 kV GSS Barmer and Dhorimanna has also been reduced. Hence, additional downstream transmission systems resulted in the utilization of the full capacity of the 400/220 kV transformer installed on the 400 kV GSS Barmer, and simultaneously it will also help to reduce the loading on the 220/132 kV transformers installed on the 400 kV GSS Barmer and 220 kV GSS Dhorimanna.

The flow of power on the EHV transmission lines in the Barmer region for the proposed option 3 is tabulated in Table 10.3. The flow of reactive power is not considered because this power will not play any role in the planning of the power system network. For power system planning, active power flow is an important key factor. From Table 10.3, it is observed that the loading on the 220 kV Barmer-Dhorimanna line has been reduced, and it is now within the permissible limits even during the scenario of high RE. Further, loading on the 132 kV Dhorimanna-Ranasar line has been reduced and it is within the permissible limits for all the scenarios of RE power generation. Hence, the creation of a 220 kV GSS in the region of Sawa, Barmer with a transmission system included in the proposal 1 has resulted in reducing the loading on the highly loaded lines and simultaneously will evacuate the RE power from the 400 kV GSS Barmer to the load centres.

Total losses of the Rajasthan transmission system during an operating scenario of high RE during the option 3 study are tabulated in Table 10.4. It is inferred that system loss has become equal to 703.488 MW. This is due to the export of the power to other states on the inter-state tie system (ISTS) lines. Furthermore, the loss has been reduced in comparison to the base case in all scenarios of RE power generation.

10.6.5 OBSERVATIONS

The following observations are made from the study discussed in the above-mentioned sections:

- It is observed that in Proposed Case 2, the 220 kV S/C Barmer-Dhaurimana and 220 kV S/C Rajwest-Dhaurimana lines continue to be overloaded under the full RE scenario.
- In terms of transmission line loadings, proposed Case 3 appears to be the most technically feasible.
- Options 1 and 3 save money on losses.

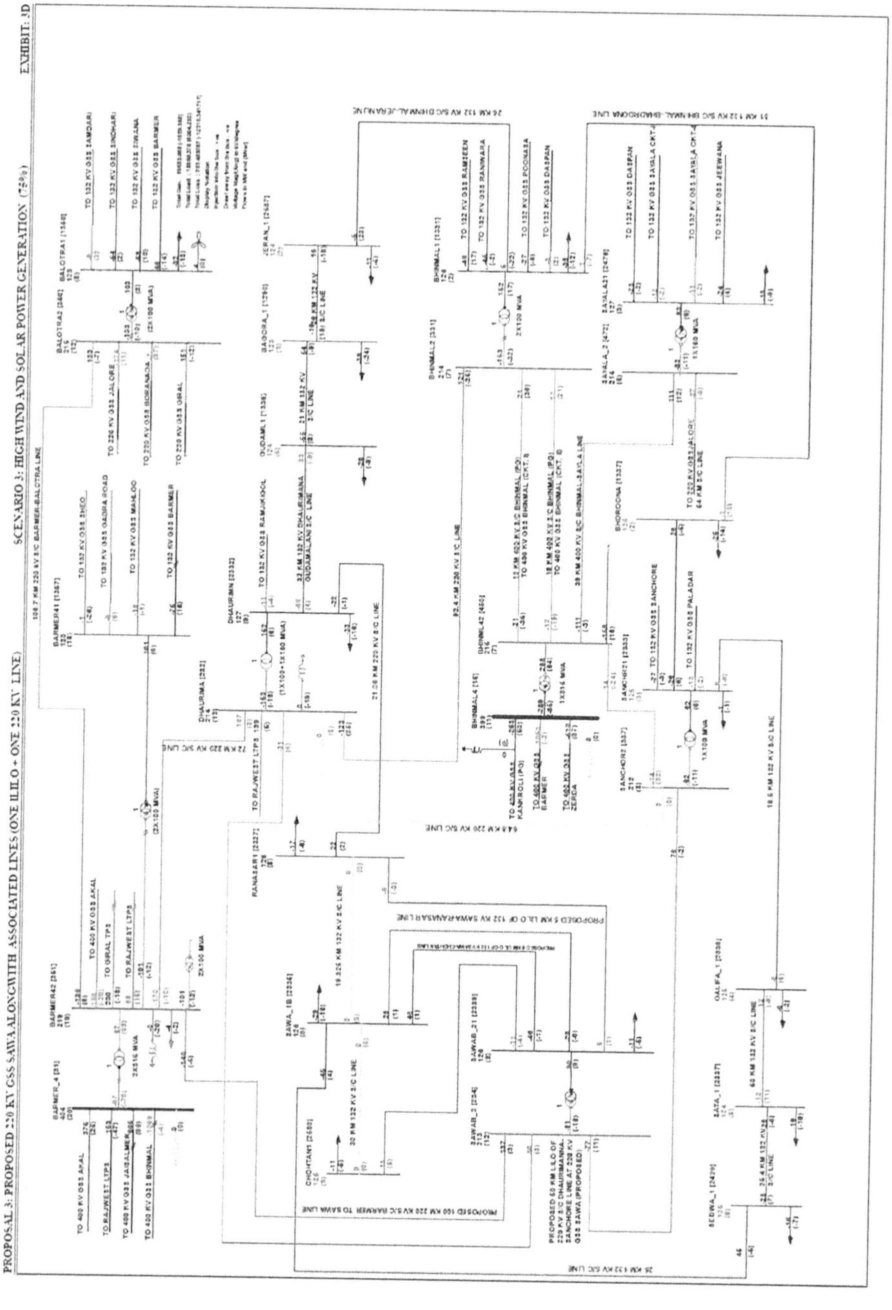

FIGURE 10.5 Power flow plots for option 3.

TABLE 10.5

Loading during Contingency Conditions of Proposed Options 1 and 3

S.No.	Transmission Lines/Transformers	Contingency 1	Contingency 2
1	220 kV S/C Barmer-Dhaurimana line	0	0
2	220 kV S/C Rajwest-Dhaurimana line	221	173
3	220 kV S/C Dhaurimana-Bhinmal line	62	95
4	220 kV S/C Dhaurimana-Sanchore line	68	0
5	220 kV D/C Barmer-Sawa line	154	255
6	220 kV S/C Dhaurimana-Sawa line	-	(−) 52
7	220 kV S/C Sawa-Sanchore line	-	90
8	260 MVA, 220/132 kV Transformer at Dhaurimana	92	130
9	132 kV S/C Dhaurimana-Ranasar line	(−) 30	5
10	160 MVA, 220/132 kV Transformer at Sawa	152	109

10.6.6 Contingency of Options 1 and 3

Options 1 and 3 are examined in the following contingency conditions:

Contingency 1: Option 1 + Outage of 220 kV Barmer-Dhaurimana line
Contingency 2: Option 3 + Outage of 220 kV Barmer-Dhaurimana line

Loadings on the important transmission lines are provided below in Table 10.5.

Observations of Proposed Options 1 and 3
- Transmission loss savings are roughly comparable in both proposed Options 1 and 3 under both zero and high solar/wind scenarios.
- The overloaded 220 kV S/C Barmer-Dhaurimana line under high RE, i.e., 221 MW, would be reduced to 173 MW in proposed option 3.
- In the case of an outage of the 220 kV S/C Barmer-Dhaurimana line under a high solar/wind scenario, overloading is observed in proposed option 1 but no overloading is observed in proposed option 3.

10.7 SHORT CIRCUIT STUDY

A short circuit study for the base without considering the proposed 220 kV GSS at Sawa and the associated transmission system was carried out, and short circuit MVA as well as short circuit current were obtained. Further, a short-circuit study is also performed for the proposed cases. It is established that the short circuit level on all buses of the existing and proposed transmission system in the Barmer region of the western parts of Rajasthan is within permissible limits. The protection system in the RVPN transmission system is effective for the short-circuit fault current of 50 kA. Here, all the fault currents are below 50 kA.

10.8   CONCLUSION

A detailed study was conducted for the planning of the new transmission system in the Barmer region of the Western parts of Rajasthan to evacuate the RE power from the wind farms to the load centers. The study was carried out for the base case of the Rajasthan transmission system without considering restructuring and considering various proposals to restructure the existing transmission network. The following conclusions are made based on the study included in this chapter:

- It is concluded that maximum loss saving is observed for the transmission system included in the proposed option 3.
- It is also concluded that in option 2, the 220 kV S/C Barmer-Dhaurimana and 220 kV S/C Rajwest-Dhaurimana lines continue to be overloaded under the full RE scenario.
- In terms of transmission line loadings, proposed option 3 appears to be the most feasible.
- In the case of an outage of the 220 kV S/C Barmer-Dhaurimana line under a high solar/wind scenario, overloading is observed in option 1, but no over-loading is observed in option 3.
- Short-circuit studies show that the fault current in all the proposed cases of new transmission systems will remain below 50 kA.

Considering the above-mentioned concluding points, the following transmission system is recommended in the Barmer region of the Rajasthan State of India to evacuate the RE power to the load centres:

- At 220 kV GSS Sawa, 1 × 160 MVA, 220/132 kV power transformer and 1 × 20/25 MVA, 132/33 kV power transformer are proposed.
- From 400 kV GSS Barmer to 220 kV GSS Sawa, a 100 km 220 kV D/C line is proposed.
- 5 km LILO of the existing 132 kV S/C Sawa (132 kV GSS)-Choutan line (proposed) at 220 kV GSS Sawa.
- 5 km LILO of the existing 132 kV S/C Sawa (132 kV GSS)-Ranasar line (Proposed) at 220 kV GSS Sawa.
- 220 kV S/C at 220 kV GSS Sawa for 50 km of Dhorimanna-Sanchore line (Proposed).

DECLARATION BY AUTHORS

The authors declare that the contents of this chapter are purely scholarly ideas of the authors and do not pertain to any department.

REFERENCES

1. S. K. Lakwal, O. P. Mahela, M. Kumar, and N. Kumar, "Optimized Approach for Restructuring of Rajasthan Transmission Network to Cater Load Demand of Rajasthan Refinery and Petrochemical Complex," *IEEE International Conference on Computing, Power and Communication Technologies (GUCON 2019)*, September 27–28, 2019, Greater Noida, India.

2. A. Kulshrestha, O. P. Mahela, M. K. Gupta, N. Gupta, N. Patel, T. Senjyu, M. S. S. Danish, and M. Khosravy, "A Hybrid Protection Scheme Using Stockwell Transform and Wigner Distribution Function for Power System Network with Solar Energy Penetration," *Energies*, vol. 13, no. 14, p. 3519, 2020. doi:10.3390/en13143519.

3. G. S. Yogee, O. P. Mahela, K. D. Kansal, B. Khan, R. Mahla, H. H. Alhelou, and P. Siano, "An Algorithm for Recognition of Fault Conditions in the Utility Grid with Renewable Energy Penetration," *Energies*, vol. 13, no. 9, p. 2383. doi:10.3390/en13092383.

4. S. R. Ola, A. Saraswat, S. K. Goyal, S. K. Jhajharia, B. Khan, O. P. Mahela, H. H. Alhelou, and P. Siano, "A Protection Scheme for Power System with Solar Energy Penetration," *Applied Sciences*, vol. 10, no. 4, p. 1516, February 2020. doi:10.3390/app10041516.

5. S. Ali, A. Bhargava, R. Singh, O. P. Mahela, B. Khan, and H. H. Alhelou, "Mitigation of Power Evacuation Constraints Associated with Transmission System of Kawai-Kalisindh-Chhabra Thermal Power Complex in Rajasthan, India," *AIMS Energy*, vol. 8, no. 3, pp. 394–420, 2020. doi:10.3934/energy.2020.3.394.

6. B. Khan, G. Agnihotri, S. E. Mubeen, and G. Naidu, "A TCSC Incorporated Power Flow Model for Embedded Transmission Usage and Loss Allocation," *AASRI Procedia*, vol. 7, pp. 45–50, 2014.

7. B. Khan, G. Agnihotri, G. Gupta, and P. Rathore, "A Power Flow Tracing based Method for Transmission Usage, Loss & Reliability Margin Allocation," *AASRI Procedia*, vol. 7, pp. 94–100, 2014.

8. B. Khan, G. Agnihotri, P. Rathore, A. Mishra, and G. Naidu, "A Cooperative Game Theory Approach for Usage and Reliability Margin Cost Allocation under Contingent Restructured Market," *International Review of Electrical Engineering*, vol. 9, no. 4, pp. 854–862, 2014.

9. B. Khan and G. Agnihotri, "A Comprehensive Review of Embedded Transmission Pricing Methods Based on Power Flow Tracing Techniques," *Chinese Journal of Engineering*, vol. 2013, Article ID 501587, 13 pages, 2013.

10. B. Khan, G. Agnihotri, and A. S. Mishra, "An Approach for Transmission Loss and Cost Allocation by Loss Allocation Index and Co-operative Game Theory," *Journal of the Institution of Engineers (India): Series B*, vol. 97, pp. 41–46, 2016.

11. B. Khan and G. Agnihotri, "A Novel Transmission Loss Allocation Method Based on Transmission Usage," *2012 IEEE Fifth Power India Conference*, 2012, pp. 1–3.

12. P. Rathore, G. Agnihotri, B. Khan, and G. Naidu, "Transmission Usage and Cost Allocation Using Shapley Value and Tracing Method: A Comparison," *Electrical and Electronics Engineering: An International Journal (ELELIJ)*, vol. 3, pp.11–29, 2014.

13. B. Khan and G. Agnihotri, "An Approach for Transmission Usage & Loss Allocation by Graph Theory," *WSEAS Transactions on Power Systems*, vol. 9, pp.44–53, 2014.

14. S. Khare, B. Khan and G. Agnihotri, "A Shapley Value Approach for Transmission Usage Cost Allocation under Contingent Restructured Market," *2015 International Conference on Futuristic Trends on Computational Analysis and Knowledge Management (ABLAZE)*, 2015, Noida, India, pp. 170–173.

15. B. Khan, G. Agnihotri, and G. Gupta, "A Multipurpose Matrices Methodology for Transmission Usage, Loss and Reliability Margin Allocation in Restructured Environment," *Electrical & Computer Engineering: An International Journal*, vol. 2, no. 3, p. 11, September 2013.

16. M. M. Tripathi, A. K. Pandey, and D. Chandra, "Power System Restructuring Models in the Indian Context," *The Electricity Journal*, vol. 29, pp. 22–27, 2016.

17. V. Popovici, "2010 Power Generation Sector Restructuring in Romania—A Critical Assessment," *Energy Policy*, vol. 39, pp. 1845–1856, 2011.

18. D. Saha and R. N. Bhattacharya, "Analysis of the Welfare Implications of Power-Sector Restructuring in West Bengal, India," *Utilities Policy*, vol. 56, pp. 62–71, 2019.

19. C. Chen, J. Gao, and J. Chen, "Behavioral Logics of Local Actors Enrolled in the Restructuring of Rural China: A Case Study of Haoqiao Village in Northern Jiangsu," *Journal of Rural Studies*, 2019. doi:10.1016/j.jrurstud.2019.01.021.
20. D. K. Sharma, O. P. Mahela, and S. Agarwal, "Design and Implementation of System Protection Scheme for Kawai-Kalisindh-Chhabra Thermal Complex in Rajasthan, India," *IEEE International Conference on Computing, Power and Communication Technologies (GUCON 2019)*, September 27–28, 2019, Greater Noida, India.
21. O. P. Mahela, B. Khan, H. H. Alhelou, and P. Siano, "Power Quality Assessment and Event Detection in Distribution Network with Wind Energy Penetration Using Stockwell Transform and Fuzzy Clustering," *IEEE Transactions on Industrial Informatics*, Early Access, January 2020. doi:10.1109/TII.2020.2971709.
22. B. Rathore, O. P. Mahela, B. Khan, H. H. Alhelou, and P. Siano, "Wavelet-Alienation-Neural Based Protection Scheme for STATCOM Compensated Transmission Line," *IEEE Transactions on Industrial Informatics*, Early Access, June 2020. doi:10.1109/TII.2020.3001063.
23. O. P. Mahela, B. Khan, H. H. Alhelou, and S. Tanwar, "Assessment of Power Quality in the Utility Grid Integrated with Wind Energy Generation," *IET Power Electronics*, January 2020. doi:10.1049/iet-pel.2019.1351.
24. S. R. Ola, A. Saraswat, S. K. Goyal, S. K. Jhajharia, B. Rathore, and O. P. Mahela, "Wigner Distribution Function and Alienation Coefficient Based Transmission Line Protection Scheme," *IET Generation, Transmission and Distribution*, vol. 14, no. 10, pp. 1842–1853, 22 May 2020, doi:10.1049/iet-gtd.2019.1414.
25. G. S. Chawda, A. G. Shaik, O. P. Mahela, S. Padmanaban, and J. B. Holm-Nielsen, "Comprehensive Review of Distributed FACTS Control Algorithms for Power Quality Enhancement in Utility Grid with Renewable Energy Penetration," *IEEE Access*, vol. 8, pp. 107614–107634, 2020, doi:10.1109/ACCESS.2020.3000931.
26. O. P. Mahela, J. Sharma, B. Kumar, B. Khan, and H. H. Alhelou, "An Algorithm for the Protection of Distribution Feeder Using Stockwell and Hilbert Transforms Supported Features," *CSEE Journal of Power and Energy Systems*, 2020. doi:10.17775/CSEEJPES.2020.00170.
27. S. R. Ola, A. Saraswat, S. K. Goyal, S. K. Jhajharia, and O. P. Mahela. "Detection and Analysis of Power System Faults in the Presence of Wind Power Generation Using Stockwell Transform Based Median," *Springer Lecture Notes in Electrical Engineering Series*, ISSN: 1876-1100, pp. 319–329, 2019. https://link.springer.com/chapter/10.1007/978-981-15-0214-9_36.
28. J. Sharma, B. Kumar, O. P. Mahela, and A. R. Garg, "Protection of Distribution Feeder Using Stockwell Transform Supported Voltage Features," *IEEE 9th Power India International Conference (PIICON 2020)*, February 28–01 March, 2020, Deenbandhu Chhotu Ram University of Science and Technology, Murthal, India. doi:10.1109/PIICON49524.2020.9113014.
29. D. Gupta, O. P. Mahela, and S. Ali, "Voltage Based Transmission Line Protection Algorithm Using Signal Processing Techniques," *2020 IEEE International Students' Conference on Electrical, Electronics and Computer Science (SCEECS 2020)*, February 22–23, 2020, MANIT Bhopal, India. doi:10.1109/SCEECS48394.2020.15.
30. R. K. Pathak, S. Agarwal, O. P. Mahela, and R. R. Choudhary, "Recognition of Faults in Grid Connected Solar Photovoltaic Farm Using Current Features Evaluated Using Stockwell Transform Based Algorithm," *2020 IEEE International Students' Conference on Electrical, Electronics and Computer Science (SCEECS 2020)*, February 22–23, 2020, MANIT Bhopal, India. doi:10.1109/SCEECS48394.2020.13.

31. S. Karmakar, G. Ahmad, O. P. Mahela, and R. R. Choudhary, "Algorithm Based on Combined Features of Stockwell Transform and Hilbert Transform for Detection of Transmission Line Faults with Dynamic Load," *First IEEE International Conference on Power, Control and Computing Technologies (ICPC2T)*, January 3–5, 2020, NIT Raipur, India. doi:10.1109/ICPC2T48082.2020.9071516.

32. S. Thukral, O. P. Mahela, and B. Kumar, "Detection of Transmission Line Faults in the Presence of Wind Energy Power Generation Source Using Stockwell's Transform," *IEEE International Conference on Issues and Challenges in Intelligent Computing Techniques (ICICT 2019)*, September 27–28, 2019, KIET Group of Institutions, Delhi-NCR, Ghaziabad, India. doi:10.1109/ICICT46931.2019.8977695.

33. O. P. Mahela, S. Agarwal, and N. K. Saini, "Power Quality Improvement In Hybrid Power System Using Synchronous Reference Frame Theory Based Distribution Static Compensator With Battery Energy Storage System," *IEEE International Conference on Computing, Power and Communication Technologies (GUCON 2019)*, September 27–28, 2019, Greater Noida, India.

34. O. P. Mahela, K. D. Kansal and S. Agarwal, "Detection of Power Quality Disturbances in Utility Grid with Wind Energy Penetration," *8th IEEE India International Conference on Power Electronics (IICPE-2018)*, December 13–14, 2018, MNIT Jaipur, India. doi:10.1109/IICPE.2018.8709578.

35. S. R. Ola, A. Saraswat, S. K. Goyal, S. K. Jhajharia, and O. P. Mahela, "A Technique Using Stockwell Transform Based Median for Detection of Power System Faults," *2018 IEEE 8th Power India International Conference (PIICON-2018)*, December 10–12, 2018, NIT Kurukshetra, India. doi:10.1109/POWERI.2018.8704459.

36. A. Meena, S. Ali, and O. P. Mahela. "Power System Oscillations Damping During Faulty Events in the Presence of Linear Load," *IEEE International Conference on Computing, Power and Communication Technologies 2018 (GUCON-2018)*, September 28–29, 2018, Galgotias University, Greater Noida, India. doi:10.1109/GUCON.2018.8675015.

37. M. Meena, O. P. Mahela, M. Kumar, and N. Kumar, "Detection and Classification of Complex Power Quality Disturbances Using Stockwell Transform and Rule Based Decision Tree," *IEEE PES International Conference on Smart Electric Drives and Power System (ICSEDPS-2018)*, June 12–13, 2018, G H Raisoni College of Engineering, Nagpur, India. doi:10.1109/ICSEDPS.2018.8536028.

38. T. Panchori, O. P. Mahela, and S. Agarwal, "Transient Performance of a Series Compensated 765 kV Transmission Line During Faulty Conditions," *2018 IEEE Students' Conference on Electrical, Electronics and Computer Science (SCEECS 2018)*, February 24–25, 2018, MANIT, Bhopal, India. doi:10.1109/SCEECS.2018.8546911.

39. B. S. Joshi, O. P. Mahela, and S. R. Ola, "Implementation of Thyristor Controlled Series Capacitor in Transmission System to Improve the Performance of Power System Network," *4th IEEE International Conference on Power, Control and Embedded Systems (ICPCES 2017)*, March 9–11, 2017, MNNIT Allahabad, India. doi:10.1109/ICPCES.2017.8117644.

40. A. Sharma, S. R. Ola, and O. P. Mahela, "Analysis of Faults on Series Compensated EHV Transmission Line in the Presence of Wind Generation," *IEEE 7th Power India International Conference on Advances in Signal Processing (PIICON 2016)*, November 25–27, 2016, Bikaner, India. doi:10.1109/POWERI.2016.8077202.

41. A. Sharma, S. R. Ola, and O. P. Mahela, "Analysis of Grid Disturbances in Power System Network in the Presence of Series Compensated 765 kV Transmission Line," *IEEE 7th Power India International Conference on Advances in Signal Processing (PIICON 2016)*, November 25–27, 2016, Bikaner, India. doi:10.1109/POWERI.2016.8077158.

42. RVPN, https://energy.rajasthan.gov.in/content/raj/energy-department/rajasthan-rajya-vidyut-prasaran-limited/en/about-us/power-map.html#, Accessed on 23.07.2020.

43. CEA, http://cea.nic.in/planning.html, Accessed on 23.07.2020.
44. T. F. Agajie, B. Khan, H. H. Alhelou, and O. P. Mahela, "Optimal Expansion Planning of Distribution System Using Grid-Based Multi-Objective Harmony Search Algorithm," *Computers & Electrical Engineering*, vol. 87, p. 106823, 2020.
45. B. Khan, H. H. Alhelou, and F. Mebrahtu. "A Holistic Analysis of Distribution System Reliability Assessment Methods with Conventional and Renewable Energy Sources," *AIMS Energy*, vol. 7, no. 4, pp. 413–429, 2019.
46. D. Anteneh and B. Khan, "Reliability Enhancement of Distribution Substation by Using Network Reconfiguration a Case Study at Debre Berhan Distribution Substation," *International Journal of Economy, Energy and Environment*, vol. 4, no. 2, pp. 33–40, 2019.
47. B. Khan, G. Agnihotri, G. Gupta, and P. Rathore, "A Power Flow Tracing based Method for Transmission Usage, Loss & Reliability Margin Allocation," *AASRI Procedia*, vol. 7, pp. 94–100, 2014.
48. B. Khan, G. Agnihotri, P. Rathore, A. Mishra, and G. Naidu, "A Cooperative Game Theory Approach for Usage and Reliability Margin Cost Allocation under Contingent Restructured Market," *International Review of Electrical Engineering*, vol. 9, no. 4, pp. 854–862, 2014.

11 Performance of Multifunctional Grid-tied Photovoltaic Inverters with Active and Harmonic Power Weight (AHPW) Control Technique

Surender Kumar Sharma and Vijayakumar Gali

CONTENTS

11.1 INTRODUCTION

With the growing demand for electricity across the globe and the exhaustion of traditional sources such as petroleum, coal, etc., many countries are looking forward to maximizing their efforts to install non-conventional energy resources. Among non-conventional energy resources, photovoltaic energy is becoming a promising alternative source to conventional energy sources due to its extensive advantages, such as being omnipresent, freely available, having lower operational and maintenance

DOI: 10.1201/9781003278030-11

costs, and being environmentally friendly. More importantly, the use of conventional sources pollutes the environment and causes global warming. More focus has to be put on environmentally friendly distributed generation technology [1,2].

Energy generation by solar-based systems is gaining greater fame as a result of lower maintenance, low operating expenses, and low transmission costs. The distributed generation based on renewable energy is connected to the local grids through proper power electronics converters. The PV plant generates DC power by photoelectric effect, which has to be converted to AC by inverters and connected to the grid through proper synchronization. The grid-connected inverters play a key role in the grid-connected system [3].

This power electronic converter has to be designed in such a way that it not only feeds the active power to the grid but also incorporates it as an active power filter. The rapid advancement and impact of digitalization have caused the widespread use of power electronics-based equipment in various industrial and domestic applications [4–13]. Hence, the power electronic devices are operated as switches for efficient control of power, and such operation affects the power quality (PQ) of three-phase systems, including single-phase electrical power system networks. Hence, harmonic mitigation and reactive power compensation have become of prime importance for suppliers and end utilities [14,15]. The grid-connected photovoltaic inverters act as a DC to AC converter to mitigate current and reactive power and increase the efficiency of the system through custom power devices like series, shunt active power filters, unified power quality conditioners (UPQC), etc [16,17].

Many researchers and practitioners believe that reference current techniques play an important role in harmonic compensation and grid synchronization. Researchers have implemented various time-domain and frequency-domain techniques to calculate the exact reference current to mitigate various odd harmonic components from the power system network [18,19]. But, very little research has been conducted on the elimination of even harmonics from power systems [20,21]. Many control strategies are available in the literature, such as unit template technique, Enhanced phase-locked loop (EPLL)-based control algorithm, instantaneous reactive power theory (IRPT), synchronous reference frame (SRF) theory, power balance theory, etc. [22,23]. These control techniques are highly affected by load conditions, supply voltage, etc. Due to poor dynamic response and poor harmonic compensation under distorted grid voltage conditions, researchers came up with adoptive control techniques such as least mean square (LMS)-based control technique, mean square error (MSE) technique, least absolute deviation (LAD) techniques, etc. [24,25]. These adaptive control techniques improve the steady-state stability and dynamic response of the system with reduced convergence speed. However, these show very sluggish response under conditions of high penetration of non-conventional energy sources and the presence of even-order harmonics in the power grid.

In order to overcome the problems of the abovementioned control algorithms, a novel active and harmonic power weight (AHPW)-based control technique is proposed in this chapter. This technique is applied to PV inverter systems where high penetration takes place. Also, this technique extracts the fundamental frequency components from highly polluted grid currents without affecting the synchronization problem. A detailed explanation is given in Section 3.

This paper is categorized into five sections. Section 11.2 depicts the architecture of a grid-connected PV system. A proposed AHPW control technique is explained in Section 11.3. A detailed simulation study is being implemented to test the proposed control algorithm in Section 11.4 and go along with the conclusions in Section 11.5.

11.2 ARCHITECTURE OF MULTIFUNCTIONAL GRID TIED PV SYSTEM

The PV system is connected to the grid supply through a single-stage inverter. In the daytime, it feeds the active power to the grid, and at night, when PV generation is off, the PV inverter acts as a shunt active power filter. A single-stage system topology is picked as shown in Figure 11.1. The solar PV solar system is associated with the grid by utilizing a single DC–AC transformation stage.

The loads could be linear, non-linear, or a combination of both types. The capacitor assumes a significant role in keeping up the energy balance across the voltage source inverter (VSI) in the single stage. The voltage across the DC-bus capacitor is exactly the same as the estimated and planned utility grid voltage. A single-stage topology can be introduced with PV solar output voltage and gives a high utilization factor. The three-phase VSI is associated with the utility grid through interfacing inductors at the point of common coupling (PCC) to limit the switching ripples of the VSI output current and evaluate its quality.

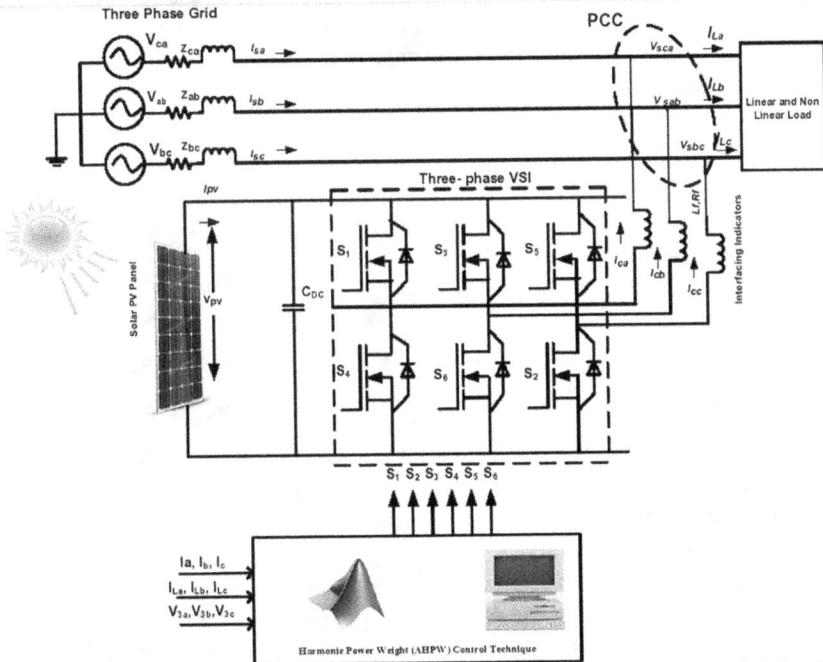

FIGURE 11.1 Designed circuit diagram for single-stage grid-connected solar photovoltaic system.

11.3   PROPOSED CONTROL TECHNIQUE

The performance of the controller for grid-tied solar PV systems is dependent on the control techniques that are used to compose the control algorithm of the controller. The active and harmonic power weight (AHPW) control technique is used to overcome the power quality issues. The AHPW control method depends upon the overall system control, for example, the generation of gating signals, the extraction of the reference signal, inductor current control, balance control of capacitor voltage, etc.

The AHPW control technique is represented in the time domain control and the frequency domain control. Traditional Fourier and quick Fourier transform algorithms, sine multiplication, and improved Fourier arrangement are all used in frequency domain calculations. Frequency domain approaches are appropriate for both single-phase as well as three-phase supply systems. Control techniques for the APF in the time domain are, for the most part, dependent on the momentary determination of remunerating orders as either current or voltage signs from deformed and harmonically contaminated voltage or current signals.

Time domain methods are, for the most part, utilized for three-phase supply systems. The single-stage topology of the grid-associated solar PV system requires different controls. The single transformation stage requires robust control for dealing with PQ issues and dynamic force movement. At this stage, an AHPW filter is utilized to separate the fundamental weight segment from the load segment, producing the reference currents and harmonically dirty load currents.

A complex vector filter (CVF)-based approach has been assumed to reduce the voltage harmonic effects and, thusly, exact estimates of unit vector designs and the amplitude of grid voltage are taken care of by using the removed basic positive sequence segments as a reference.

$$V_{s,abc} = \sqrt{\frac{2}{3}\left(v_a^2 + v_b^2 + v_c^2\right)} \tag{11.1}$$

$$S_{pa} = \frac{v_a}{V_s}; S_{pb} = \frac{v_b}{V_s}; S_{pc} = \frac{v_c}{V_s} \tag{11.2}$$

11.3.1   ACTIVE AND HARMONIC POWER WEIGHT (AHPW) ALGORITHM FOR VSI

In order to give a proficient solution to the imbued conundrum between steady-state lack of appropriate adjustments and exceptions in the classical least mean square (LMS) and least absolute difference (LAD)-based algorithms, the AHPW algorithm is presented. However, its presentation is seen to be debased in the presence of impulsive noise.

Usually, the optimum solution obtained by adaptive algorithms is limited to quadratic error-based cost work, but AHPW doesn't give a decent result through this cost work. Consequently, AHPW requires the convex amalgamation for different cost work, which is made up of two error norms.

$$Z(m) \triangleq \alpha(m) \times E\left\{e^2(m)\right\} + \{1 - \alpha(m)\} \times E\{e(m)\} \tag{11.3}$$

where e (m) shows is the output error signal and α (m) to the mixing boundary having a range [0, 1], this cost work turns into an error norm of returns to of LAD control for $\alpha = 0$ and error norm LMS control for $\alpha = 1$.

The selection is important for obtaining the momentary representation of LAD and LMS control, and along these lines, the issue can be moderated related to LMS control. The error signal implies significant work in obtaining the correct solution. The desired response of the error signal is q (m) and the filter output r (m). Besides, the traditional error for signal working is written regarding input vector x (n) and weight vector w (m) as,

$$p(m) = q(m) - r(m)$$

$$= q(m) - w^T(m) \times x(m) \tag{11.4}$$

Now, separating the cost function as for w (m) of the update weight equation for the AHPW algorithm is obtained as

$$r(m+1) = r(m) + \{2(m)e(m)\} + \{1-(m)\}sign\{e(m)\}y(m) \tag{11.5}$$

The significant and adequate state for parameter μ is composed as the conjunction investigation of mean as

$$0 < \mu < \frac{2}{\left(2 + [1-\lambda]\sqrt{2l? \ \sigma_n^2 l^2}\right)N\sigma_y^2} \tag{11.6}$$

where σ_y^2 and σ_n^2 are the sound powers for input and estimation. N shows the length of the filter is N and λ is chosen with index term p as a combined parameter. In this way, adaptive mixing parameter λ (p) is assessed through the given probability appropriation:

$$\lambda\{p\} = prob\{D(p) > d_0 u D(p) < -D_0\} \tag{11.7}$$

where D_0 is the positive number and D (p) is the even distribution required for the signal model within the sight of impulse aggravations for the state $d_0 \sim N(0, \sigma_s)$ as

$$\beta(p) = 2qrft\left[\frac{|f(p)|}{\sigma_s}\right] \tag{11.8}$$

where $qrft$ (.) represents the complementary function, σ_s means the approximation of $D(p)$ within the sight of exception. This AHPW calculation is estimation more concentrated.

For something else, the LMS calculation is preferred for a quicker combination according to the rationale that appears in Figure 11.2. Similarly, the advantages of AHPW calculation for controls with minimal system complexity are similar. The weight improvement of the AHPW calculation is introduced as

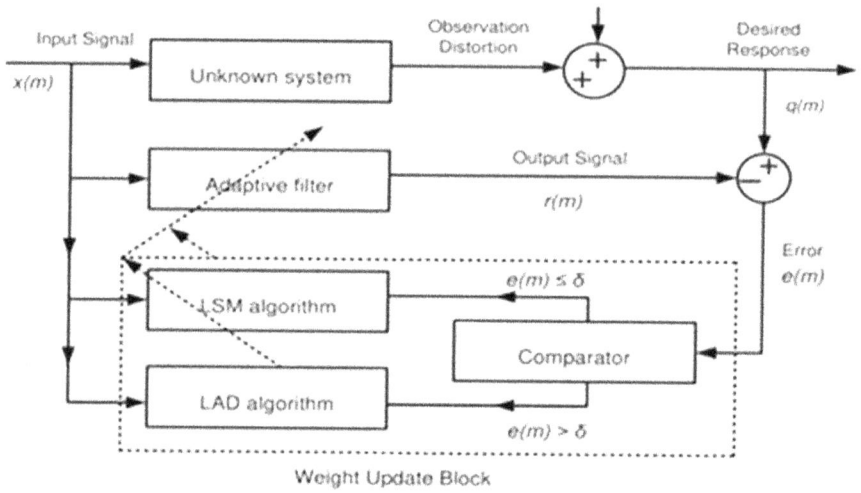

FIGURE 11.2   Block diagram of AHPW algorithm.

$$w_c(m+1) = \begin{cases} w_c(m) + \beta \cdot \text{sign}\{e_a(m)\} \cdot t(m) & \text{if } |e_a(m)| > \delta \\ w_c(m) + \beta \cdot e_a(m) \cdot t(m) & \text{if } |e_a(m)| < \delta \end{cases} \quad (11.9)$$

In this way, the weight update condition is introduced as GSPV parameters for all stages and can be composed as

$$w_a(m+1) = \begin{cases} w_a(m) + \mu \cdot \text{sign}\{e_a(m)\} \cdot u_{pa}(m) & \text{if } |e_a(m)| > \delta \\ w_a(m) + \mu \cdot e_a(m) \cdot u_{pa}(m) & \text{if } |e_a(m)| < \delta \end{cases} \quad (11.10)$$

So also, the updated weight conditions for the rest of the stages can be composed as Eq. (11.10). Furthermore, since the updated load of the considerable number of stages is acquiring the balanced source current wave forms in this way, the standard of all the updated load parts has been taken for additional handling as

$$t_{abc} = \frac{t_a + t_b + t_c}{3} \quad (11.11)$$

This voltage is calculated by a voltage transducer and compared with the DC capacitor. Proportional–integral (PI) is a regulator and the analysed output and output segment as

$$t_{DC}(m+1) = \{K_p \times V_{DC}(m)\} + \left\{ K_1 \sum_{m=0}^{k} V_{DC}(m) \right\} \quad (11.12)$$

where K_p and K_I are the acting boundaries of the DC-bus PI controller.

$$wpv = \frac{2 \times Vpv \times I_{pv}}{3 \times V_i} \tag{11.13}$$

Subsequently, net active weight segment is the overall condition for is composed as

$$p_N = p_{abc} + p_{DC} - p_{PV} \tag{11.14}$$

Finally, the reference currents are determined by multiplying by the net active weight segment and respective unit vector layout as

$$i_{ta} = v_{ra} \times z_a; \quad i_{tb} = v_{rb} \times z_b; \quad i_c = v_{rc} \times z_c; \tag{11.15}$$

balanced the reference current with the real origin current and found the error in each stage.

11.3.2 COMPLEX VECTOR FILTER-BASED TECHNIQUE

A complex vector filter strategy in a three-phase system has been embraced for removing the basic positive sequence element by the polluted grid voltages, utilizing the CVF technique as

$$w_{s\alpha\beta}(s) = w_{s\alpha}(s) + jw_{s\beta}(s) \tag{11.16}$$

This filter has better harmonic weakening and dynamic reaction capacity [25]. It's required in a three-phase PV connected grid system and this filter work replaced the $s = s - j\omega_0$ as

$$Q(s) = Q'(s - j\omega_0) = \frac{\omega_c}{s - j\omega_0 + \omega_c} \tag{11.17}$$

where $\omega_0 = 2\pi f_o$ is the essential angular frequency of source voltages having $f_o = 50\,\text{Hz}$ and ω_c represents the angular cut-off recurrence. The reaction of this transfer work is to block other harmonic segments. Regardless, this filter provides a solidarity addition for critical frequency segments while blocking all other harmonic parts (Figure 11.3). Presently, this filter is assessed by taking the Laplace reverse of Eq. (11.17) as

$$H(t) = \omega_k e^{j\omega_k t - \omega_k t} = \omega_k e^{-\omega_k t}\{\cos(\omega_k t) + j\sin(\omega_k t)\} \tag{11.18}$$

Along these lines, a prudent exertion is needed to choose the estimation of ω_c during complex vector filter usage. The usage of first-order complex vector filter is executed by utilizing Eq. (11.17) as $1/(s - j\omega_c)$ as shown [24]

$$\frac{1}{s - j\omega_c} = \frac{s + j\omega_c}{s^2 + \omega_c^2} = \frac{s}{s^2 + \omega_c^2} + j\frac{\omega_c}{s^2 + \omega_c^2} \tag{11.19}$$

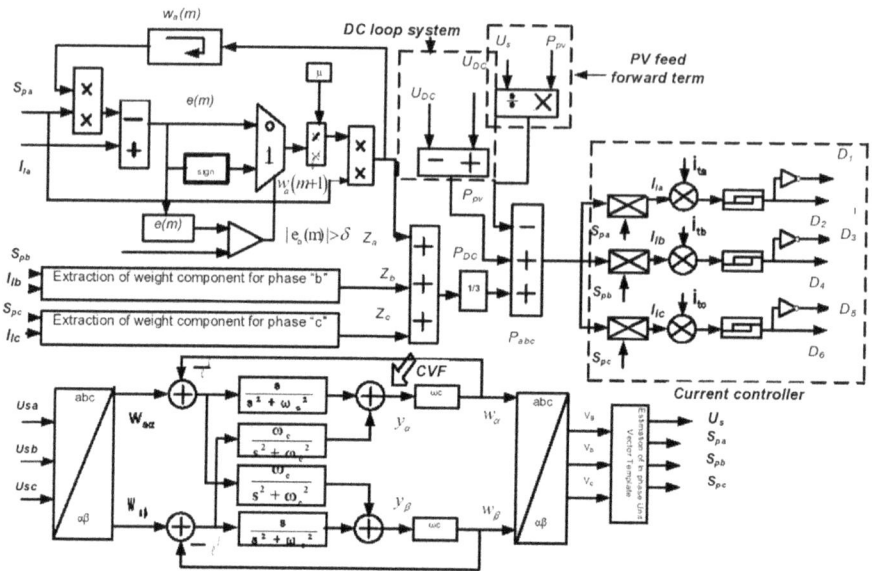

FIGURE 11.3 Control diagram for single-stage grid-connected solar photovoltaic system.

The scalar usage of first-order CVF has been introduced in terms of Eqs. (11.20) and (11.21). The fanciful piece of the CVF transfer function is dependent on these terms. Additionally, the dual output of the CVF filter is written as

$$z_\alpha(s) = \left(\frac{s}{s^2+\omega_c^2}\right) \times z_{s\alpha}(s) - \left(\frac{\omega_c}{s^2+\omega_c^2}\right) \times z_{s\beta}(s) \tag{11.20}$$

$$z_\beta(s) = \left(\frac{\omega_c}{s^2+\omega_c^2}\right) \times z_{s\alpha}(s) - \left(\frac{s}{s^2+\omega_c^2}\right) \times z_{s\beta}(s) \tag{11.21}$$

11.4 SIMULATION RESULTS AND DISCUSSIONS

The execution of the proposed AHPW control technique is analysed during the presence of different types of load and grid supply conditions. Two types of non-linear loads are used to create different types of harmonics, i.e., even and odd harmonics, and inject them into the grid side. Hence, the robustness and compensation capability of the proposed control technique can be tested. MATLAB software is used to implement the proposed multifunction of the PV inverter system. We compared the performance of the proposed AHPW control technique with that of the LMS and LAD control techniques. The PV system parameters are tabulated in Appendix A.

11.4.1 PERFORMANCE ANALYSIS WITH NON-LINEAR LOAD WITH PV INTEGRATION

The multifunctional PV system execution parameters of three-phase supply currents (I_{Sabc}), supply voltages (V_{Sabc}), compensating currents (I_{Cabc}) and load currents (I_{Labc})

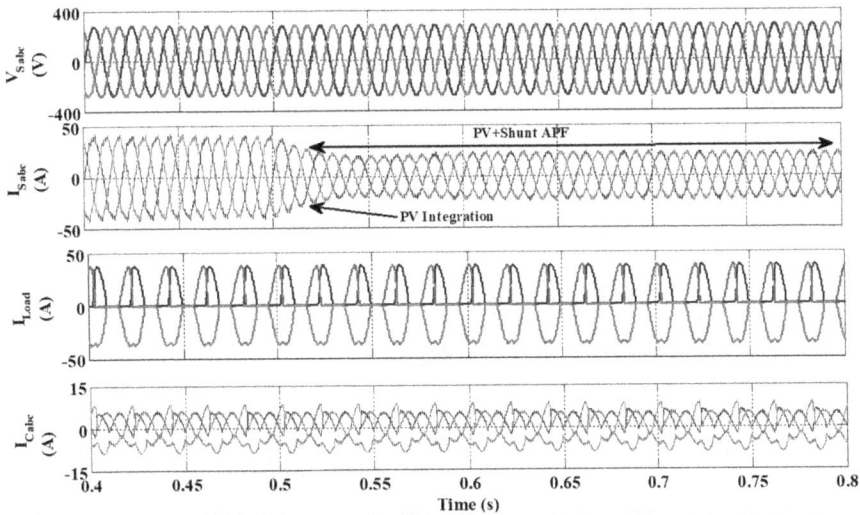

FIGURE 11.4 Multifunctional performance of the grid-tied photovoltaic system.

are shown in Figure 11.4. The PV power integration starts at $t = 0.5$ s; the PV array supplies the power to the grid and fulfills the demand of supply.

At the same time, the PV inverter plays an important role in injecting active power generated by PV into the grid. At the same time, it acts as a shunt APF to eliminate harmonics from the grid. The compensation capability of the proposed AHPW is tested by connecting a three-phase diode half-wave rectifier with an R-L load. This type of load injects more dominant, even harmonic content into the system. Before the adjustment, the total harmonic distortion (THD) of phase-a grid current was 68.51%, as shown in Figure 11.5a. The AHPW extracts the fundamental segments of current from the load currents, and these fundamental load currents will be processed to produce the reference currents. The generated reference currents will be compared with the actual source currents and produce error signals. These error signals are further processed for the generation of pulse width modulation signals by a hysteresis current controller. According to IEEE 519 standards, the phase-a grid current THD became 3.13%, as shown in Figure 11.5c. It tends to be resolved that the AHPW control algorithm is effectively working in PV synchronization along with even-order harmonic mitigation.

11.4.2 Performance Analysis during Adverse Condition of Grid Supply

The grid voltage will also experience distortions due to the high perforation of RES into the grid, high power non-linear loads connected near to the PCC, etc. To test the toughness of the proposed control algorithm, they made the grid forcefully polluted with high odd-order harmonics.

As shown in Figure 7.1a, the THD of grid voltages was recorded as 12.32%. In Figure 11.6, the distorted grid voltages, source currents, load currents, and compensating currents are shown. The proposed AHPW control algorithm uses a CVF filter

(a)

(b)

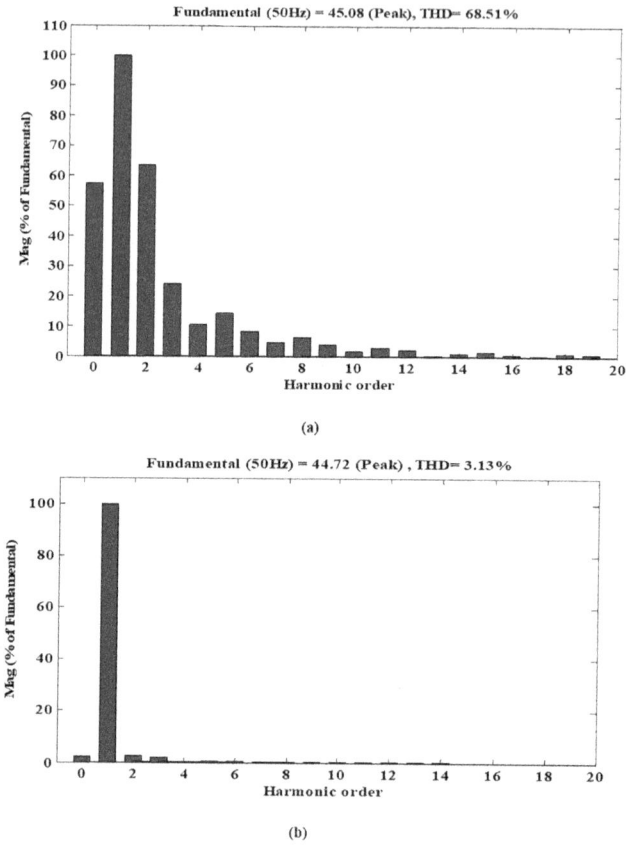

FIGURE 11.5   Harmonic spectrum of phase-a grid current (a) without compensation (b) with shunt APF.

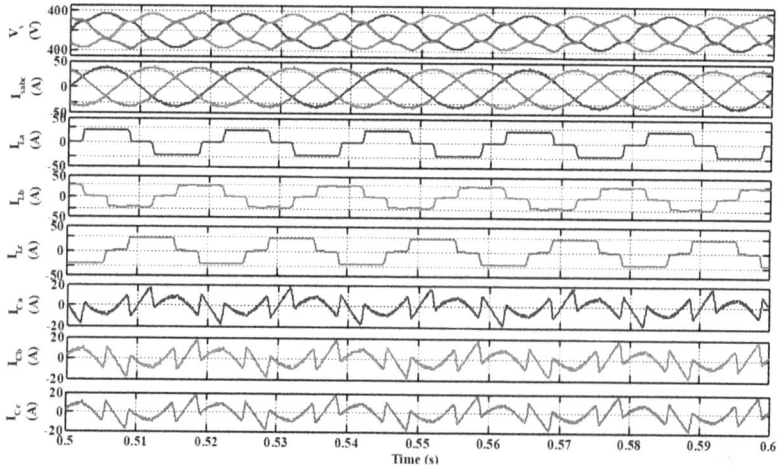

FIGURE 11.6   Performance of AHPW-CVF control technique.

to extract the fundamental frequency components from the distorted grid voltages. The reference currents generated by the AHPW control algorithm are the same as explained in Section 4.1. Figure 11.7b and c show the THD of phase-a grid current after and before compensation. Before compensation, THD was at 27.21%, and it was reduced to 3.78% after compensation. It is observed that the proposed control controller is robust in compensation under adverse conditions of grid voltages.

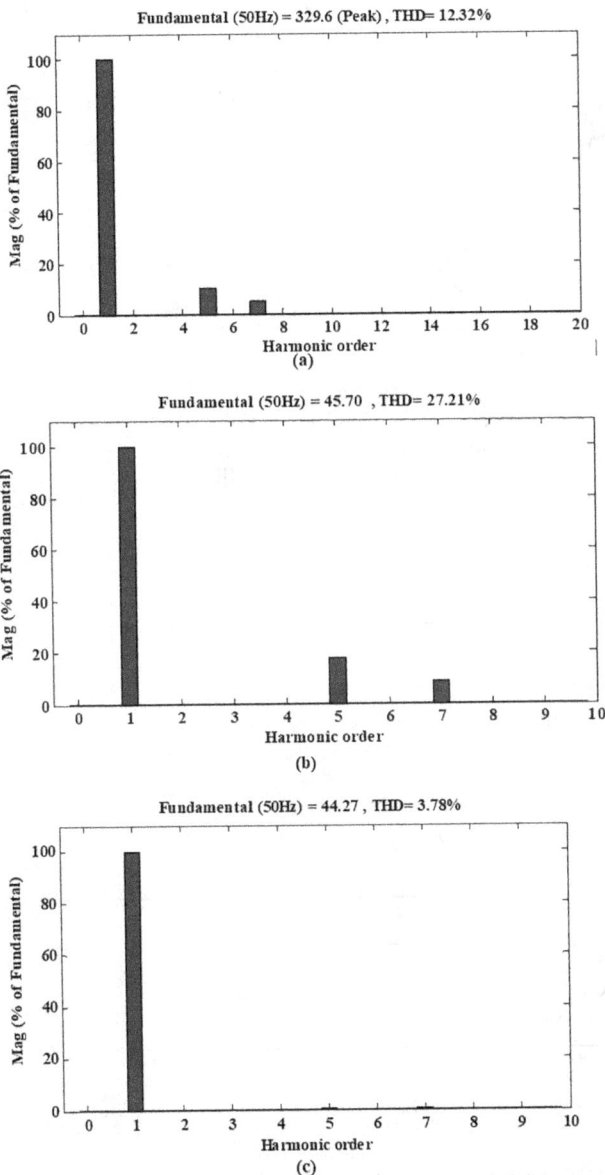

Fundamental (50Hz) = 329.6 (Peak) , THD= 12.32%

(a)

Fundamental (50Hz) = 45.70 , THD= 27.21%

(b)

Fundamental (50Hz) = 44.27 , THD= 3.78%

(c)

FIGURE 11.7 Harmonic spectrum. (a) Grid supply voltage. (b) Phase-a grid current with harmonic. (c) Phase-a grid current after shunt APF on.

11.4.3   COMPARATIVE PERFORMANCE STUDY UNDER LOAD PERTURBATION

In this section, the proposed AHPW control technique is checked under load perturbation conditions in comparison with popular LMS and LAD control algorithms.

When the load is changed at 0.9 and 1.4 s, the performance parameters of grid currents and load currents are shown in Figure 11.8a. In addition, the DC-bus voltage tracking capability using LMS, LAD, and AHPW controller techniques is shown in Figure 11.8b. It is the factor δ in the AHPW controller that makes it fast at tracking

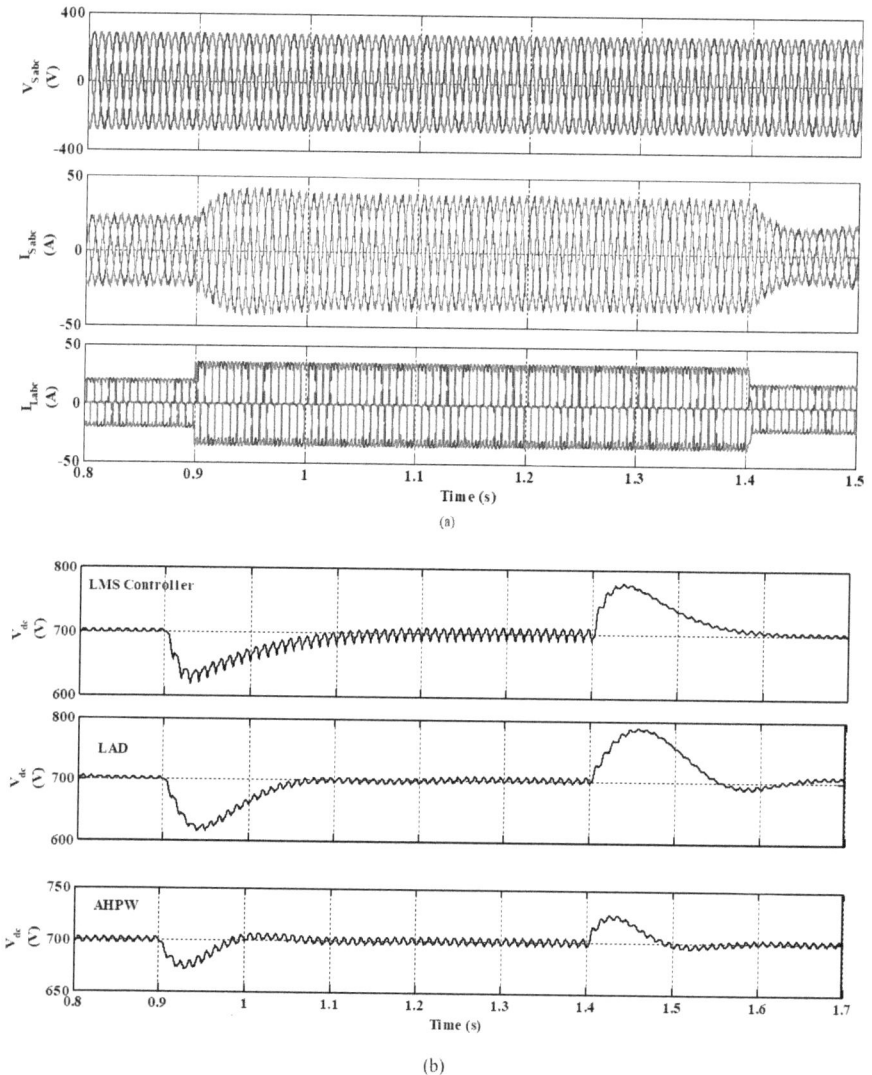

FIGURE 11.8 Performance parameters under load perturbation (a) grid voltages, grid currents and load currents (b) DC-link voltage stabilization using LMA, LAD and AHPW control algorithms.

TABLE 11.1

Performance Comparison of LMS, LAD and APHW Control Algorithms

Control Algorithm for DC-Voltage Regulation	The Steady-State Condition of the Load					Transient Condition of the Load			
	%Accuracy of DC Voltage	%THD			Settling Time	Overshoot	Undershoot	Response Time	Ripple Voltage
		I_a	I_b	I_c	(s)	(V)	(V)	()	(V)
LMS	94.13	4.55	4.67	4.73	0.9	70	65	2.27	20
LAD	95.68	4.56	4.51	4.23	0.4	65	55	1.32	10
AHPW	98.87	3.13	3.22	3.41	0.03	20	15	0.02	4

reference DC-bus voltage, lowering the settling time during load perturbation and switching parameters at a faster rate, which enhances the harmonic mitigation and synchronization capability. A fair comparison of LMS, LAD, and AHPW controller algorithms is shown in Table 11.1.

11.5 CONCLUSIONS

In this chapter, a novel AHPW-based control technique is developed for eliminating the even-order and odd-order harmonics from power system network and synchronization problems. Even harmonics create EMI problems in power electronic devices. Hence, the temperature of the device will rise, which leads to damage to the whole circuitry. This AHPW calculates the active power generation by a PV plant and feeds it to the grid without any phase delay. In addition, it extracts the harmonically weighted components in the supply and processes the reference current generation. The proposed PV system with AHPW is implemented using MATLAB software to check its performance under various conditions of PV, load, and grid power supply. The simulation results confirm that the proposed control algorithm is able to feed the active power to the grid in synchronism without phase delay and compensates for even low-order harmonics; hence, there are no EMI problems. A fair performance comparison was conducted between LMS, LAD, and APHW control techniques under load perturbations. It is observed that APHW has shown good tracking capability, better synchronism, less overshoot, undershoot, and high convergence speed compared with LMS and LAD control techniques.

REFERENCES

1. Panwar, N. L., Kaushik, S. C., and Kothari, S. (2011). Role of renewable energy sources in environmental protection: A review. *Renewable and Sustainable Energy Reviews*, 15(3), 1513–1524.
2. Carrasco, J. M., Franquelo, L. G., Bialasiewicz, J. T., Galván, E., PortilloGuisado, R. C., Prats, M. M., and Moreno-Alfonso, N. (2006). Power-electronic systems for the grid integration of renewable energy sources: A survey. *IEEE Transactions on Industrial Electronics*, 53(4), 1002–1016.

3. Singh, B., Chandra, A., and Al-Haddad, K. (2014). *Power Quality: Problems and Mitigation Techniques*. John Wiley & Sons, New York.
4. Mahela, O. P., Khan, B., Alhelou, H. H., and Siano, P. (November 2020). Power quality assessment and event detection in distribution network with wind energy penetration using stockwell transform and fuzzy clustering. *IEEE Transactions on Industrial Informatics*, 16(11), 6922–6932.
5. Mahela, O. P., Khan, B., Alhelou, H. H., and Tanwar, S. (14 October 2020). Assessment of power quality in the utility grid integrated with wind energy generation. *IET Power Electronics*, 13(13), 2917–2925.
6. Mahela, O. P., et al. Recognition of power quality issues associated with grid integrated solar photovoltaic plant in experimental framework. *IEEE Systems Journal*. doi:10.1109/JSYST.2020.3027203.
7. Mahela, O. P., Shaik, A. G., Khan, B., Mahla, R., and Alhelou, H. H. (2020). Recognition of complex power quality disturbances using S-transform based ruled decision tree. *IEEE Access*, 8, 173530–173547.
8. Mahela, O. P., Khan, B., Alhelou, H. H., Tanwar, S., and Padmanaban, S. (2021). Harmonic mitigation and power quality improvement in utility grid with solar energy penetration using distribution static compensator. *IET Power Electron*, 14, 912–922.
9. Pachauri, R. K., et al. (2020). Impact of partial shading on various PV array configurations and different modeling approaches: A comprehensive review. *IEEE Access*, 8, 181375–181403.
10. Pachauri, R. K., Mahela, O. P., Khan, B., Kumar, A., Agarwal, S., Alhelou, H. H., and Bai, J. (2021). Development of arduino assisted data acquisition system for solar photovoltaic array characterization under partial shading conditions. *Computers & Electrical Engineering*, 92, 107175.
11. Pachauri, R. K., et al. (2021). Shade dispersion methodologies for performance improvement of classical total cross-tied photovoltaic array configuration under partial shading conditions. *IET Renewable Power Generation*, 15, 1796–1811.
12. Khan, B., Agnihotri, G., Rathore, P., Mishra, A., and Naidu, G. (2014). A cooperative game theory approach for usage and reliability margin cost allocation under contingent restructured market. *International Review of Electrical Engineering*, 9(4), 854–862.
13. Khan, B., Agnihotri, G., and Mishra, A. S. (2016). An approach for transmission loss and cost allocation by loss allocation index and co-operative game theory. *Journal of The Institution of Engineers (India): Series B*, 97, 41–46.
14. Gali, V., Gupta, N., and Gupta, R. A. (2021). Experimental investigations on single-phase shunt APF to mitigate current harmonics and switching frequency problems under distorted supply voltage. *IETE Journal of Research*, 67, 333–353.
15. Kocewiak, Ł. H., Hjerrild, J., and Bak, C. L. (2010, September). The impact of harmonics calculation methods on power quality assessment in wind farms. In *Proceedings of 14th International Conference on Harmonics and Quality of Power-ICHQP 2010* (pp. 1–9). IEEE.
16. Fabricio, E. L. L., Júnior, S. C. S., Jacobina, C. B., and de Rossiter Correa, M. B. (2017). Analysis of main topologies of shunt active power filters applied to four-wire systems. *IEEE Transactions on Power Electronics*, 33(3), 2100–2112.
17. Marini, A., Ghazizadeh, M. S., Mortazavi, S. S., and Piegari, L. (2019). A harmonic power market framework for compensation management of DER based active power filters in microgrids. *International Journal of Electrical Power & Energy Systems*, 113, 916–931.
18. Gali, V., Gupta, N., and Gupta, R. A. (2017, June). Mitigation of power quality problems using shunt active power filters: A comprehensive review. In *2017 12th IEEE Conference on Industrial Electronics and Applications (ICIEA)* (pp. 1100–1105). IEEE.

19. Rajagopal, R., Palanisamy, K., and Paramasivam, S. (2018, July). A technical review on control strategies for active power filters. In *2018 International Conference on Emerging Trends and Innovations In Engineering And Technological Research (ICETIETR)* (pp. 1–6). IEEE.
20. Xu, J., Qian, Q., Zhang, B., and Xie, S. (2019). Harmonics and stability analysis of single-phase grid-connected inverters in distributed power generation systems considering phase-locked loop impact. *IEEE Transactions on Sustainable Energy*, 10(3), 1470–1480.
21. Liu, Y. and Heydt, G. T. (2005). Power system even harmonics and power quality indices. *Electric Power Components and Systems*, 33(8), 833–844.
22. Dey, P. and Mekhilef, S. (2014, May). Synchronous reference frame based control technique for shunt hybrid active power filter under non-ideal voltage. In *2014 IEEE Innovative Smart Grid Technologies-Asia (ISGT ASIA)* (pp. 481–486). IEEE.
23. Ruan, X., Wang, X., Pan, D., Yang, D., Li, W., and Bao, C. (2018). *Control Techniques for LCL-Type Grid-Connected Inverters*. Springer, Singapore.
24. Chaudhary, P. and Rizwan, M. (2017, February). A three phase grid connected SPV power generating system using EPLL based control technique. In *2017 Second International Conference on Electrical, Computer and Communication Technologies (ICECCT)* (pp. 1–6). IEEE.
25. Singh, S., Kewat, S., Singh, B., and Panigrahi, B. K. (2020). Enhanced momentum LMS LMS-based control technique for grid-tied solar system. *IET Power Electronics*, 13, 2767–2774.

APPENDIX A: THE SIMULATION PARAMETERS

Parameters	Symbol	Simulation
Grid supply	$V_{s,abc}$	400 V RMS (50 Hz) (L-L)
Smoothing inductor	L_{ca}, L_{cb}, L_{cc}	5 mH
DC-bus electrolytic capacitor	C_{dc}	1,500 µF
PV side DC-bus voltage	V_{dc}	700 V
Non-linear load	Resistance	12 Ω
Resistance	Inductance	25 mH
Inductor inductance		
Two panels of 250 W are connected in parallel total of 500 W		
$P_{max}=$	**250 W**	
$I_{sc}=$	8.95 A	
$I_{mpp}=$	8.35 A	
$V_{mpp}=$	29.95 V	

12 Valuation of Dynamic VAR Support in Deregulated Power System

Pankaj Mishra and T. Ghose

CONTENTS

12.1 INTRODUCTION

The consideration of technical aspects in framing the pricing mechanism for dynamic dispatch of reactive power is very important for voltage stability as well as economic operation and should not be ignored because of the extremely low running cost of VAR generation. The localized behavior, the dependency of node voltages and transmission line losses, the type of sources, the response of sources and the cost of generation are some of the major factors associated with reactive power, which influence the pricing mechanism for reactive power support. These make it the same voltage. In addition, the dissimilar operational importance of static and dynamic reactive power complicates its nature even as a commodity to price. Moreover, a competitive power market raises the expectations of customers to receive quality and uninter-rupted power. These are the reasons that draw the attention of several researchers towards reactive power management and its pricing aspects. The regulatory authori-ties, along with researchers, find it necessary to have a separate pricing mechanism for this service, which should suit the restructured power system. Several technical and economic criteria have been analysed to achieve a proper pricing technique. Some researchers also arrived at some convincing and feasible techniques. Among the proposals so far, the exact cost of reactive power in an electrical market is still a debatable issue for two main reasons. One is that the running cost for reactive power generation is very low, and the other is the incomparable establishment cost of VAR sources, as the variation in the range of their costs is not so small. A convinc-ing costing of reactive power produced by generators and other network devices has been developed, and, based on the loadability curve of generators, a cost component called "opportunity cost" for generators has been proposed in the literature [1]. Still, the volatility in VAR pricing and management in the market operation of deregulated power systems is a major concern for the system operators.

12.1.1 Deregulated Power System & Ancillary System Requirement

In the deregulated power system, the generation, transmission, and distribution of electrical energy are three separate entities. With the presence of many generation companies (GENCOs), the generation market has become fully competitive in sell-ing their produced energy. Similarly, the distribution companies (DISCOMs) came into existence for the purpose of retailing electrical energy. On the other hand, due to geographical constraints, the operation of a transmission system remains a regu-lated monopoly whose function is to allow open, nondiscriminatory, and compa-rable access to all suppliers and consumers of electrical energy [2–11]. To carry out this function, the Independent System Operator (ISO) is framed as a separate entity. Along with supporting the main transactions of energy, the ISO also arranges or procures some services termed ancillary services (AS), which are required to main-tain the reliability and security of energy supply. Although the listing of AS is still under discussion, AS is broadly classified as frequency control, network control, and system restart AS. These services are required to maintain acceptable frequency and voltage levels in the system, provide a reserve to ensure secure operation of the system and handle emergency situations such as restarting the system after a black-out. In addition, in market operations, these services are important for the support of commercial transactions. The ISO must arrange AS either on its own or with other power system utilities. More specifically, in the power market, the ISO must purchase these services from the utilities, and this entails the requirement of an AS market.

12.1.2 Reactive Power as an Ancillary Service

‖‖‖Reactive power management is one of the ASs that requires special attention in a restructured power system. The fast-growing demand and transmission network are escalating the complexity of power system operation. Further, the competition among the utilities to maximize profit gives additional stress to the operators. In this scenario, voltage instability is a major concern for the system operator. Since the voltage profile of nodes in a power system is directly affected by the reactive power status of the system, it is essential to search for a common mechanism to deal with both reactive power balance and voltage stability simultaneously. The compatibility of reactive power management with the restructured power system and its marketing environment anticipates a separate marketing provision of the service for secure and reliable operation of the system. Prior to restructuring, the reactive power supply and consumption were controlled by the system operator as a component of a real power transaction, and no separate costing of the service was required. However, because sole control over generation and distribution is no longer in the hands of a single operator, and maintaining system stability is the responsibility of ISO, defining such services separately becomes a requirement of system operation. As mentioned above, ISO must procure reactive power to maintain the voltage profile of the system. This again demands a separate marketing of reactive power and defining its cost to develop a pricing mechanism. There are several research studies available that define the pricing and management of reactive power. Some theories developed for defining the cost of reactive power include opportunity cost, cost of loss, marginal price of reactive power, and costing based on the depreciation cost of static reactive power sources. The localized nature of reactive power, which restricts its transmission over a long distance due to associated high loss, also limits the number of suppliers of reactive power. The strong dependency of voltage stability on reactive power urges an improved solution to its pricing.

12.1.3 REACTIVE POWER PRICING APPROACHES

Considering all aspects of reactive power, including its localized behavior, performance requirements, and reactive power loss, an appropriate pricing scheme is discussed and proposed by the author in [12]. However, to achieve a perfect market structure, the spot pricing technique using Lagrange's function gives a way of pricing reactive power, and some researchers find it efficient to price reactive power along with active power by minimizing the cost of active power for real-time pricing [13]. The same concept of spot pricing is extended with modifications like the inclusion of the cost of capacitor placement in the objective function [14]. Further developments made in this concept focus on the security aspects of the power system, and several constraints are considered while determining the marginal cost of reactive power. Line losses, reactive and active power reserve, voltage stability margin, reactive power flow tracing, sensitivity of generator and other VAR sources for incremental demand, and sensitivity of bus voltage for reactive power mismatch are some of the important factors which are incorporated in optimization of either cost or loss of the system [15–19]. However, most of these techniques use the cost of active power production of a generator, which has a very low portion of the true cost of providing reactive power service. The implicit and explicit costs of the generator [1] are also used in the generator's cost function for spot pricing, and the same costs of network devices are used as the cost function [19–22]. Some authors use the same concept in developing new payment functions to suit the competitive market environment for generators and network reactive power devices [23,24]. The effect of reactive power on real power revenue is another approach in this regard. How much real power shipment price is saved by injecting reactive power is the basis for such a methodology. In [25], the locational marginal price is used as a price signal for reactive power. There are a few research papers available that present reactive power pricing based on cost. However, due to the negligible running cost and almost nil fuel cost of producing reactive power, the pricing based on cost still requires more in-depth analysis.

Maintaining the voltage profile under normal and contingency conditions, minimizing congestion of real power flows, and minimizing real power losses are the three basic requirements of reactive power in a system. Another approach to designing a reactive power pricing mechanism is by deciding its value and importance in meeting these requirements. A value-based procurement method is suggested in [26]. In this method, the value of the VAR support of each generator with respect to the load bus is determined by calculating the utilization factor for the VAR supplied. However, the implicit and explicit cost functions discussed in [1] are used here for the cost of reactive power supplied by generators. The author in [27] included the maximization of voltage stability along with the minimization of system congestion and total payment for the procurement of reactive power. Optimizing the reactive power reserve to increase the voltage stability is considered for VAR scheduling in [28], where the author proves that maximizing the system's reactive power reserve improves the voltage stability of the system. The importance of voltage stability in deciding the reactive power compensation is shown [29]. The authors use a new voltage stability index, the bus index, for regulating the stability and deciding the vulnerability of buses to voltage collapse.

Moreover, the unbundling of high demand growth and natural limitations on extending the transmission system force the power network to operate under stressed conditions. The high MW flow requires proper dynamic power compensation to avoid voltage instability. The value of reactive power can be precisely defined. Dynamic reactive power plays a vital role in maintaining system stability and security during the real-time operation of a power system. The valuation of dynamic VAR sources based on their effect on system security and stability has been carried out [30–32]. Another important factor is transmission line loss and congestion, which should be considered for proper reactive power management and voltage stability. Apart from considering the topology of a power system for voltage stability, the authors in [33] find it necessary to include the MW loss in the voltage stability analysis for a stressed power system. Thus, along with voltage stability, real power loss and congestion are the other two important factors associated with reactive power management. This generates the necessity of a pricing scheme for reactive power which should incorporate these factors.

12.2 REACTIVE POWER & VOLTAGE STABILITY

12.2.1 REACTIVE POWER MANAGEMENT

Controlling reactive power generation is required to facilitate commercial transactions of real power across transmission networks. When the magnitudes of the load impedance and line impedance are equal, the transmitted MW power is maximum.

An increase in real power demand causes an increase in current but a decrease in load bus voltage. Therefore, the MW received at the load bus is again a function of nonlinear changes in current and voltage. This is a dynamic reactive power management and, therefore, the voltage is only the means of keeping the desired MW flow. Controlling voltage is affected by the nature of the transmission system, which itself is a nonlinear consumer of reactive power depending on system loading. In addition, the system's reactive power requirement varies with load and generation patterns. Besides this, controlling reactive power flow can reduce losses and congestion on the transmission system. These facts result in a dynamic reactive power requirement. Furthermore, any contingency can have the compounding effect of reducing the reactive supply and, at the same time, reconfiguring flows such that the system is consuming additional reactive power. In such situations, at least a portion of the reactive power must be capable of responding quickly to changing reactive power demands and maintaining acceptable voltages throughout the system.

In a system, the reactive power and voltage control are managed by a dynamic process like real-power AGC. The secondary voltage control is achieved by controlling reactive power flow and supply using dynamic VAR sources. For this reason, synchronous condensers, STATCOM, and SVC are VAR sources with the ability to adjust their VAR supply continuously, making them more valuable for the transmission system operators. Table 12.1 shows the speed, voltage support, and costs for the different sources of dynamic reactive power [34].

12.2.2   LINE LOSSES AND VOLTAGE STABILITY

Here it is worth discussing the phenomenon of voltage instability in a constrained situation. During heavily loaded conditions, the series loss in the transmission line accelerates the downfall of voltage. When the voltage decreases gradually, the industrial motor loads are thrown off by undervoltage relays at lower voltages, which leads to enhanced load bus voltage. In such situations, it is traditional to boost the distribution voltage by a transformer tap change, even though the transmission voltages of upstream nodes remain critical. This increases the bus loads again by reconnecting the industrial loads, which results in the deterioration of the reactive power position. The series reactive loss may increase tremendously, thus depressing the load bus voltage of upstream nodes to an alarming level. In addition to this, the remote generator, along with local generators, tries to push more reactive power through the line, which further increases the series reactive loss, resulting in voltage decline. Thus, in a heavily loaded line and depressed bus voltage, the reactive power demand increases, and the transmission line series reactive loss is quite substantial, which results in voltage collapse. Apart from other reasons for the decline in voltage in transmission systems, like sudden failure of generating units near load centres and outages of a heavily loaded transmission line, the transmission line losses have an appreciable effect on the progress of a system from a normal or marginal voltage stability point to the voltage collapse point [35].

Let us consider a simple two-bus radial network. For long transmission lines, the X/R ratio is very high, so the line is purely reactive. Reactance X also includes the series reactance of transformers in the line. The generator terminal voltage $E \angle 0$ is

TABLE 12.1
Characteristics of Voltage Control Equipments

| Equipment Type | Speed of Response | Voltage Support | | | Costs | | |
		Ability	Availability	Disruption	Capital (per kvar)	Operating	Opportunity
Generator	Fast	Excellent, additional short term capacity	Low	Low	Difficult to separate	High	Yes
Synchronous condenser	Fast	Excellent, additional short term capacity	Low	Low	$30–35	High	No
Static var compensator	Fast	Poor, drops with V^2	High	Low	$45–50	Moderate	No
STATCOM	Fast	Fair, drops with V	High	Low	$50–55	Moderate	No
Distributed generation	Fast	Fair, drops with V	Low	Low	Difficult to separate	High	Yes

the e.m.f. at the highly regulated bus which keeps the voltage at E. Let us consider S a constant MVA load, at voltage $V \angle \delta$ is drawn at the load bus, and Q_L is the series reactive loss occurring in line.

Then,

$$|Q_L| = |I^2 \times X| \tag{12.1}$$

where

$$|I| = \left| \frac{S}{V} \right| \tag{12.2}$$

Therefore,

$$|Q_L| = \left| \frac{S}{V} \right|^2 \times |X| \tag{12.3}$$

The change in series reactive loss for a change in voltage is

$$\left| \frac{\partial Q_L}{\partial V} \right| = \frac{-2}{|V|^3} \times |S|^2 \times |X|$$

Assuming $|X|$ to be constant and for constant MVA load, we get.

$$\left| \frac{\partial Q_L}{\partial V} \right| \propto \frac{1}{|V|^3} \tag{12.4}$$

Thus, the change in series reactive loss increases sharply with any drop in transmission voltage.

Generally, for long transmission lines, the X/R ratio is considered very high, and the value of resistance for the line is neglected for the analysis of most of the theoretical concepts. This assumption does not affect the validity of the result to a very large extent. However, in the case of a stressed power system, the resistive losses in the line cannot be neglected as the magnitude of the current is very high. Taking resistance into account, the series loss occurring in the transmission line may have an approximate relation as developed above. As in Eq. (12.4), for any instant of operation, assuming MVA load (S) and line impedance to be constant, the change in MVA loss with voltage can be represented as

$$\left| \frac{\partial S_L}{\partial V} \right| \propto \frac{1}{|V|^3} \tag{12.5}$$

i.e. change in MVA loss increases sharply with any drop in voltage.

12.3 VOLTAGE STABILITY SIGNAL

A two-bus equivalent model, having series impedance connecting two buses, can be used for voltage stability analysis by deriving the series impedance value from total generation and demand at any operating time period. Using this postulate, the authors have derived a voltage stability signal representing the voltage condition of the whole system at any time period [36]. The two-bus equivalent of a complex network has a generator bus, where all the generators of the network are clubbed together, having a fixed terminal voltage, i.e., the generator bus is considered an infinite bus. The second bus is considered an equivalent load-bus feeding the total demand of the network, and the voltage of this bus is considered V_{syst}, the voltage stability signal. The mathematical expression for V_{syst} is given by Eq. (12.6).

$$V_{syst} = \left| \frac{S_{DT} \times \sqrt{Z_{eq}}}{\sqrt{S_{LT}}} \right| \tag{12.6}$$

Where,
S_{DT} = Total MVA demand in the system
S_{LT} = Total MVA loss in the system
Z_{eq} = Series impedance of the line connecting the two buses

As per Eq. (12.6), the voltage of the equivalent load bus (V_{syst}) can be represented in terms of the total losses occurring in the system corresponding to the total load during that period. The impedance seen by the generator is also included in the equation. Since the voltage collapse is a problem of the stressed power system, it is quite reasonable to choose the network condition of a heavily loaded power system. A brief justification is also presented here below. The signal thus derived gives an appropriate indication of the system voltage stability condition caused by the reactive power mismatch in the system.

Let γ be the percentage of loss with respect to total demand occurring in the whole network such that $S_{LT} = \gamma \times S_{DT}$.

$$\dot{V}_{syst} = f(\gamma, V_{syst}) \tag{12.7}$$

Equation (12.7) represents a first-order differential equation. The solution to this equation, obtained by using the MATLAB toolbox ode45, is graphically shown in Figure 12.1. The limiting values of E and γ, for the solution, are chosen as 1.0 and (0.01, 1), respectively. The decreasing nature of the graph reflects the fact that the system voltage profile deteriorates with an increase in transmission losses. Figure 12.1 gives the range of V_{syst} according to the series of transmission losses occurring in the system. Under normal operating conditions, the transmission line series loss remains in the range of 5%–7%. V_{syst} is accordingly normalized for practical application.

12.4 VALUATION OF DYNAMIC REACTIVE POWER

Controlling reactive power generation is required to facilitate commercial transactions of real power across transmission networks. An increment in real power

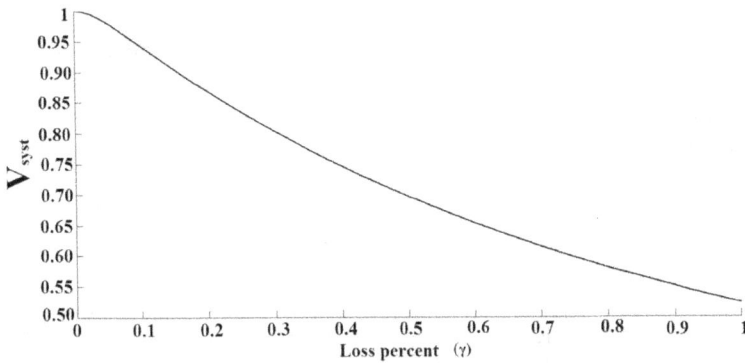

FIGURE 12.1 *V*syst vs. transmission line loss as percent of total demand.

demand results in an increase in current but, at the same time, a decrement in load bus voltage. Therefore, the MW received at the load bus is again a function of nonlinear changes in current and voltage. Dynamic reactive power management, therefore, voltage, is the only means of keeping the desired MW flow. Further, any contingency can have the compounding effect of reducing the reactive supply and, at the same time, reconfiguring flows such that the system is consuming additional reactive power. In such situations, at least a portion of the reactive power must be capable of responding quickly to changing reactive power demands to maintain acceptable voltages throughout the system. In this section, the proposed scheme is framed and explained. The first cost of the VAR supplied or drawn has been formulated, and then the formulation for valuation of the VAR is carried out. Finally, the concept has been illustrated by using two bus test cases.

Nomenclature
$k =$ Operating period,
$P_{Cij} =$ MW loss between branch connecting nodes i and j,
$P_{Tl} =$ Change in MW loss between branch connecting nodes i and j between two consecutive operating periods,
$P_L =$ Change in total system MW loss between two consecutive operating periods,
$P_{LT} =$ Total system MW loss,
$Q_{LT} =$ Total system MVAR loss,
$V_{syst} =$ System voltage stability signal,
$\lambda_i =$ MW marginal cost at any period at the ith bus.

12.4.1 COSTING

The importance of dynamic reactive power in power system operation is vital in transmitting active power. Static reactive power management is generally done in the planning and scheduling process, and the need for dynamic reactive power is fulfilled by the reactive power reserve kept during scheduling. All such reactive power support is priced either on the basis of a fixed cost for capital cost recovery or on the basis of long-term or short-term bilateral contracts. As there is very little or negligible running cost of reactive power, the dynamic reactive power is also priced along with the static reactive power pricing. The main disadvantage of pricing dynamic reactive power in this manner is that it does not provide a proper pricing signal and the dynamic reactive power is not valued according to its importance in the system's voltage stability.

Let us consider a multi-bus network where N is the number of branches incidental to the ith bus. The total real power injected from ith bus varies with each transaction. This variation depends on the loading of the lines and losses in the line. If the loss is high, then expected power could not be injected from the said bus to satisfy a particular transaction, i.e., a high MW loss in the lines may bring the line close to its loading limit, which may lead to a violation of the committed transaction by GENCO and/or TRANSCO. The change in line transaction limit can be determined either by comparing the difference in branch MW capacity or by comparing the branch MW loss for each transaction. The saved MW loss shows the possibility of more MW shipments; therefore, the increased MW loss with more loading implies a loss of the MW transaction capability of the line. With an injection of reactive power, the reduced loss or increased transaction capacity shows a saving in cost or an increase in revenue. Thus, defining the change in transaction limit as observed for ith bus

$$P_{TI} = \sum_{j=0}^{N} P_{Cij}^{k} - \sum_{j=0}^{N} P_{Cij}^{k-1} \tag{12.8}$$

The change in total transmission line loss as

$$P_L = P_{LT}^{k} - P_{LT}^{k-1} \tag{12.9}$$

where k is the period of operation and $k-1$ is the previous period from where any change in loading/generation like reactive power injection/drawl or an increase in load resulted in k^{th} period of operation.

Taking λ_i as the marginal cost of real power at the kth period, the revenue lost due to reactive power mismatch for the ith bus is defined as

$$\text{Revenue lost} = \lambda_i \left[(P_{TI}) + (P_L) \right] \tag{12.10}$$

Eq. (12.10) gives the cost of loss incurred due to reactive power deficiency or excess at a node. Because the proposed mechanism is designed with stressed power conditions in mind, i.e., a high requirement for dynamic reactive power, it is worthwhile to define real power loss in terms of reactive power loss that occurred between two stages of real-time operation. Thus, including the influence of MVAR loss in the system on the loss of MW revenue, i.e. $\dfrac{\partial P_l}{\partial Q_l}$, the cost directly includes the effect of VAR on the system.

Thus, the revenue loss containing the effect of reactive power loss in the system is,

$$\text{cost}(Q) = \lambda_i \left\{ \frac{\left[(P_{TI}) + (P_L) \right]}{\left(Q_{LT}^{k} - Q_{LT}^{k-1} \right)} \right\} \tag{12.11}$$

In Eq. (12.11), the ratio of MW losses to the change in MVAR losses is defined for stressed power system conditions. Wherein, on a practical operational basis, it is assumed that any change in voltage level and real power loss is also associated with a change in reactive power loss, i.e., for the stressed condition, the denominator remains nonzero.

Equation (12.11) gives the cost of dynamic reactive power, which is actually a quantification of the value of reactive power in context with its technical importance.

The direct effect of a shortage or excess of reactive power is seen in the voltage profile of the system. Hence, the quantified value of VAR is justifiable for the shortage of reactive power in the system, which results both in power losses and a reduction in the power handling capacity of the transmission line.

12.4.2 VALUATION

The shortage of reactive power has a direct effect on the bus voltage, and this effect is exaggerated by an increase in transmission line loss. In this work, it is proposed that the pricing of reactive power be carried out on the basis of its contribution to the system voltage stability. The value of the reactive power delivered to the system is determined by its effect on the system voltage stability. The V_{syst} discussed above has been used here to measure the system voltage stability condition.

It can be easily shown that the reactive power injected by the source is valuable if it pushes the system away from the voltage stability limit. In addition, reactive power injected from a distant source has the effect of increasing transmission line losses and burdening the power system, which results in the system's being more susceptible to voltage instability. The injected reactive power should not have any value in such cases. The valuation is implemented by multiplying the cost with V_{syst} as given in Eq. (12.12).

Eq. (12.12) thus gives a system quality-based pricing structure of dynamic reactive power. The price of reactive power at any operating stage is:

$$\left|\left(V_{\text{syst}}^{k} - V_{\text{syst}}^{k-1}\right)\right| \times \text{Cost}(Q) \tag{12.12}$$

In Eq. (12.12), the absolute value of the system voltage difference is used to quantify the price of VAR according to the system voltage condition. It may be asked here why one should get any payment for degrading the system's condition. This may be explained because of the very low numerical value of prices at any stage with a good voltage profile; the cost of generating VAR will always be higher than the amount it gets for VAR.

12.4.3 THE PRICING MECHANISM

Requirement: For each operating time period, a record of the transmission line's MW transaction limit, total MW and MVAR loss, and V_{syst} is to be maintained.

The steps involved in determining the price of VAR supplied or consumed at any kth operating period are as follows:

I. Data collection: V_{syst}, MW transaction limit of lines, and total MW & MVAR losses between $(k-1)$th and kth operating period.

II. Determine the change in MW transaction limit of originating branches for each bus. Repeat this till all the buses are covered and each branch should be covered only once. Use Eq. (12.8)

III. Determine the change in total MW & MVAR loss between $(k-1)$th and kth operating period.

IV. Find the MW marginal cost of all buses for the kth operating period.

V. Evaluate the cost of reactive power Cost(Q) using Eq. (12.11).

VI. Find the change in V_{syst} between $(k-1)$th and kth operating period.

VII. Calculate the price of Q at each bus for the kth operating period using Eq. (12.12).

Illustration with an Example

The proposed concept of costing and pricing of dynamic reactive power is presented with a simple two-bus example in the following section.

Let us create a base case showing the poor voltage condition of the load bus shown in Figure 12.2. The line loss is 29.27 MVA. The loading of the line and buses 1 and 2 are also shown. The effect of injecting reactive power is then checked, and the cost of reactive power injected is determined as discussed above. For simulation, Power World Simulator software is used. The two-bus OPF case has been used to explain the proposed dynamic reactive power cost and pricing. Figure 12.3 shows the case of an improved voltage condition where reactive power is locally injected on bus 2.

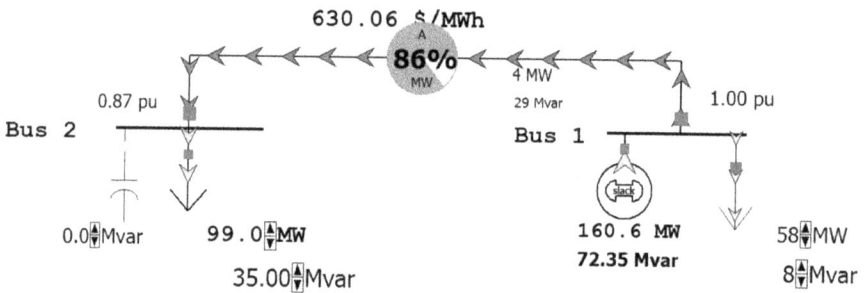

FIGURE 12.2   Poor voltage condition at load bus as base case.

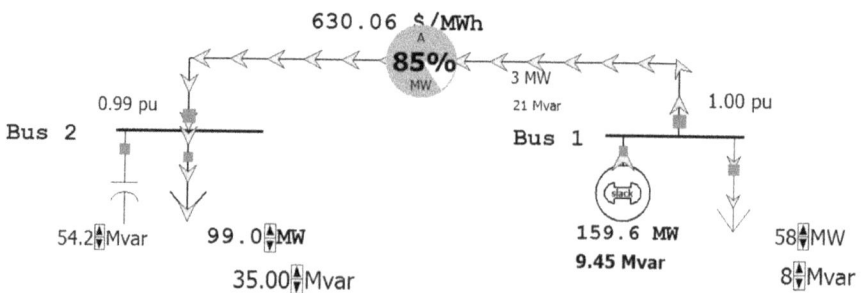

FIGURE 12.3   Case showing voltage of load bus corrected.

Referring to Figure 12.3, from the improved values of parameters, it is apparent that when the reactive power demand at bus no. 2 is supplied by the generator at bus no. 1, the line connecting the two buses gets stressed. When the reactive demand is supplied locally, it saves the MW loss in line and, also, due to the reduced reactive demand in line, the bus voltage improves. The cost of reactive power supplied by the source at bus no. 2 for improving the bus voltage from 0.87 to 0.99 p.u. is calculated hereunder as per the proposal. Here, the change in MW loss is 1 MW, and the change in line capacity is +1%.

According to the OPF result, the marginal price of active power during this period is 630.06 \$/MWh, so the loss in revenue at 0.87 p.u. voltage is $630.06 \times 1 = 630.06$ \$/h and hence cost of reactive power for the same period will be $(630.06 \times 1)/(29-21) = 78.757$ \$/MVArh. The change in reactive power loss is used in the denominator of the cost function to give the cost of saving per unit reactive power loss, which in turn defines the cost of voltage support because the more the MVAR loss, the more the system approaches the voltage stability limit. Now the value of reactive power is determined by the change in V_{syst}. Hence, the price of supplying reactive power to improve the voltage condition will be the multiplication of two, that is, Cost × Change in V_{syst}. Thus, the amount of money paid to the supplier of reactive power at bus 2 is

$$\left[(78.757\$) \times |0.6514 - 0.5165|\right] \times (54.2 \text{MVAR}) = 575.787\$$$

This gives the price of the VAR supplied by the source to improve the system's condition with respect to MW loss and voltage profile.

12.5   DISCUSSION ON RESULTS OF THE PROPOSED POSTULATE

The proposed pricing scheme has been verified using the IEEE 30 bus test system. The power flow data of the system is referred to in [37–39]. Buses 1, 2, 13, 22, 23, and 27 are the PV buses, which include the dynamic VAR sources of the system. The minimum and maximum voltages on buses are taken as 0.95 and 1.06 p.u., respectively. Starting from a base case, the price of VAR for the next operating stages has been calculated using the proposed scheme. The effect on the VAR price with VAR injection from the demand side is shown and compared with the price determined by an existing method using OPF. The VAR prices on all buses are also determined and plotted with the VAR demand increment. The results are shown with the help of different graphs and discussed in the following paragraphs.

Starting from a base case, $(k-1)$th operating period, the VAR demand at bus no. 30 is increased in steps up to loading prior to the divergence of Newton Raphson's load flow solution, which is taken as a kth operating period. Following the steps mentioned above, the price of VAR for all buses has been calculated for the kth operating period. At this critical load, 2 MVAR of reactive power is injected into the network at bus no. 28 to observe the effect on the voltage profile of the system and thereby calculate the price of VAR for this $(k+1)$th operating period. The proposed price at the kth stage is more than that at $(k+1)$th stage for each bus, and that is because the kth stage is a stressed system condition reflecting point closed to voltage collapse and reactive power generation. The 28th bus relieves stress. That is why we are saying the $(k+1)$th stage. Further, the prices at the kth stage are higher, which indicates that the contribution of reactive power should have a higher value when the voltage profile is poor. This is expected as the dynamic power supplied/drawn by each bus has an effect on the voltage condition of the system. The $(k+1)$th stage has lower values of price, thus discouraging the dynamic VAR sources from changing the VAR generation.

12.5.1 COMPARISON OF PROPOSED SCHEME WITH OPF BASED METHOD

The VAR price obtained from a method using OPF and that of the proposed scheme is compared in the graphs of Figures 12.4 and 12.5.

Figure 12.4 shows the graphs for the system condition changing from the base case to the kth operating period. The graphs in Figure 12.4 show voltage at bus no. 30, the overall voltage indicator V_{syst}, VAR price calculated up to a kth operating period using OPF and proposed VAR price with respect to change in reactive demand at bus no. 30. The nature of the VAR price determined using the proposed mechanism is the same as that of using OPF. However, the value of the proposed VAR price is very low as compared to that of determining it using OPF. This is because the proposed method only includes the revenue which can be earned from MW lost in the line, which is less as compared to the incremental cost for MW of power generated or demanded.

Figure 12.5 shows three graphs representing the VAR price for each bus, calculated using the OPF-based method and the proposed method. The third graph shows the voltage at each bus; all three graphs are drawn for the kth stage when the voltage is prior to collapse. Comparing the graphs showing the proposed price at each bus with the graph showing the voltages on each bus, the variation in VAR price is in accordance with the bus voltages. This verifies the effectiveness of the proposed scheme in reflecting the VAR price according to the voltage condition of the bus.

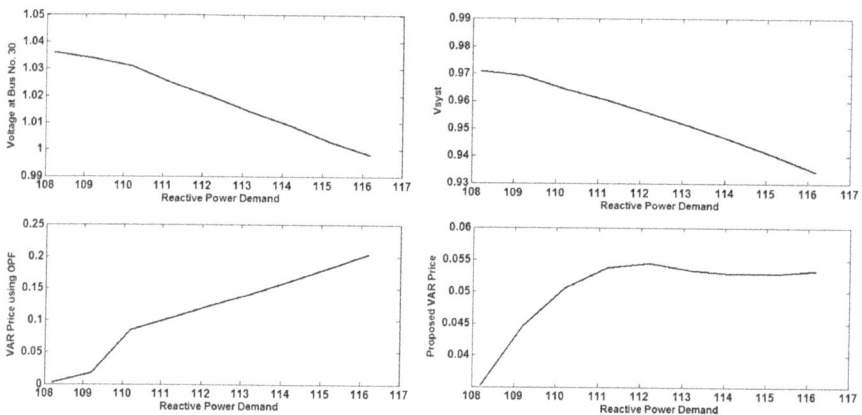

FIGURE 12.4   Effect of VAR load change at bus no 30 on proposed VAR price at a PQ bus.

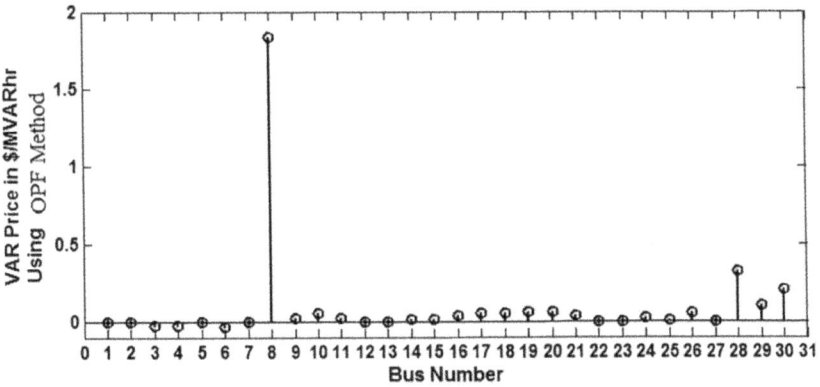

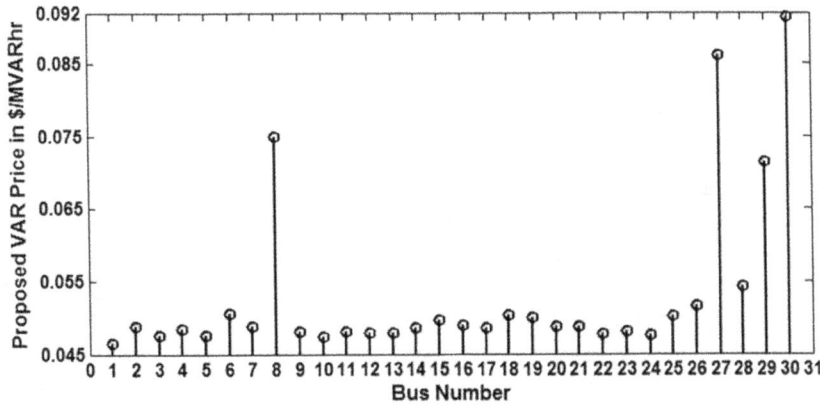

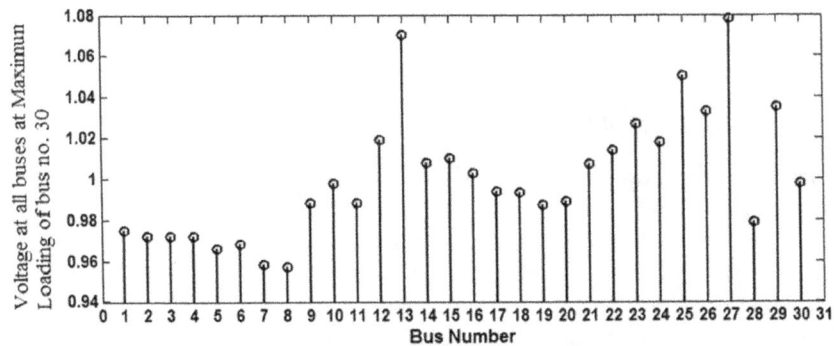

FIGURE 12.5 Comparison of proposed VAR price with VAR price calculated using OPF at all buses.

12.5.2 Locational Effect of VAR & the Proposed Pricing Scheme

Figure 12.6 illustrates the locational effect of VAR in the proposed pricing scheme, which also shows the higher prices for the reactive power sources that help to restore V_{syst}. Graphs in Figure 12.6 compare the proposed VAR price at PV bus 27 to other

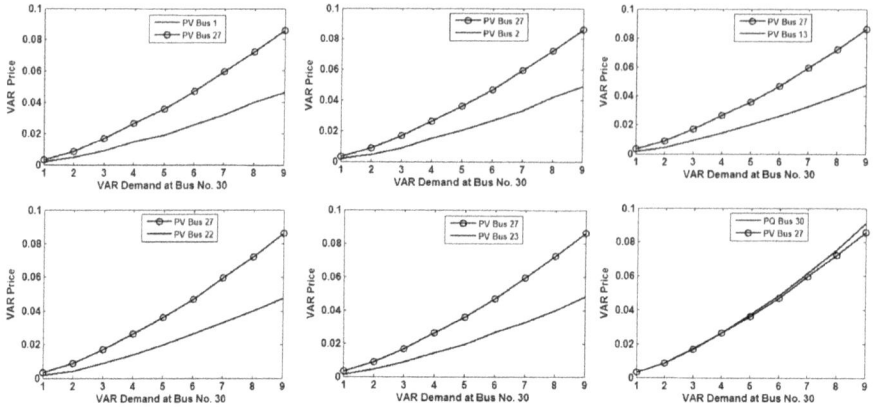

FIGURE 12.6 Locational effect of VAR on proposed pricing scheme.

PV buses, namely 1, 2, 13, 22, and 23, and the last graph in Figure 12.6 compares the VAR price at PV bus 27 to the PQ bus 30, where VAR demand varies. In the first five graphs, the VAR price at PV bus number 27 is higher than the other PV buses, which is because the VAR demand is required at bus number 30, which is nearer to PV bus number 27. The last graph in the series depicts the higher VAR price of bus number 30 as compared to bus number 27 at higher loading. This demonstrates the importance of load variation at bus no. 30 on the system voltage condition.

Thus, we may conclude that the numerical values of the proposed VAR price indicate the actual system condition with respect to voltage stability. Along with this, the proposed mechanism also takes care of the locational effect of reactive power management, load power factor, and intensity of the change that occurred in the system.

12.6 CONCLUSION

Dynamic reactive power supported by generators and other sources has a significant role, particularly when the system condition is taking the system toward a voltage collapse point. The importance of using the system voltage signal, i.e., V_{syst}, is critically examined. The pricing equation includes the V_{syst} in a bid to encourage critical sources as required by the prevailing situation to support voltage stability. The amount of pricing value for a source automatically shows the sensitive location of the reactive power source and, thus, the high value of pricing encourages the service provider to inject more reactive power. Different cases are created to show the dynamic nature of pricing that depends on the variation in benefit value and V_{syst}. The locational effect of a VAR source on the voltage stability of the system is incorporated into pricing by the real-time variation of V_{syst} with reactive power generation from any source. The result also shows that the revenue earned is different for sources located at different places, even though they generate the same amount of reactive power. It is also observed that the technical aspect of reactive power requires more attention while framing its pricing mechanism. The scheme is an approximate approach for pricing the dynamic reactive power and can be made effective for its practical use by considering other factors affecting the voltage and MW revenue loss in a system and considering the other running cost components of the VAR sources.

REFERENCES

1. J. W. Lamont and J. Fu, "Cost analysis of reactive power support," in IEEE Transactions on Power Systems, vol. 14, no. 3, pp. 890–898, Aug. 1999.
2. B. Khan, G. Agnihotri, S. E. Mubeen, and G. Naidu, "A TCSC Incorporated Power Flow Model for Embedded Transmission Usage and Loss Allocation," *AASRI Procedia*, vol. 7, pp. 45–50, 2014.
3. B. Khan, G. Agnihotri, G. Gupta, and P. Rathore, "A Power Flow Tracing based Method for Transmission Usage, Loss & Reliability Margin Allocation," *AASRI Procedia*, vol. 7, pp. 94–100, 2014.
4. B. Khan, G. Agnihotri, P. Rathore, A. Mishra, and G. Naidu, "A Cooperative Game Theory Approach for Usage and Reliability Margin Cost Allocation under Contingent Restructured Market," *International Review of Electrical Engineering*, vol. 9, no. 4, pp. 854–862, 2014.
5. B. Khan and G. Agnihotri, "A Comprehensive Review of Embedded Transmission Pricing Methods Based on Power Flow Tracing Techniques," *Chinese Journal of Engineering*, vol. 2013, Article ID 501587, 13 pages, 2013.
6. B. Khan, G. Agnihotri, and A. S. Mishra, "An Approach for Transmission Loss and Cost Allocation by Loss Allocation Index and Co-operative Game Theory," *Journal of the Institution of Engineers (India): Series B*, vol. 97, pp. 41–46, 2016.
7. B. Khan and G. Agnihotri, "A Novel Transmission Loss Allocation Method Based on Transmission Usage," *2012 IEEE Fifth Power India Conference*, 2012, pp. 1–3.
8. P. Rathore, G. Agnihotri, B. Khan, and G. Naidu, "Transmission Usage and Cost Allocation Using Shapley Value and Tracing Method: A Comparison," *Electrical and Electronics Engineering: An International Journal (ELELIJ)*, vol. 3, pp. 11–29, 2014.
9. B. Khan and G. Agnihotri, "An Approach for Transmission Usage & Loss Allocation by Graph Theory," *WSEAS Transactions on Power Systems*, vol. 9, pp. 44–53, 2014.
10. S. Khare, B. Khan, and G. Agnihotri, "A Shapley Value Approach for Transmission Usage Cost Allocation under Contingent Restructured Market," *2015 International Conference on Futuristic Trends on Computational Analysis and Knowledge Management (ABLAZE)*, Noida, 2015, pp. 170–173.
11. B. Khan, G. Agnihotri, and G. Gupta, "A Multipurpose Matrices Methodology for Transmission Usage, Loss and Reliability Margin Allocation in Restructured Environment," *Electrical & Computer Engineering: An International Journal*, vol. 2, no. 3, p. 11, September 2013.
12. Shangyou Hao and A. Papalexopoulos, "Reactive power pricing and management," in IEEE Transactions on Power Systems, vol. 12, no. 1, pp. 95–104, Feb. 1997.
13. M. L. Baughman and S. N. Siddiqi, "Real-time pricing of reactive power: theory and case study results," in IEEE Transactions on Power Systems, vol. 6, no. 1, pp. 23–29, Feb. 1991.
14. D. Chattopadhyay, K. Bhattacharya and J. Parikh, "Optimal reactive power planning and its spot-pricing: an integrated approach," in IEEE Transactions on Power Systems, vol. 10, no. 4, pp. 2014–2020, Nov. 1995.
15. G. M. Huang and H. Zhang, "Pricing of generators reactive power delivery and voltage control in the unbundled environment," 2000 Power Engineering Society Summer Meeting (Cat. No.00CH37134), 2000, pp. 2121–2126 vol. 4.

16. J. Barquin Gil, T. G. San Roman, J. J. Alba Rios and P. Sanchez Martin, "Reactive power pricing: a conceptual framework for remuneration and charging procedures," in IEEE Transactions on Power Systems, vol. 15, no. 2, pp. 483–489, May 2000.

17. D. Chattopadhyay, B. B. Chakrabarti and E. G. Read, "Pricing for voltage stability," PICA 2001. Innovative Computing for Power - Electric Energy Meets the Market. 22nd IEEE Power Engineering Society. International Conference on Power Industry Computer Applications (Cat. No.01CH37195), 2001, pp. 235–240.

18. V. M. Dona and A. N. Paredes, "Reactive power pricing in competitive electric markets using the transmission losses function," 2001 IEEE Porto Power Tech Proceedings (Cat. No.01EX502), 2001, pp. 6. vol.1.

19. V. L. Paucar and M. J. Rider, "Reactive power pricing in deregulated electrical markets using a methodology based on the theory of marginal costs," LESCOPE 01. 2001 Large Engineering Systems Conference on Power Engineering. Conference Proceedings. Theme: Powering Beyond 2001 (Cat. No.01ex490), 2001, pp. 7–11.

20. Y. Dai, Y. X. Ni, F. S. Wen and Z. X. Han, "Analysis of reactive power pricing under deregulation," 2000 Power Engineering Society Summer Meeting (Cat. No.00CH37134), 2000, pp. 2162–2167 vol. 4.

21. Shangyou Hao, "A reactive power management proposal for transmission operators," in IEEE Transactions on Power Systems, vol. 18, no. 4, pp. 1374v1381, Nov. 2003.

22. Rider, M.J.; Paucar, V.L.: 'Application of a nonlinear reactive power pricing model for competitive electric markets', IEE Proceedings - Generation, Transmission and Distribution, 2004, 151, (3), p. 407–414.

23. J. Zhong and K. Bhattacharya, "Toward a competitive market for reactive power," in IEEE Transactions on Power Systems, vol. 17, no. 4, pp. 1206–1215, Nov. 2002.

24. J. Zhong, "A pricing mechanism for network reactive power devices in competitive market," 2006 IEEE Power India Conference, 2006, pp. 6.

25. T. A. Vaskovskaya, "Market price signals for customers for compensation of reactive power," 11th International Conference on the European Energy Market (EEM14), 2014, pp. 1–4.

26. S. K. Parida, S. N. Singh, and S. C. Srivastava, "VAR Cost Allocation by Using a Value-Based Approach," *IET Generation, Transmission & Distribution*, vol. 3, no. 9, pp. 872–884, 2009.

27. A. Kargarian, M. Raoofat, and M. Mohammadi, "VAR Provision in Electricity Markets Considering Voltage Stability and Transmission Congestion," *Electric Power Components and Systems*, vol. 39, pp. 1212–1226, 2011.

28. Q. Sun, J. Zhang, B. Li, and Y. Song, "A method to improve reactive reserve management with respect to voltage stability," 2015 IEEE Power & Energy Society General Meeting, 2015, pp. 1–6.

29. D. O. Dike and S. M. Mahajan, "Voltage Stability Index-Based VAR Compensation Scheme," *Electrical Power and Energy Systems*, vol. 73, pp. 734–742, 2015.

30. W. Xu, Y. Zhang, L. C. P. da Silva, P. Kundur and A. A. Warrack, "Valuation of dynamic reactive power support services for transmission access," in IEEE Transactions on Power Systems, vol. 16, no. 4, pp. 719–728, Nov. 2001.

31. A. Pirayesh, M. Vakilian, R. Feuillet, and N. HadjSaid, "A Conceptual Structure for Value-Based Pricing of Dynamic VAR Support," *2005 IEEE/PES Transmission and Distribution Conference & Exhibition*, Asia and Pacific Dalian, China.

32. X. R. Li, C. W. Yu, and W. H. Chen, "A Novel Value Based VAR Procurement Scheme in Electricity Markets," *Electrical Power and Energy Systems*, vol. 43, pp. 910–914, 2012.

33. J. W. Simpson-Porco, F. Dorfler, and F. Bullo, "Voltage Collapse in Complex Power Grids," *Nature Communications*, vol. 7, p. 10790, 2016. doi:10.1038/ncomms10790.

34. Principles of Efficient and Reliable Reactive Power Supply and Consumption, Federal Energy Regulatory Commission Staff Report Docket No. AD05-1-000 February 4, 2005, Washington, DC.

35. A. Chakrabarti, D. P. Kothari, A. K. Mukopadhyay, and A. De, *An Introduction to Reactive Power Control and Voltage Stability in Power Transmission System*, PHI Learning Pvt. Ltd. ISBN-978-81-203-4050-3, 2010.

36. P. Mishra and T. Ghose, "A Direct Method for Assessment of Overall Voltage Condition of Power System," *Electrical Power and Energy Systems*, vol. 81, pp. 232–238, 2016.

37. R. D. Zimmerman, C. E. Murillo-S_anchez, and R. J. Thomas, "Matpower: Steady-State Operations, Planning and Analysis Tools for Power Systems Research and Education," *IEEE Transactions on Power Systems*, vol. 26, no. 1, pp. 12–19, February 2011. doi:10.1109/TPWRS.2010.2051168.

38. C. E. Murillo-S_anchez, R. D. Zimmerman, C. L. Anderson, and R. J. Thomas, "Secure Planning and Operations of Systems with Stochastic Sources, Energy Storage and Active Demand," *IEEE Transactions on Smart Grid*, vol. 4, no. 4, pp. 2220–2229, December 2013. doi:10.1109/TSG.2013.2281001.

39. L. L. Lai, *Power System Restructuring and Deregulation: Trading, Performance and Information Technology*, 2002, Print ISBN: 9780471495000. doi:10.1002/0470846119.

13 Smart Grid Cyber Security Threats and Solutions

Vasundhara Mahajan, Neeraj Kumar Singh,
Praveen Kumar Gupta, Atul Kumar Yadav,
and Soumya Mudgal

CONTENTS

DOI: 10.1201/9781003278030-13

13.1 INTRODUCTION

The power grid is the world's largest interconnected machine, consisting of power plants, transmission lines, substations and consumers. In the traditional electrical system, electrical energy is distributed from generating stations to the consumers by increasing and decreasing voltage at different levels during transmission. Traditional grid was built in the 1890s and improved slowly over the decades with upgradation in technology. Today, the grid consists of thousands of generating stations connected with more than 100,000 km of high-voltage transmission lines. This engineering marvel has now reached its capacity. Newer grid technology has emerged in the past few years. The new grid overcomes the limitations of the traditional grid and is built to handle the digital and computerized equipment and technology. It automates and manages the increased complexity to provide a reliable power supply. Nowadays, along with the increased use of renewable energy in the grid, an intelligent system is needed to accommodate the distributed power generation.

13.1.1 SCHEMATIC VIEW OF SMART GRID

Typically, a smart grid consists of a microgrid, smart meters, renewable energy sources and hybrid electric vehicle. Figure 13.1 shows a schematic view of the smart grid. It is divided into three categories named as distribution side, transmission side and generation side. Generation side focuses on the conventional and non-conventional electricity sources, transmission side is used to transfer power and distribution side takes care of customers for efficient power supply. Both energy and communication data are exchanged by two-way network. A different area of the smart grid consists of different smart equipment. A brief description of each side of the grid is given as follows:

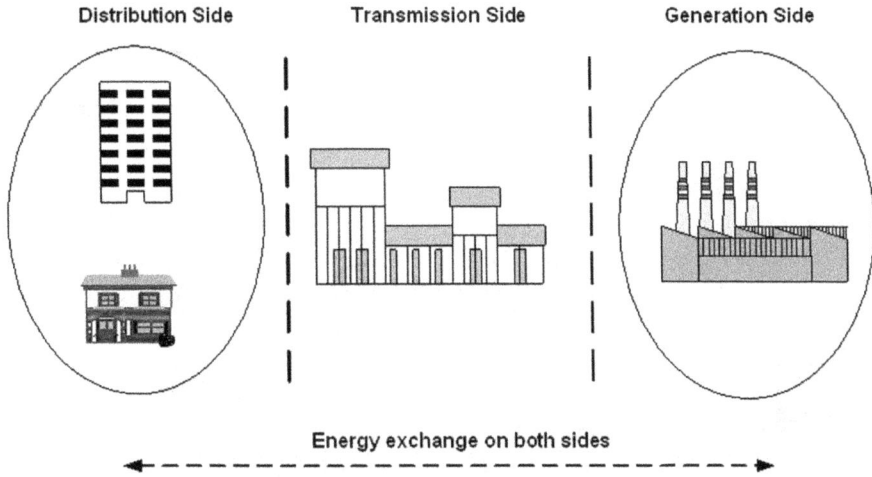

FIGURE 13.1 Schematic diagram of smart grid.

1. **Distribution Side:** This side includes the customers and local substations. These substations are built with modern technologies to integrate and monitor the power flow from generation to distribution. These substations report to the main control centre about the distribution side power consumption and the local power generation (e.g. solar roof top power). Customers' homes are equipped with the smart meters which periodically send the power consumption report to the substations. Further, the appliances installed at the customers' ends are connected to the smart meters in order to have a smart energy monitoring system.

2. **Transmission Side:** This side of the power system mainly focuses on the power transfer from the generation side to the distribution side. The transmission side is also enabled with the bidirectional communication network to have real-time monitoring of the entire grid. Multiple communication technologies are used for an improved grid monitoring.

3. **Generation Side:** The generation side mainly focuses on power generation which includes traditional power generation and renewable power generation. This side is also equipped with modern communication technologies to have a real-time energy supply status.

13.1.2 SMART GRID COMMUNICATION MODEL

The smart grid is an approach that ensures safe, real-time monitoring with the reliable quality power supply to all the customers. The grid also provides two-way communication between the customers and utilities. Figure 13.2 shows the architecture of a smart grid as conceptualized by the National Institute of Standards and Technology. The smart grid framework includes several entities such as market, customers, utility providers, generating units, transmission units and distribution units. Each entity is directly or indirectly associated with each other for proper operation of smart grid. Concurrent

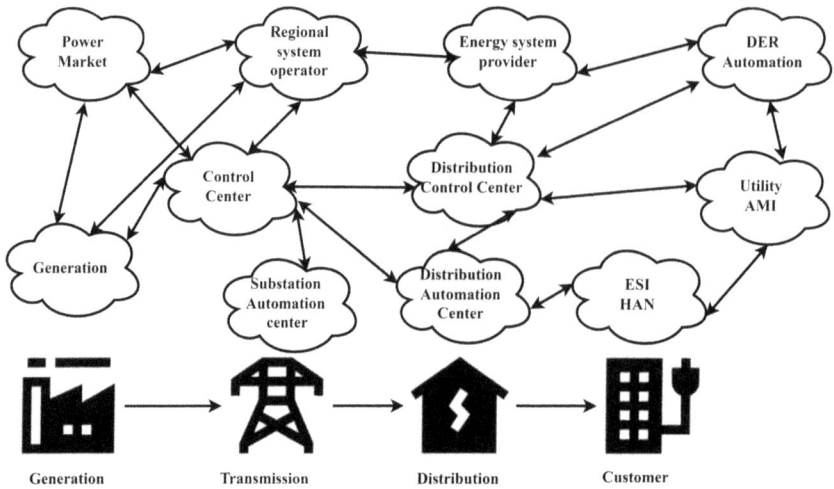

FIGURE 13.2 Smart grid complex working architecture.

to the overall framework, the communication model of the smart grid is also conceptualized as the information collection and the management. Table 13.1 highlights the difference between smart grid and traditional grid. The role of the operator includes the monitoring of the overall smart grid systems and various policy implementations. Utility providers are the middle persons between the customers and the energy market, thus helping in to have multiple utility providers in a single energy market.

TABLE 13.1
Basic Difference between Power Grid and Smart Grid

Characteristics	Power Grid	Smart Grid
1. Customer participation	Static policies are present irrespective of the real-time energy consumption.	Dynamic policies are expected to be developed considering the real-time power consumptions from customers.
2. Communication facility	Only power line communication (PLC) present.	Bidirectional communication facility present.
3. Distributed energy generation	Few distributed generations such as solar and wind are considered.	All types of distributed generation (small or medium) are considered.
4. Inclusion of storage devices	Not present.	Different distributed energy storage devices can be used in different situation.
5. Real-time communication with the customers	Not present.	Using bidirectional communication, the real-time communication with the customers can be established.
6. Security	No proper implementation is present to prevent energy theft and other security breaches.	Adequate measures and policies are taken into consideration to avoid any theft and security breaches.

Today any disruption in the electricity such as fault has a domino effect that can lead to failure of banking, industry, communication and other associated sectors. This cascaded failure leads to loss of billions of revenues. The progressive features of smart grid will add reliance to the power network and thus will make the grid prepared in advance for any situation, leading to early detection and preparedness. Smart grid technology can detect and isolate outage before it leads to a large blackout. Two-way communication feature provides automatic rerouting of the equipment and power during power failure.

In addition to the above properties, the smart grid also accommodates small distributed power generation or "customer-owned generation", which supports the customer to transfer the extra generated power to the grid. This helps in promoting the new renewable energy generation. In a way, the smart grid is a solution to replace or modernize the exiting power grid infrastructure. It is a way to provide an efficient, reliable and customer-centred grid network.

13.1.3 SMART GRID VISION

According to the department of energy, the smart grid is envisioned to fulfil these objectives:
1. **Operational Efficiency:**
    The efficiency of the smart grid involves the following:
    * Integration of distributed generation units
    * Improvement of resource utilization
    * Real-time energy monitoring
2. **Active Participation of the Customer:**
    In traditional grid customer do not participate actively in the energy trading process. But in the smart grid, customers participate in all the trading process. Active participation of the customer helps in reduction in electricity outage.
3. **Energy Efficiency:**
    Energy efficiency is mainly focused on the reliable energy distribution and includes:
    * Reduction in energy loss
    * Reduction in imbalance between demand and supply
    * Reliable power supply
4. **Green Energy:** Involves the reduction in the carbon emission in the generation, transmission and distribution.

13.1.4 VULNERABILITIES IN SMART GRID NETWORK

The smart grid is a Cyber-Physical System integrated with an electrical network supported by a communication network. To support the entire network various parties are involved in the network. As a result, the smart grid must have an appropriate security measure. Energy security or grid security is a national security concern. A smart grid consists of millions of smart components that support two-way communication. Communication between them is usually through a wireless medium that enables the grid network to perform smarter facilities.

Advanced Metering Infrastructure (AMI) and other communication devices in the grid have become more prone to cyber-attacks with the introduction of the new communication technologies. These devices have introduced new vulnerable points in the network. These susceptible points become access points for the attackers. This can hamper the grid availability and confidentiality. In the past few decades, utility companies are facing Denial of Service (DoS) attacks very frequently. According to researchers, the future smart grid will face cyber-attacks using attack vectors on AMI, communication networks or other smart meters. With the expansion of the electrical grid, endpoints (e.g. smart meter, smart devices) will also grow. Thus smart grid protection will be infeasible without the deployment of the proper security infrastructure.

13.2 ARCHITECTURE OF SMART GRID

The idea of the need for a "smart" grid arose from the use of AMI. The aim was to design a network with improved management and increased energy efficiency at the demand side. The grid would be equipped with smart protection schemes to make it reliable and secure from malicious activities as well as natural disasters [1]. Several models for a smart grid across different locations have been suggested in the literature [1–3]. Due to the differences in all these models, it became essential to have a standardized model based on which all the future standards of the smart grid would be set. Therefore, the IEEE P2030 standards were set that made the smart grid a complex and large "system of systems" [4]. Under IEEE P2030, the components of the smart grid could interoperate between power, communication and information technology systems. An efficient communication platform between the systems was established for effective data transfer. The smart grid infrastructure is broadly classified into three systems, as shown in Figure 13.3. For better security smart protection system must be present, for efficient management, the best management algorithms must be implemented and for longer life smart infrastructure must be included.

In the smart infrastructure system of a smart grid, bidirectional communication is supported. The data flow in the grid can be in the form of information or energy. This system is further divided into three subsystems, as shown in Figure 13.4. The smart energy subsystem forms the physical layer of the smart grid architecture. The smart information and smart communication subsystem together form the cyber layer.

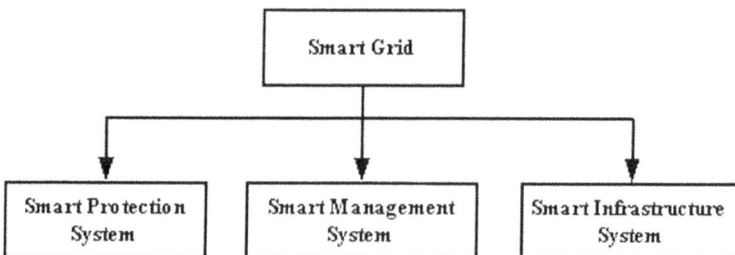

FIGURE 13.3 Smart grid subsystems.

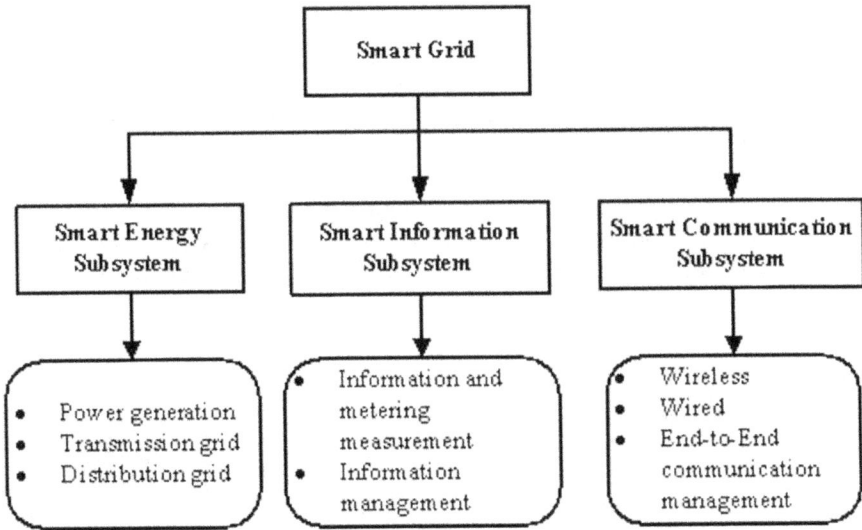

FIGURE 13.4   Smart infrastructure system classification.

13.2.1   Smart Energy Subsystem

This section deals in detail about subparts of smart grid and their roles.

13.2.1.1   Bulk Power Generation

Electricity in a grid is generated from traditional sources like coal and natural gas or from renewable energy sources like solar or wind energy sources. In a smart power generation, the energy flows in two directions. The flow of energy in a smart grid, from the generation side to demand side is same as that of conventional grid. Addition to this, the smart grid supports energy flow from customer's side to the grid, like in the case of energy generation by a rooftop solar panel. This key feature of the smart grid is called distributed generation [5].

13.2.1.2   Transmission Grid

The development of the transmission network in a smart grid is driven by two factors:
- **Infrastructure Challenges:** involves solving problems related to fast ageing of components or ever-increasing load demands.
- **Innovative Technology:** involves the incorporation of advance power electronics, new materials and communication technology.

The smart transmission system is built up of three interactive components [6]:

Smart Control Centre
    The smart control centres enable efficient monitoring, analysis and visualization of the transmission network.

Smart Power Transmission Network

The smart power transmission network is a modification to the existing infrastructure of the network. Newer materials and technology along with signal processing improves the reliability and power quality of the network.

Smart Substation

The smart substations also work on high voltages. They have modifications in the control, measuring and monitoring equipment. The substation undergoes automation, self-healing and digitalization. Smart substations also have increased operator security.

13.2.1.3   Distribution Grid

The smart distribution system serves users more efficiently as compared to the traditional grid distribution network. The integration of distributed generators in the smart grid increases the feasibility of the power generation in the grid. Contrary to this, the control of the power flow becomes complicated. The smart distribution system acts like a smart power router [7]. Several units of payloads are formed by dividing electricity. To each unit, a header and a footer are attached, making them an electric energy packet. When the smart router receives a packet, it sorts the packet as per the load address and forwards it to the corresponding loads. The power regulation becomes easy by controlling the quantity of electric energy packets transferred.

13.2.2   SMART INFORMATION SUBSYSTEM

In a smart grid, a smart information subsystem supports information generation, analysis, modelling, optimization and integration. The following subsections explain the parts of the subsystem.

13.2.2.1   Information Metering and Measuring

Smart Metering

Smart metering is crucial for a smart information subsystem as it gathers information from the devices at users' end. Automatic Meter Reading (AMR) and Automatic Meter Infrastructure (AMI) are used for this [8]. AMR is used to collect the consumption and status data of the user automatically. This data is transferred to the central database for analysis and billing. The function of AMI is similar to AMR, the only difference being that AMI is capable of two-way communication. The information gets available at real-time. Similar to AMI, Smart Meters record the power consumption data of the user at pre-defined intervals (hourly or less). This data is sent back to the service provider on a regular basis.

Smart Monitoring and Measurement

Measurement and monitoring the status of the grid is important in a smart grid. Two major approaches followed here are sensors and phasor management units.

Sensors

Sensor networks are used for monitoring and measurement purposes in a smart grid [9]. To detect mechanical failures in a power grid, sensor networks are embedded in transmission lines to access electrical and mechanical conditions in real-time, diagnose faults and determine counter-measures for fault clearing in the transmission line [10]. Wireless sensor network provide a low-cost and feasible communication platform for diagnosis and remote system monitoring [11].

Phasor Management Units

Phasor measurement units (PMUs) help in creating a reliable smart grid infrastructure [12]. They determine the health of the system by measuring the electrical waves in the grid. The readings of PMU are taken at dispersed locations in the network and get synchronized using GPS radio clock. With multiple PMUs, the alternating current readings at different locations in the grid are compared and this sampled data is used by the system operators.

13.2.2.2 Information Management

Due to metering, monitoring and sensing, a large amount of data and information is generated. Therefore, information management in the grid is essential. It includes the following tasks:

Data Modelling

The aim of data modelling is to create displayable, transferable, persistent, compatible and editable data representation to use within the smart grid. This includes all the components of the grid from generation to consumer levels. The data exchange is only meaningful when the information can be used for both components involved [5].

Information Analysis, Integration and Optimization

Information analysis supports interpretation, processing and correlation of observations as the monitoring and metering systems in smart grid generates a large amount of data. Information integration aims to merge information from disparate sources with typographical, contextual and conceptual representations. This includes cross-correlation and verification of information for designation and validity of sources. Information optimization improves the effectiveness of information. Due to large-scale sensing, monitoring and measurement, the data size of the smart grid is very large. This data may also consist of useless or redundant data. Therefore, optimization is essential.

13.2.3 SMART COMMUNICATION SUBSYSTEM

This part of the smart grid is responsible for information transfer and communication connectivity among components of the grid. The different types of network components in smart grid are:

- **Expertise Bus:** connects control centre, generators and markets.
- **Wide Area Network (WAN):** connects geographically distant location.

- **Field Area Network (FAN):** connects electronic devices controlling transformers and circuit breakers.
- **Premises Network:** include both customer and utility network within the user domain.

13.2.3.1 Wireless Technologies
They offer low-cost installation, mobility and are more suitable for remote locations. The following technologies are used:

- Wireless Mesh Network
- Cellular Communication System
- Cognitive Radio
- Wireless Communication based on 802.14.4
- Satellite Communication
- Microwave or free-space optical communication

13.2.3.2 Wired Technologies
The important technologies in wired communications are:

- Fibre-optic communication
- Powerline communication

13.2.3.3 End-to-End Communication Management
For an electric utility to deploy various communication systems, a common management platform using Transmission Control Protocol/Internet Protocol (TCP/IP) technology for maintaining end-to-end communications is used. This is an easy solution for problems due to incompatible lower layer technology. The smart grid architecture of the smart infrastructure systems is shown in Figure 13.5. The energy flow in the physical layer is represented by a solid line and the information flow in the cyber layer is represented by a dotted line. Both cyber and physical layer connects each part of smart grid network for better and efficient operation.

13.2.4 Smart Management System
Smart management systems are important to manage several functions of a smart grid. They are of two types, as seen in Figure 13.6.

13.2.4.1 Management Objectives
This category of the management system focuses on the different objectives essential for the smooth functioning of a smart grid [13,14].

1. Minimization of energy losses.
2. The shaping of the demand profile to match the demand with supply.
3. Price stabilization.
4. Optimization of utility and cost for an individual user, multiple users or electric industry.
5. Emission control.

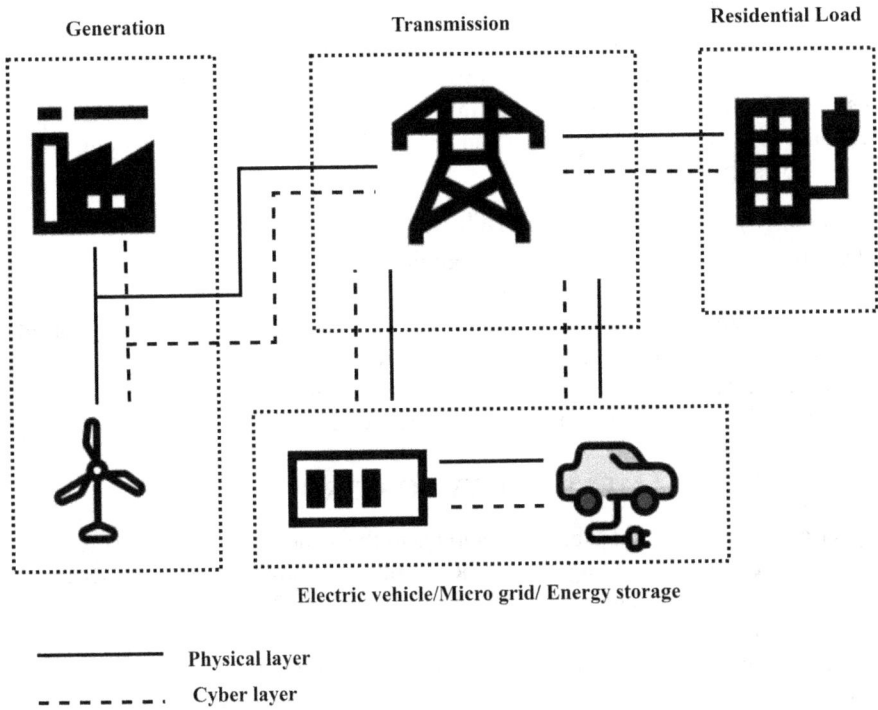

FIGURE 13.5 Cyber-physical architecture of smart grid.

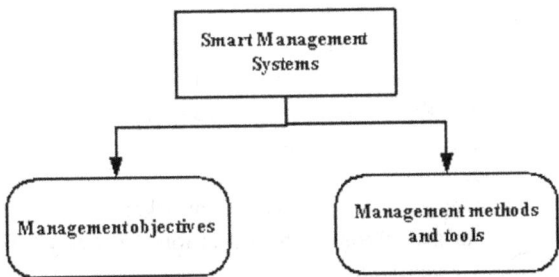

FIGURE 13.6 Smart grid managerial.

13.2.4.2 Management Methods and Tools

This category of a smart management system discusses various methods and tools useful for the above-stated management objectives. These include [13,15]:

1. Optimization methods.
2. Machine Learning.
3. Auction.
4. Game Theory.

13.2.4.3   Smart Protection System

In a smart grid, the smart protection system serves two purposes. It prevents the system in case of compromise in the infrastructure of the smart grid due to natural disasters, mechanical failures or user input errors. The protection system also protects the grid against cyber-attacks and spies. The smart protection system can be classified as shown in Figure 13.7.

13.2.4.4   System Reliability and Failure Protection

The future of the smart grid expects more integration of distributed generators and renewable energy sources in the network. The reliability of the grid fluctuated due to the uncertainty of these sources. Therefore, alternate innovative technology needs to be incorporated into the grid for improving the architecture [1]. In order to avoid malfunction, failure protection is essential. A protection system also needs to function after failure in the grid. They function in failure identification, detection and recovery.

13.3   PHYSICAL LAYER SECURITY CONCERNS

Cyber-attack on AMI and substation belongs to the same category as AMI is a sub-part of substation for metering. So in other words if the attack on AMI is successful, it means the attack affected the substation. The role of AMI is to link communication channels between customers, electric companies and substations. It sends the updated price information and monitor energy consumption. The possible physical layer attacks on the substation-based AMI system are as follows:

- **Denial of Service Attack (DoS):** In this type of attack, the intruders make the system/network unavailable for commands and signals from the operator such that the system performs unhealthy and can cause damage. Most of the attack consists of DoS category.
- **Stealing Attack:** This is done to acquire maximum information about the customers. Through the same Virtual Private Network (VPN) it might be possible that the customer curial information gets leaked to intruders, which then causes difficulties.
- **Maloperation through Gain Access:** The attacker after gaining the access to the network can give wrong commands forcing the system to perform wrong which may lead to damage.

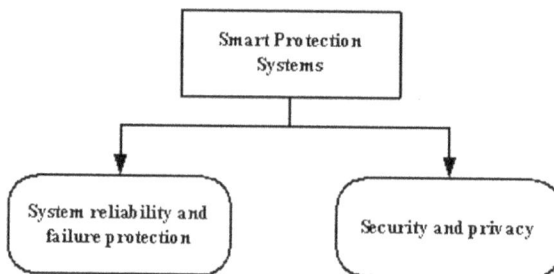

FIGURE 13.7   Smart grid protection system classification.

13.4 POSSIBLE SOLUTIONS TO TACKLE PHYSICAL LAYER SECURITY CONCERN

This section contributes to the latest technique used to detect cyber security issues in smart grid.

13.4.1 CONSUMPTION PATTERN BASED ELECTRICITY THEFT (CPET) DETECTION ALGORITHM

Every consumer linked with the power grid has its own fixed pattern of energy consumption. The smart meters send consumption data on regular intervals for monitoring. The load consumption pattern is recorded and stored for further studies. Many techniques are used to detect electric theft using consumption patterns like game theory, classification-based, data mining and machine learning. In general, each customer/industry has its own pattern of energy consumption; any change from the pattern can be a sign of cyber intrusion.

The major problem associate with CPET is lack of theft sample with customers and industry, also zero-day attack data cannot be obtained from historical data. So a few detection techniques for CPET are discussed below:

- **Mixing Data Mining with Support Vector Machine (SVM) [16]:** In this method, the operator calculates and visualizes 2–3 year energy consumption of customer and detects the malicious activities when there is abrupt change in load profile.
- **K-mean Fuzzy System [17]:** Using this method the cluster is formed with the historical data of the customers. Those customers whose pattern is out and far from the cluster main point is considered as malicious customers/points.
- **Non-intrusive Load Monitoring (NILM) [18,19]:** Although the customers load details are available to the power operators, using NILM the operator can analyse the appliance used by the customers; in case the variation is very high it means the customer is doing frauds.

13.4.1.1 CPET Threat Model

The aim of intruders is to manipulate the pattern and possibly pay less than the real actual bill. So the attackers can have three options to do so:

- **Physical Manipulation:** This is most common way to reduce the energy bill. The malicious customer can tamper the meter in such a way that it starts showing less reading than the original.
- **Cyber Manipulation:** This type of manipulation is used where smart meters are connected through communication link. The meter is hacked to reduce the billing and to access gain to the control centre for more severe attack.
- **Customer privacy:** Some time customer privacy and theft issue is related to the meters. This may cause social issues in the society.

13.4.1.2   Steps for CPET Detection Algorithm

Any CPET detection algorithm has the steps as shown in Figure 13.8. The description is as follows:

- **Training of the Algorithm:** This phase is essential because each customer/consumer has its own pattern of energy consumption. So any algorithm must know the energy pattern of individual customer/consumer such that it can detect the abnormalities in the pattern.
- **Data Reduction Technique:** It must be noted that to study a single customer energy behaviour data of 3–4 years is required. This would need large storage, which can make the algorithm sluggish against the detection process. To overcome this, the algorithm must also focus on the data reduction technique to handle large data simultaneously without adding burden on the algorithm.
- **Implementation:** This step is an application phase where the algorithm is applied to the news data which has to be tested for the possible detection of malicious activities. The complete steps can be easily understood from Figure 13.8.

13.4.2   Anomaly Detection System for Smart Grid Substation

Electric power grid networks are flexible and complex cyber-physical systems. The physical layer consists of conventional/modern power plants, transmission and distribution networks. And the cyber layer consists of communication infrastructure used for the exchange of data throughout the electrical power grid. For proper operation of the power grid, the physical systems rely on cyber infrastructure communication. Due to extensive use of communication infrastructure the power grid network comes under "critical infrastructure". The most basic unit of this critical infrastructure is power substations which play a vital role between generation, transmission and distribution. It consists of advanced information technology and automated systems with an energy management system installed for proper monitoring and control. The power outages through substation can be caused by equipment failure, climatic conditions and human errors. Much research has been

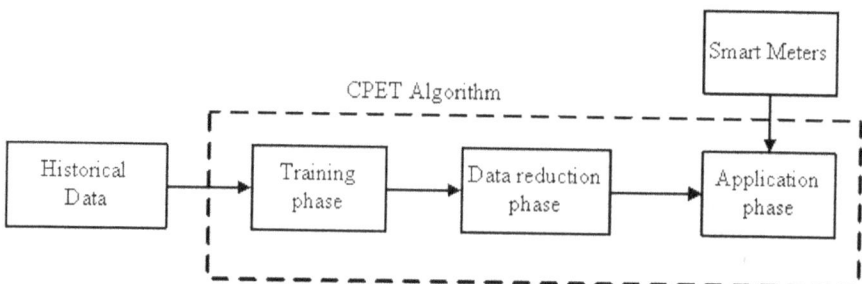

FIGURE 13.8   Consumption Pattern Based Electricity Theft (CPET) detection algorithm steps.

conducted on the previously stated issues and the new concept of self-healing substation. But due to the presence of a large cyber network in the substation, the power grid is likely to be vulnerable to cyber intrusion. Like the stuxnet cyber intrusion on industrial Supervisory Control and Data Acquisition (SCADA) system infected approximately 90,000 computers worldwide [20]. So in order to mitigate cyber intrusion in power substation an engineer/researcher must know about the vulnerabilities regarding substation.

13.4.2.1 Cyber Vulnerabilities of Substation

13.4.2.1.1 Traditional Industrial Communication Protocol

The industrial communication protocol is a key element for data transfer between different elements of power grid substation. These protocols can only be modified by operators for the proper and smooth working scenario. Most of the industrial protocol did not incorporate the condition of cyber intrusion, making it unsuitable for the twenty-first century. Some important protocols like Sampled measured value (SMV) and Generic object oriented substation event (GOOSE) use a multicast scheme for data communication [21]. Using this protocol, the message containing critical information may suffer cyber vulnerabilities like group centre trust control and group access control. The reader may think to implement an encryption method to enhance the multicast scheme, but it will affect the performance of SMV and GOOSE. Both these protocols work within the time frame of 4 milliseconds which does not allow encryption and decryption methods. Most of the latest protocols focused only on the DoS attack. To prevent a different type of cyber intrusion in power substation new Intrusion Detection Techniques are required. The IEC 61850-based substation cyber automated system needs more upgradation against cyber intrusion.

13.4.2.1.2 Remote Access Control Point

Some of the substation components are located in remote areas to gather information using Remote Terminal Units (RTU). For transferring data VPN is used. These RTU are not well secured and are poorly configured using an old firewall and low level of cryptography making them more vulnerable to cyber intrusion. In most of the critical infrastructure failure, RTU had been used to breach the grid network [22,23]. With an increase in digitalization, the managers/operators must enhance the security features of RTU by using new firewall policies, provision of changing passwords frequently and enhance light cryptography.

13.4.2.1.3 Built-in Web Server with Default Password

A power substation consists of a large number of Intelligent Electronic Devices (IED). So it is hard for the operator to remember a new password for all IED, in this case, all IEDs must have a default password or same password to decrease the reaction time of the substation. Each substation has its own web server for operation making the system vulnerable. The substation manager must continuously check the unauthorized access to this network.

13.4.2.2 Cyber Intrusion and Detection Scenario in Substation

Substation security threats can be divided into two parts as stated below:

- **Physical Layer:** It consists of complete hardware component issues like malfunctioning and defective equipment.
- **Cyber Layer**: It consists of linkage between cyber and physical layers like sensors, firewall, communication and software. Cyber intrusion is successful only if the physical layer gets affected through cyber layer, so cyber layer can be considered as cyber-physical systems.

As illustrated in Figure 13.9, the substation should use advanced firewall policies for each connection attached to it. Nowadays the threat of cyber-attack is higher due to digitalization, and for maintenance, it required remote access. Different cyber-physical attack scenarios for the substation are discussed in the next sections.

13.4.2.3 Minor Substation Attack (MSA) and Prevention

In Minor Substation Attack, a single substation is targeted for cyber intrusion, forcing system towards unhealthy operations [24,25]. There are four possibilities of cyber intrusion in substation as is stated below and shown in Figure 13.9:

1. Through RTUs and IEDs of generation units the attacker may manipulate data for unhealthy operations.
2. A false customer from distribution network can be used to attack the substation and may cause data loss.

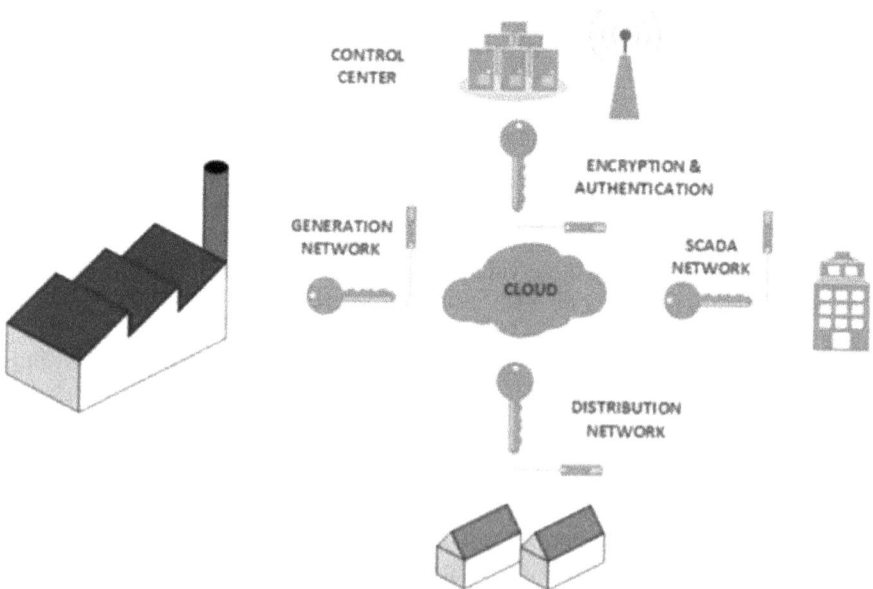

FIGURE 13.9   Possible ways intrusion in substation.

3. In case if control centre is under threat, then it is quite easy to manipulate substation communication network.
4. Through SCADA VPN network.

The easiest way to hack any VPN is to send a false email to the operator. This email may consist of fake websites or malware software which may create a mirror image of a substation computer screen on attacker helping to gather all credentials of the original operator. Man In The Middle Attack is possible through an unknown operator who is using a pen drive that consists of malware software for intrusion [26,27]. This malware will find a path to the communication network through which intruders may control the system. Gaining access to external devices may cause substation blackout.

13.4.2.4 Multiple Substation Attack Scenario

Each substation belongs to the power grid to play a different and vital role for electric power transfer. The priority for the cybersecurity level must be fixed according to the level of power the substation handles. For large economic losses just like that which happened in Ukraine Power Grid, more than one substation can be targeted. Targeting more substations simultaneously can trigger cascade failure. So by giving priority numbering to different substation it enhances the system by two-folds:

1. Cost of cyber security on priority basis can be reduced.
2. Cascade failure can be prevented in a better way.

13.4.2.5 Graph-Based Attack Tree Evaluation

A graph-based attack tree has been used in computer science and ICT to analyse potential cyber threats [28,29]. The most popular is the node-based system design and security analysis. Figure 13.10 shows the node-based cyber intrusion attack scheme. The attacker can choose any path to enter the VPN of substation and get full access externally. From social sites or through IEDs/RTUs, the intrusion can have many more paths depending on the system under study. So operators must evaluate all possible drawbacks and should eliminate it as shown in Figure 13.10.

13.4.2.6 MATLAB Example for Cyber Intrusion Prevention

This section deals with MATLAB simulation to prevent cyber intrusion in the IEEE-33 bus test system [30]. The system is modelled with a graph-based approach. This will have 33 nodes because each bus will be treated as a node in graph formation. To give priority to the nodes, fuzzy logic is implemented which has three parameters as stated in Table 13.2. Using these parameters, the fuzzy rules are designed for the prevention of cyber intrusion indicated in Table 13.3. The fuzzy membership function details are given in Table 13.4.

The system is simulated for two cases:

- **Case A:** Dynamic Host Configuration Protocol (Commonly used ordinary protocol)
- **Case B:** Fuzzy logic-based data protection

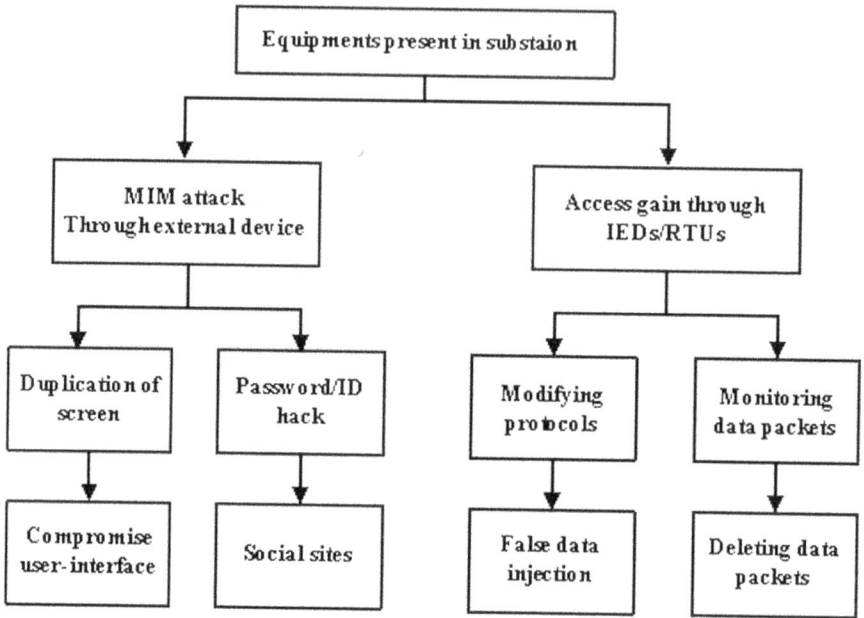

FIGURE 13.10   Attack tree evaluation.

TABLE 13.2
Input-Output Rule Parameters

Parameters	To Generate Fuzzy Rule 3 Variation Are Chosen		
Priority	1st	2nd	3rd
Node	L	M	H
Degree	L	M	H

H, High; L, Low; M, Medium.

The data loss rate plot as shown in Figure 13.11 indicates that the fuzzy logic-based system reduces the data loss significantly. For case A, it is found to be 411.337 Mbps, and for fuzzy logic-based protection system, it is 167 Mbps. Cyber intrusion is high for ordinary protocol, so protocol must always be updated periodically.

13.5   CYBER LAYER SECURITY CONCERN

The chance of cyber intrusion becomes high if intruders manage to gain access on cyber layer and physical layer simultaneously. This section deals with different cyber layer intrusions in detail.

TABLE 13.3

Fuzzy Rule Base for Cyber Intrusion Priority Identification

Sr. No.	IF	AND	THEN
	Node	Degree	Priority
1	L	L	1st
2	M	M	1st
3	H	H	2nd
4	L	L	1st
5	M	M	2nd
6	H	H	3rd
7	L	L	2nd
8	M	M	3rd
9	H	H	3rd

H, High; L, Low; M, Medium.

TABLE 13.4

Input-Output Membership Function Range

System	Classification	Input 2 (Gaussian)	Input 1 (Triangular)
IEEE-33 bus	Low	0–5–11	0–7
	Medium	11–16–21	7–17
	High	21–27–33	7–33

13.5.1 ATTACK ON SCADA SYSTEMS

The modern power system depends on SCADA for the smooth and healthy operation of the electric grid. The aim of SCADA is to perform some specific tasks such as monitoring, control and alarming for every section of the electric grid at one central place known as "Control Centre". To affect the performance of electric grid through SCADA, intruders must gain access to the following subsystems attached to SCADA [31–33]:

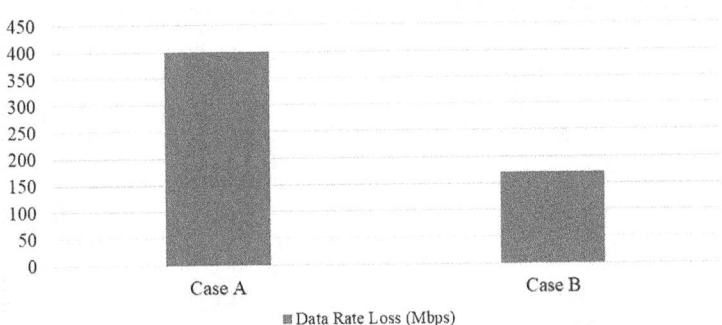

FIGURE 13.11 Comparison between ordinary protocol and fuzzy-based protocol.

- **SCADA Protocols and Servers:** SCADA is an important link between the control centre and RTU for monitoring and control of operation/process. SCADA server continuously transfers the data of RTUs to the operator for real-time updates. For normal fluctuation, SCADA is self-sufficient to detect the variation and give alarm against it. For cyber intrusion, the intruder must try to gain access to the SCADA server. Gaining access to the SCADA server means complete control of the system.
- **Intrusion in Energy Management Systems (EMS):** Most of the SCADA servers have high security features against intrusion. In such case intruders may try to gain access through EMS. EMS is capable of performing state estimation, generation control, fault identification/location and economical operation. So attackers can do maximum damage even if it gets access to EMS.
- **Human Machine Interface (HMI):** HMI plays a vital role by connecting operator, EMS and SCADA on a single platform for easy of monitoring and operation. It is very difficult to get access to HMI so in most case intruders avoid to hack/invade HMI systems.

The type of SCADA used in the electric grid is stated in Table 13.5. The low values in the table indicate the weak points, which must be improved to tackle cyber intrusion.

13.5.2 ATTACK RELATED TO WIDE AREA NETWORK (WAN)

Every modern system and infrastructure relies on real-time data communication for monitoring and control effectively and efficiently. With an increase in number of communication infrastructure WAN attacks are increasing in number. The WAN is used for the following purposes:

- Protection control during maloperation.
- Providing SCADA to the connected system.
- Human Machine Interface.
- Data storage and analysis.

TABLE 13.5
Type of SCADA System Used in the Electric Grid

Network	Distribution SCADA	Transmission SCADA	AMI SCADA
Domain	Distribution side	Transmission side	Substation side
Application	Distribution automation, fault detection	EMS, control function	Energy monitoring
Availability	High	High	Medium
Confidentiality	Low	Low	High
Integrity	Medium	High	Medium
Protocols	IEC 61850 DNP3	IEC 61850 DNP3 ModBus	ZigBee

So attackers try to get access to the WAN system for high level of attack and damage. WAN attack can be divided into the following [34]:

- **External WAN Attacks:** External attack focuses on integrity, confidentiality and availability. Manipulation of existing data or injecting new data is a subpart of external WAN attack.
- **Internal WAN Attack:** This is subdivided into two types which are application level attack and user level attack. In application attack the attacker targets the system which works on day-to-day data and automation. The most common day-to-day data monitoring system is SCADA. In user level attack the intruders try to exploit the users' authentication paths and invade the access for complete control. The chance of user level attack is very low due to trained operators. The impact of application level attack is more compared to user level.

13.6 POSSIBLE SOLUTIONS TO TACKLE CYBER LAYER SECURITY CONCERN

The smart grid is an advanced electric network that generates, transmits and provides electricity to consumers with the support of ICT, RTUs and IEDs. Presence of electronic devices and communication network makes the system vulnerable against cyber intrusion. In these sections, different methods to tackle cyber layer intrusion are discussed.

13.6.1 WEIGHT TRUST EVALUATION METRICS FOR CYBER LAYER

The cyber layer of the smart grid consists of information and communication technology used for data transfer. Both the physical layer and cyber layer work in parallel as shown in Figure 13.12. The weight trust evaluation method is more attractive and its implementation on a flexible network is very easy. The malicious sensors are detected if the weight of the sensor/RTUs/IEDs is far away from the pre-defined value. For weight trust evaluation in the electric grid following steps are followed:

- Step 1: All the sensors are modelled as node using graph theory approach.
- Step 2: Determine the type of communication, i.e. either system is following single hop or multi hop communication scheme for data transfer.
- Step3: Define a pre-weight value of each node under normal operation. In many cases it is taken as 1.

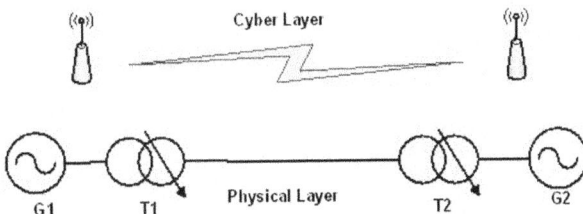

FIGURE 13.12 Cyber layer parallel to physical layer.

- Step 4: Define the trust rule for data communication between nodes. In [35], the author defines the trust rule using the formula stated below:

$$T_i(t) = \left[\Delta M \middle/ \left(\dot{X}_i^t - \dot{X}_i^{t-1} \right) \right]$$

where T_i denotes the trust value of the node, ΔM denotes difference of upper and lower bound of sensor measurements. Accordingly, the trust category can be defined as:

$$T_i = \begin{cases} > 1 \text{ and} < 0 & \text{Trusted} \\ \infty & \text{Highly trusted} : x_{t+1} = x_t \\ > 0 \text{ but} < 1 & \text{Malicious node} \end{cases}$$

So any node which has trust weight value between 0 and 1 is trusted node.

This method is flexible and can be implemented in any system consisting of RTUs/IEDs. Using this, an overall matrix can be formed as stated in Table 13.6 for easy analysis of the security score. It can be noticed that if the system score is 0, it means there are any malicious nodes, and if the score is 1, it means the system is operating correctly. Table 13.6 highlights the score evaluated for system status.

13.6.2 GRAPH THEORY-BASED EVALUATION TECHNIQUE

Graphs and attack trees have previously been used to model the electric grid and perform security analysis against cyber intrusion [36–38]. This method is used to study potential threats and possible attack scenario which can affect the system operation. Using graph theory approach the critical nodes and its feature can be highlighted which helps the researcher to understand different possible causes. Figure 13.13 shows the graphical model of the IEEE-9 bus system. Each generator and bus is denoted by node and attached through two links, one is a physical layer link and the other is the cyber layer link. Here G indicates generator and B indicates bus. In this technique, the behaviour of each node is monitored and any deviation from the predefined range is considered as malicious behaviour. The priority to the node is given according to the importance of node as shown in Figure 13.14.

There is a very marginal difference between the weight trust-based method and the graph theory approach. Both follow the evaluation of states of neighbouring nodes

TABLE 13.6
Overall System Evaluation Matrix

Cyber Layer	Physical Layer	System Score	System Status
0	1	0	Hacked
0.5	1	0	Cyber layer hacked
1	0	0	Physical layer fault/hacked
1	0.5	0	Physical layer fault/hacked
1	1	1	Healthy

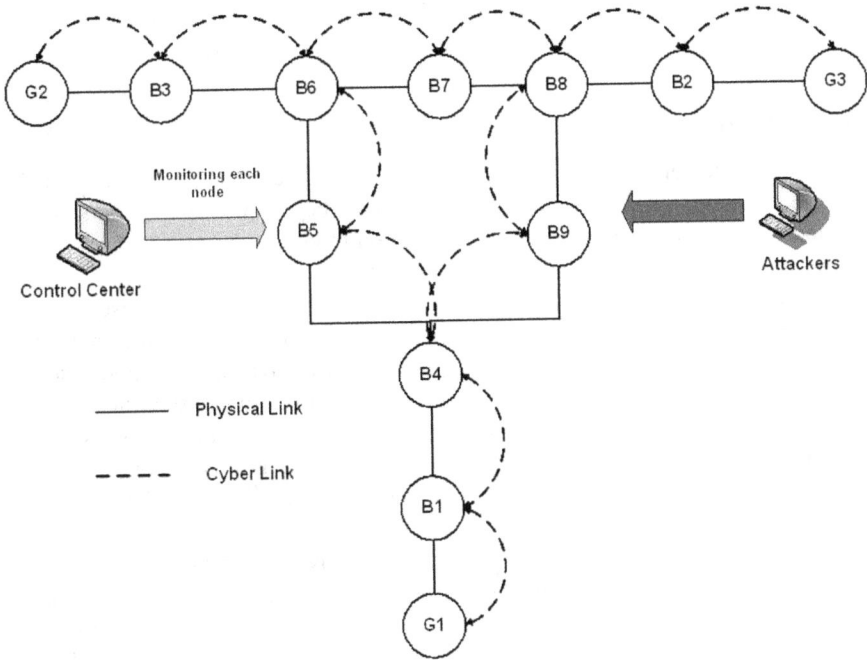

FIGURE 13.13 IEEE-9 test system modelling using graph theory.

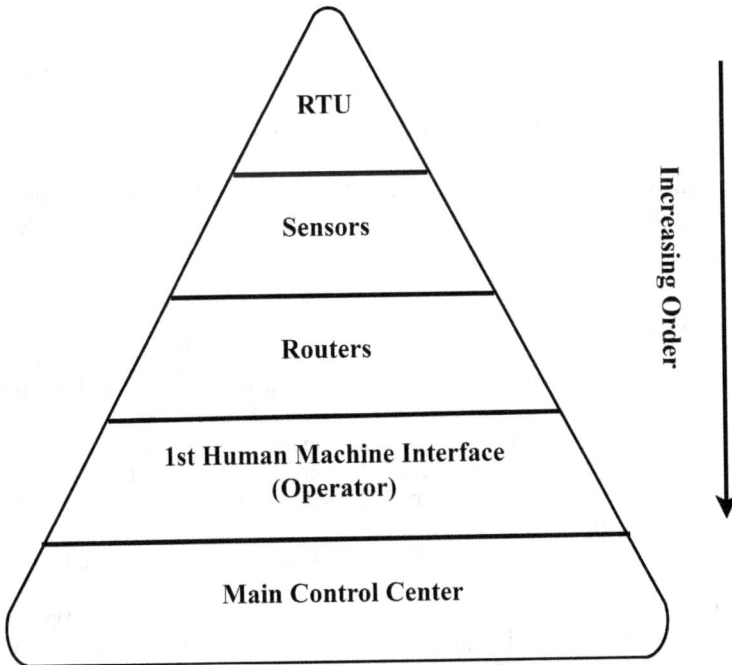

FIGURE 13.14 Priority order for system attached to smart grid network.

in terms of trust or condition. In the weight trust method if the nodes become non-malicious, it can be updated but it is a disadvantage in the graph theory approach.

13.7 CONCLUSION

The present chapter discusses the impact of cyber-attacks on smart grid and the methods to detect and prevent them. The chapter begins with an introduction to the smart grid. The bidirectional communication model of the grid is discussed in detail and its advantages over the conventional grid are compared. The vision of smart grid, i.e. improved efficiency, active customer participation and use of green energy, is discussed in detail. This is followed by the infrastructure of a smart grid based upon IEEE P2030 standards. The entire grid is divided into three systems: Infrastructure, Management and Protection. The infrastructure system of the smart grid gets further classified into Energy, Information and Communication subsystems. The energy subsystem also called the physical layer of the grid is responsible for the generation, transmission and distribution of energy in the grid. The information and communication layer together forms the cyber layer of the grid. Smart management system regulates the power flow efficiently, especially when distributed generators come into the picture. Since a smart grid has several communicating components, it becomes vulnerable to cyber-attacks. The smart protection system serves two purposes: prevention of grid from cyber-attacks and protection from mechanical damages. This chapter discusses cyber protection in the physical and cyber layer of the smart grid in detail. The physical layer of the smart grid is generally exposed to DoS attacks, stealing attacks or maloperation through gain access. The CPET detection algorithm provides a solution to these attacks. The techniques to this involve Mix data mining with SVM, K-mean fuzzy system and NILM. The algorithm for CPET has also been discussed in detail in the chapter. The vulnerabilities in the substations, the cyber-physical intrusion and detection of minor as well as multiple attacks have been discussed for both physical and cyber layer of the grid. This involves a solution by using MATLAB and graph theory. The security concerns for the cyber layer involves attacks on SCADA and WAN. The solution to these concerns is by weight trust evaluation of metrics for cyber layer or by an evaluation based on graph theory.

REFERENCES

1. F. Rahimi and A. Ipakchi, "Demand response as a market resource under the smart grid paradigm," *IEEE Transactions on Smart Grid*, vol. 1, pp. 82–88, 2010.
2. A. H. Mohsenian-Rad, V. W. Wong, J. Jatskevich, R. Schober, and A. Leon-Garcia, "Autonomous demand-side management based on game-theoretic energy consumption scheduling for the future smart grid," *IEEE Transactions on Smart Grid*, vol. 1, pp. 320–331, 2010.
3. A. P. A. Ling, S. Kokichi, and M. Masao, "The Japanese smart grid initiatives, investments, and collaborations," *arXiv preprint arXiv:1208.5394*, 2012.
4. D. G. Photovoltaics and E. Storage, "IEEE guide for smart grid interoperability of energy technology and information technology operation with the electric power system (EPS), end-use applications, and loads," in *IEEE Std 2030–2011*, pp.1–126, 10 September 2011.

5. H. Zareipour, K. Bhattacharya, and C. Canizares, "Distributed generation: Current status and challenges," in *Annual North American Power Symposium (NAPS)*, 2004, pp. 1–8.

6. F. Li, W. Qiao, H. Sun, H. Wan, J. Wang, Y. Xia, et al., "Smart transmission grid: Vision and framework," *IEEE Transactions on Smart Grid*, vol. 1, pp. 168–177, 2010.

7. T. Takuno, M. Koyama, and T. Hikihara, "In-home power distribution systems by circuit switching and power packet dispatching," in *2010 First IEEE International Conference on Smart Grid Communications*, 2010, pp. 427–430.

8. D. W. Rieken and M. R. Walker II, "Ultra low frequency power-line communications using a resonator circuit," *IEEE Transactions on Smart Grid*, vol. 2, pp. 41–50, 2011.

9. I. F. Akyildiz, W. Su, Y. Sankarasubramaniam, and E. Cayirci, "A survey on sensor networks," *IEEE Communications Magazine*, vol. 40, pp. 102–114, 2002.

10. R. A. Len, V. Vittal, and G. Manimaran, "Application of sensor network for secure electric energy infrastructure," *IEEE Transactions on Power Delivery*, vol. 22, pp. 1021–1028, 2007.

11. X. Fang, S. Misra, G. Xue, and D. Yang, "Smart grid—The new and improved power grid: A survey," *IEEE Communications Surveys & Tutorials*, vol. 14, pp. 944–980, 2011.

12. A. Armenia and J. H. Chow, "A flexible phasor data concentrator design leveraging existing software technologies," *IEEE Transactions on Smart Grid*, vol. 1, pp. 73–81, 2010.

13. A.-H. Mohsenian-Rad and A. Leon-Garcia, "Optimal residential load control with price prediction in real-time electricity pricing environments," *IEEE Transactions on Smart Grid*, vol. 1, pp. 120–133, 2010.

14. S. Bu, F. R. Yu, and P. X. Liu, "Stochastic unit commitment in smart grid communications," in *2011 IEEE Conference on Computer Communications Workshops (INFOCOM WKSHPS)*, 2011, pp. 307–312.

15. K. Moslehi and R. Kumar, "A reliability perspective of the smart grid," *IEEE Transactions on Smart Grid*, vol. 1, pp. 57–64, 2010.

16. A. Jindal, A. Dua, K. Kaur, M. Singh, N. Kumar, and S. Mishra, "Decision tree and SVM-based data analytics for theft detection in smart grid," *IEEE Transactions on Industrial Informatics*, vol. 12, pp. 1005–1016, 2016.

17. J. Devika, M. Nayak, D. Dash, and K. D. Sa, "Energy efficiency in densely distributed wireless sensor network using fuzzy K-mean," in *2017 2nd International Conference on Man and Machine Interfacing (MAMI)*, 2017, pp. 1–6.

18. S. Makonin and F. Popowich, "Nonintrusive load monitoring (NILM) performance evaluation," *Energy Efficiency*, vol. 8, pp. 809–814, 2015.

19. A. Zoha, A. Gluhak, M. A. Imran, and S. Rajasegarar, "Non-intrusive load monitoring approaches for disaggregated energy sensing: A survey," *Sensors*, vol. 12, pp. 16838–16866, 2012.

20. E. Weintraub, "Security risk scoring incorporating computers' environment," *(IJACSA) International Journal of Advanced Computer Science and Applications*, vol. 7, no.4, pp. 183–189, 2016.

21. J. Horalek, J. Matyska, and V. Sobeslav, "Communication protocols in substation automation and IEC 61850 based proposal," in *2013 IEEE 14th International Symposium on Computational Intelligence and Informatics (CINTI)*, 2013, pp. 321–326.

22. S. K. Venkatachary, J. Prasad, and R. Samikannu, "Cybersecurity and cyber terrorism-in energy sector–A review," *Journal of Cyber Security Technology*, vol. 2, pp. 111–130, 2018.

23. D. U. Case, "Analysis of the cyber attack on the Ukrainian power grid," *Electricity Information Sharing and Analysis Center (E-ISAC)*, vol. 388, pp. 1–23, 2016.

24. G. Dondossola, J. Szanto, M. Masera, and I. Nai Fovino, "Effects of intentional threats to power substation control systems," *International Journal of Critical Infrastructures*, vol. 4, pp. 129–143, 2008.

25. E. Tebekaemi, E. Colbert, and D. Wijesekera, "Detecting data manipulation attacks on the substation interlocking function using direct power feedback," in *International Conference on Critical Infrastructure Protection*, 2017, pp. 45–62.

26. U. Meyer and S. Wetzel, "A man-in-the-middle attack on UMTS," in *Proceedings of the 3rd ACM workshop on Wireless security*, 2004, pp. 90–97.

27. G. N. Nayak and S. G. Samaddar, "Different flavours of man-in-the-middle attack, consequences and feasible solutions," in *2010 3rd International Conference on Computer Science and Information Technology*, 2010, pp. 491–495.

28. A. Tajer, S. Kar, H. V. Poor, and S. Cui, "Distributed joint cyber attack detection and state recovery in smart grids," in *2011 IEEE International Conference on Smart Grid Communications (SmartGridComm)*, 2011, pp. 202–207.

29. H. A. Dawood, "Graph theory and cyber security," in *2014 3rd International Conference on Advanced Computer Science Applications and Technologies*, 2014, pp. 90–96.

30. N. K. Singh and V. Mahajan, "Fuzzy logic for reducing data loss during cyber intrusion in smart grid wireless network," in *2019 IEEE Student Conference on Research and Development (SCOReD)*, 2019, pp. 192–197.

31. B. Zhu, A. Joseph, and S. Sastry, "A taxonomy of cyber attacks on SCADA systems," in *2011 International Conference on Internet of Things and 4th International Conference on Cyber, Physical and Social Computing*, 2011, pp. 380–388.

32. C. W. Ten, C.-C. Liu, and M. Govindarasu, "Vulnerability assessment of cybersecurity for SCADA systems using attack trees," in *2007 IEEE Power Engineering Society General Meeting*, 2007, pp. 1–8.

33. C.-W. Ten, C.-C. Liu, and G. Manimaran, "Vulnerability assessment of cybersecurity for SCADA systems," *IEEE Transactions on Power Systems*, vol. 23, pp. 1836–1846, 2008.

34. P. Robertson, C. Gordon, and S. Loo, "Implementing security for critical infrastructure wide-area networks," in *Proceedings of the Power and Energy Automation Conference*, Spokane, WA, 2013, pp. 26–28.

35. N. K. Singh and V. Mahajan, "Detection of cyber cascade failure in smart grid substation using advance grey wolf optimization," *Journal of Interdisciplinary Mathematics*, vol. 23, pp. 69–79, 2020.

36. V. Shivraj, M. Rajan, and P. Balamuralidhar, "A graph theory based generic risk assessment framework for internet of things (IoT)," in *2017 IEEE International Conference on Advanced Networks and Telecommunications Systems (ANTS)*, 2017, pp. 1–6.

37. A. Srivastava, T. Morris, T. Ernster, C. Vellaithurai, S. Pan, and U. Adhikari, "Modeling cyber-physical vulnerability of the smart grid with incomplete information," *IEEE Transactions on Smart Grid*, vol. 4, pp. 235–244, 2013.

38. P. Mishra et al., "VMShield: Memory introspection-based malware detection to secure cloud-based services against stealthy attacks," *IEEE Transactions on Industrial Informatics*. doi:10.1109/TII.2020.3048791.

14 Review of Congestion Management in Deregulated Power System

Vasundhara Mahajan, Vijaykumar K. Prajapati, and Soumya Mudagal

CONTENTS

14.1 INTRODUCTION

In vertically integrated structure, there was only one utility that provided electricity to the entire region and single authority controlled different operations such as generation, transmission and distribution. The electricity price had been decided by considering the aggregated cost incurred in operation rather than considering the separate costs incurred in different operations. This has resulted in operational and economical monopoly, and the customer had no choice for purchasing electricity other than this single entity. Because of these reasons, the power utilities around the world have been transformed towards deregulation. In deregulated power system,

DOI: 10.1201/9781003278030-14

the operations are separated from each other which has induced competition among the participants and resulted in better and more efficient services to the customer at lower rates and increased operational efficiency [1]. The main entities of the deregulated power system are transmission companies (TRANSCOs), generating companies (GENCOs), distribution companies (DISCOs), power energy exchange (PEX), ancillary service providers, independent system operator (ISO) and retail service providers. The ISO maintains the reliability and security of the power system network. In vertical integrated structure, power and money flows are unidirectional and information flow is bidirectional, whereas in deregulated power system power flow is unidirectional and information and money flows are bidirectional [2]. In deregulated power system, a variety of challenges exist such as maintaining (Figure 14.1) reliability and security of power system, choosing the best auction strategy for the electricity, relieving of congestion, locational pricing issues, mitigation of market power, market efficiency and market equilibrium [3]. Among these, the congestion management is considered to be one of the important issues due to its impact such as generation-load mismatch and system constraints violation on the system performance. Congestion occurs due to limited capacity of transmission lines, increased load demand and acute competition among GENCOs. It creates the divergence in the price and puts power system at high risk by violating the system constraints [4]. As per the load requirement, if higher amount of power flows through the transmission line beyond its physical limits, then this is called congestion, and hence it restricts the

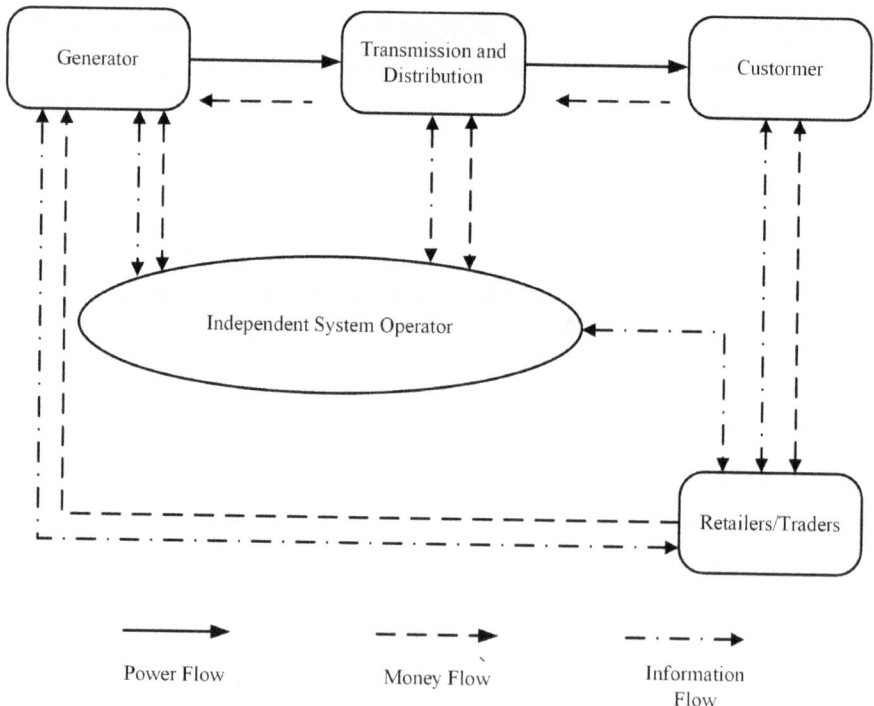

FIGURE 14.1 Understanding of deregulated power system [2].

power transfer. Under this condition, market participants can maintain energy prices at higher levels than competitive rate to earn more profit. It decreases the competition and can raise price. Due to these, it is required to control and restrict the congestion situation and is done through congestion management. Congestion management involves a set of procedures or actions taken to relieve the overloading without violating the system constraints. Congestion management is carried out by preparing a proper economic scheduling that matches generation and loads according to the load demand at that time [5,6].

The process of congestion management is as follows (Figures 14.2 and 14.3):

1. Read system data such as branch data, generator data, load demand history and different system constraints to perform optimal power flow studies.
2. Evaluate the demand and supply bids submitted by DISCOs and GENCOs, respectively, for day ahead schedule.

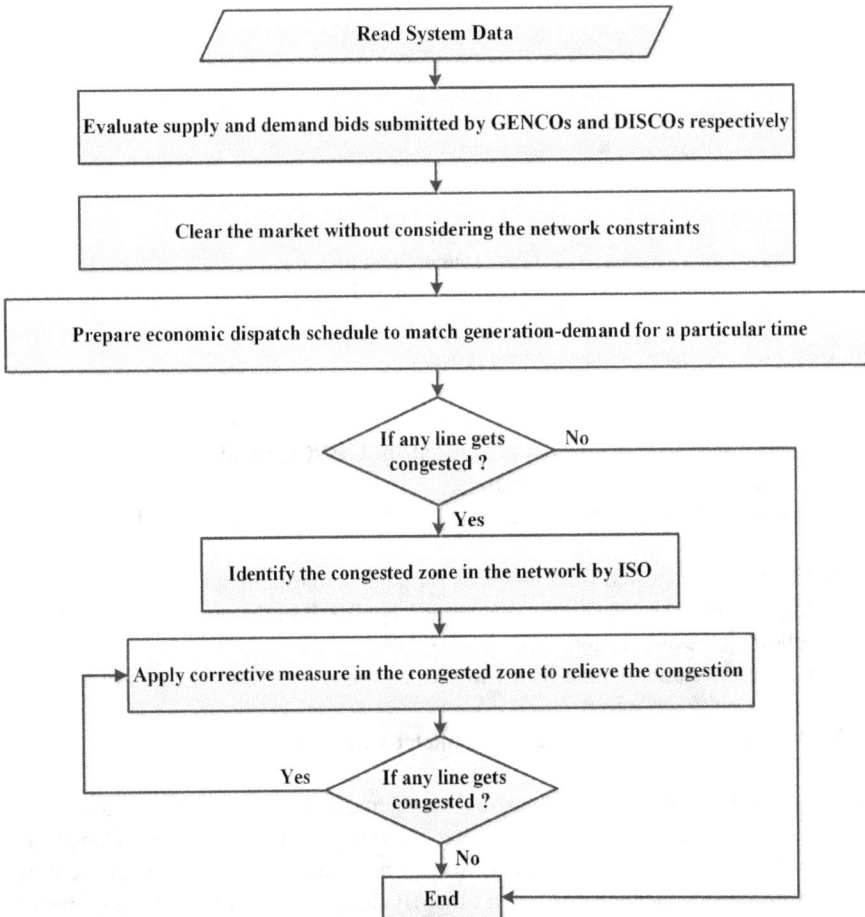

FIGURE 14.2 Congestion management process.

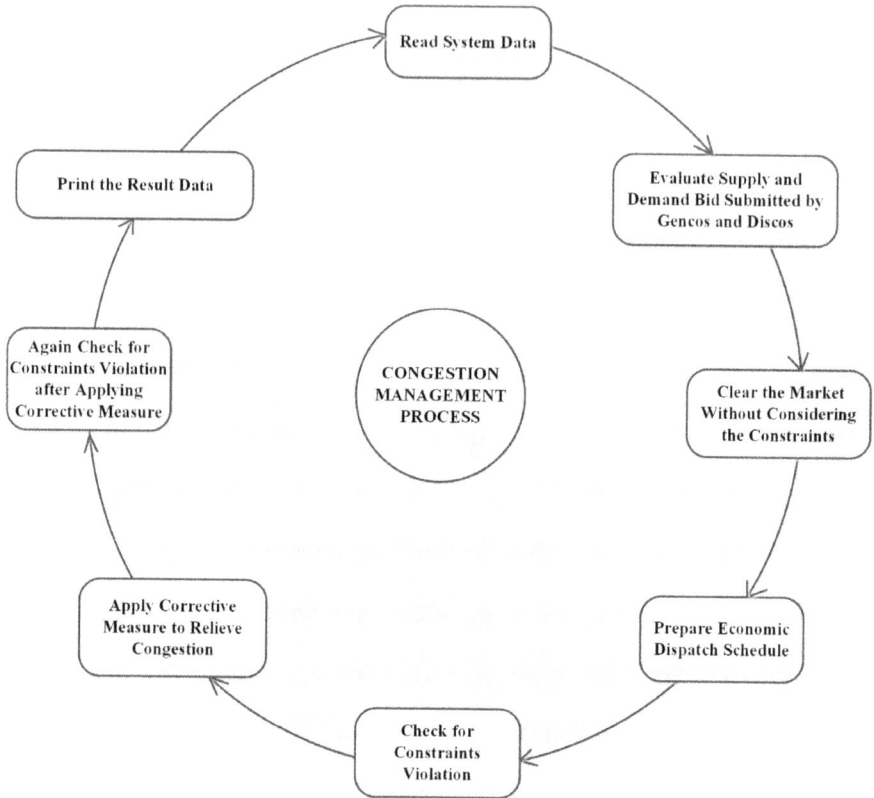

FIGURE 14.3 Congestion management process.

3. Clear the market using bid submitted by GENCOs and DISCOs without considering the system constraints.
4. Prepare the economic dispatch schedule to match generation load for a particular time.
5. Check for congested lines or zones for a particular demand on network.
6. Implement the congestion management approach in the congested zone for relieving the congestion.
7. Check the status of line flow limit violation, if persists, then go to step 6; otherwise, go to step 8.
8. Print the details of the system parameter which are now within limits.

In deregulated power system, congestion management problem and its locational pricing issues have been solved by employing optimal power flow method. Congestion in the network affects the value of locational marginal pricing (LMP) particularly at congested nodes; hence, evaluation of LMP along with its different component is required. In the following section, the use of optimal power flow and LMP in congestion management is discussed.

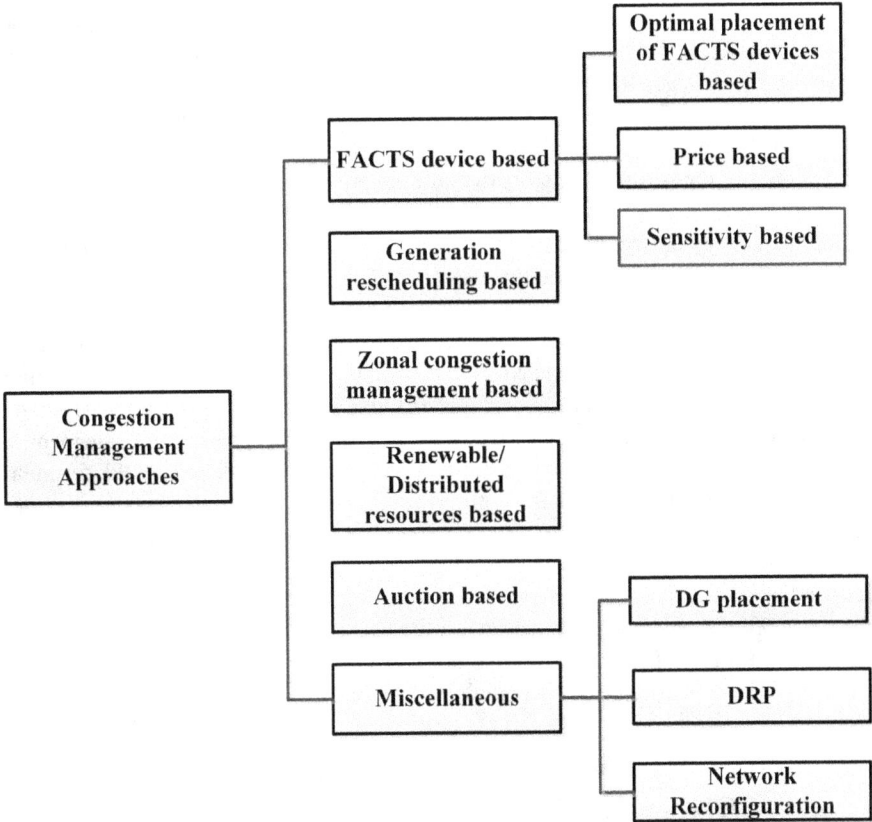

FIGURE 14.4   Optimal power flow.

This chapter is organized as follows. Sections 14.2 and 14.3 represent the use of optimal power flow and LMP in congestion management respectively. Section 14.4 discusses the integration of FACTS devices in congestion management problem formulation. In Section 14.5, the comprehensive review of various approaches for congestion management has been presented. Section 14.6 gives the conclusion.

14.2   OPTIMAL POWER FLOW IN CONGESTION MANAGEMENT

Optimal Power Flow (OPF) is a powerful, versatile and flexible tool to control the power flow in the system within constraints and operating limits of the system. It seeks to optimize the control variable of the objective function either by maximizing or minimizing the objective function value. In OPF formulation, state of the power system is represented by state variables which are continuous in nature. Bus voltage angle and magnitude and reactive and real power injections are the state variables. The other variables are controlled variables or designed variables or decision variables, and they are continuous or discrete in nature. The control variables are selected according to nature of problems. Real and reactive power generation, transformer tap ratio and switching shunt capacitor are the controlled variables. The nature of the OPF problem, depending upon type of the objective functions and constraints, may be a linear, non-linear, convex, non-convex, mixed integer linear, mixed integer non-linear programming and large-scale optimization. In OPF formulation, two types of constraints are presented in literature:

i. Equality constraints: Equality constraints are also known as balanced constraints, and power flow equations are treated as equality constraints.
ii. Inequality constraints: Minimum and maximum limits of reactive power, real power and voltage are inequalities constraints.

Solution of each OPF problem requires distinct mathematical and computational techniques. Nowadays, OPF problems are commonly used in power systems like unit commitment, economic dispatch, real and reactive power pricing issues, congestion management, voltage control and loss reduction [7,8]. In congestion management, a wide variety of single or multi-objective functions has been developed in OPF formulation. The solution of OPF gives the optimal values of design variables with respect to the objective functions. Congestion management's most common objective function is minimizing the cost of generation. Quadratic cost curves or piecewise linear sections are used to approximate the generation cost function. Maximization of social welfare, voltage profile improvement, voltage stability improvement, improvement in transient stability and minimization of system losses are the single-objective functions. Multi-objective functions are formed by combining two or more objective functions [9–16].

14.3 LOCATIONAL MARGINAL PRICING (LMP) IN CONGESTION MANAGEMENT

Generator marginal cost is a cost to serve the next incremental of load in the system that is economically dispatched. LMP is the cost of supplying the next incremental of electrical energy at a specific bus considering the generator marginal cost and the operating limits of the transmission system. LMP consists of mainly three components: (i) Energy component: It depends upon the marginal cost for generator operation. (ii) Loss component: It depends on total load on the system. (iii) Congestion component: It is considered as cost of delivery of energy at particular bus [17].

The value of the LMP may not be the same for all buses as the loss component and congestion component may be different for different bus. It depends on the sensitivities of the selected reference. At reference bus, LMP is always equal to energy component as the loss component and congestion component are always zero. For equality constraints, LMP at a node is a dual variable comprising power balance equation. It provides information to sellers and buyers to find the actual cost of supplying energy to a specific location [18].

LMP is derived from the Lagrange function of the objective function by considering all equality and inequality constraints. The value of Lagrange multipliers is obtained by applying Karush-Kuhn-Tucker (KKT) condition to the Lagrange function [19]. The value of Lagrange multiplier associated with equalities constraints gives the LMP. It is also known as nodal price or spot price and is given by

$$\rho_i = \lambda + \lambda_{L_i} + \lambda_{C_i} \tag{14.1}$$

where λ, λ_{L_i} and λ_{C_i} are the energy, loss and congestion component of LMP, respectively. With increase in load demand, the total losses will increase; hence, price at particular location will increase.

Similarly, the spot price (LMP) for bus j can be written similar to:

$$\rho_j = \lambda + \lambda_{L_i} + \lambda_{C_i} \tag{14.2}$$

The difference in spot prices between two buses i and j is given as:

$$\Delta\rho_{ij} = \left(\lambda_{L_i} - \lambda_{L_j}\right) + \left(\lambda_{C_i} - \lambda_{C_j}\right) \tag{14.3}$$

The difference in price between two nodes depends on the congestion and losses within the network, also known as congestion rent. The total congestion rent (TCC) is calculated as

$$TCC = \sum_{ij=1}^{N_L} \Delta\rho_{ij} P_{ij} \tag{14.4}$$

14.4 FLEXIBLE A.C. TRANSMISSION SYSTEM (FACTS) DEVICES IN CONGESTION MANAGEMENT

Flexible A.C. Transmission Systems (FACTS) are the power electronics converter-based devices used in transmission system to control the energy flow and allow the transmission line to operate at its full capacity. FACTS devices are classified into three categories: (i) Series Controllers: Series controller reduces the overloading of the line and increases the power transfer capacity. Thyristor Controlled Phase Angle Regulator (TCPAR), Thyristor Controlled Series Capacitor (TCSC) and Static Synchronous Series Capacitor (SSSC) come under this category. (ii) Shunt Controllers: The shunt controllers such as Static Synchronous Compensator (STATCOM) and Static Var Capacitor (SVC) are installed to compensate voltage by injecting reactive power at the buses. (iii) Combined Series-Shunt Controller: By employing the combined series-shunt controllers such as Unified Power Flow Controller (UPFC), the power flow and voltage can be controlled [20].

Installing FACTS devices is a viable solution for congestion management compared to building a new transmission line. It controls the line flow by injecting certain amount of active and reactive power at the node resulting in increased loadability of the line and reduces the price of the electricity. It can be modelled as power injection models and is treated as PQ load. The symmetrical structure of the admittance matrix is retained in the power injection model. This makes possible the integration of FACTS devices into existing power system analytic tools. To manage the congestion in the system, optimal location and setting of parameters of FACTS devices are desirable [21].

14.5 DIFFERENT APPROACHES FOR CONGESTION MANAGEMENT

Literature suggests the congestion management approaches to be categorized as (Figure 14.5):

    a. FACTS devices based
- Sensitivity based
- Optimal placement of FACTS device based
- Price based

    b. Generation rescheduling based
    c. Zonal-based congestion management
    d. Renewable/Distributed resources based
    e. Auction based
    f. Miscellaneous approaches

14.5.1 FACTS DEVICE-BASED APPROACH

The increased load demand on the power system and the limited capacity of transmission line flow require the installation of new transmission lines. However, right of way, huge capital investment and environmental issues restrict the installation of new transmission lines. An alternate solution is the placement of FACTS devices on the transmission line as the power flow capacity of the line gets improved. The literature presents different algorithms and approaches for optimal sizing and location of FACTS devices using various optimization techniques to optimize multi- and single-objective functions of congestion management. FACTS devices can also be placed on the transmission line by using sensitivity analysis. In sensitivity analysis, impact of independent variable on dependent variable is determined under certain assumption. This technique is used within specific boundaries that depend on one or more input variables. Different sensitivity indices are used either to control the power flow or to locate FACTS devices in the transmission lines for relieving the congestion. The effect of FACTS devices on LMP can also be assessed. LMP is an effective indicator to assess the congestion and degree of congestion. LMP is the cost of delivering the energy at a particular bus.

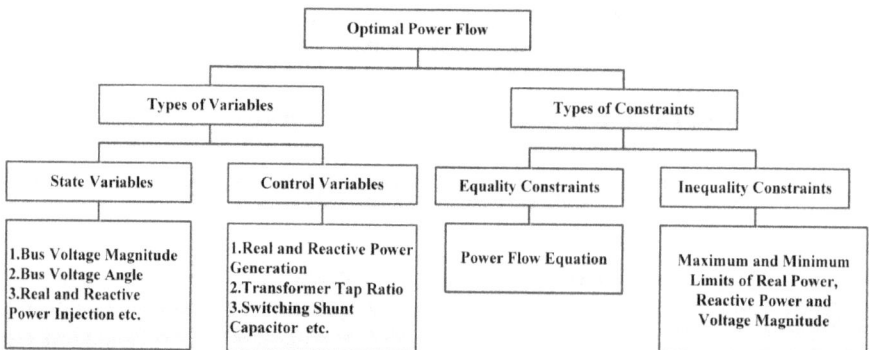

FIGURE 14.5 Congestion management approaches.

The placement of FACTS device, rescheduling of generators and forming of zones in the network will affect the LMP issues. Different approaches have been presented in the literature to minimize the LMP and relieve the congestion.

14.5.1.1 Sensitivity-Based Approach

The optimal placement of TCPAR and TCSC for congestion relief has been determined by sensitivity-based approach using "real power line flow index" [22,23]. Verma et al. [24] present sensitivity-based approach which reduces the "system losses" and the congestion by optimally placing the UPFC in the line. The optimal location of UPFC is determined using real power flow performance index and loss sensitivity. The major drawback of this method is that it does not capture the non-linearity associated with the system even though it will locate the FACTS devices optimally in the line. Lee [25] presented the optimal placement of TCSC by considering combination of the "line flow sensitivity" and "shadow price" to control power flow through line.

Alvarado [26] discussed the congestion management strategies in deregulated electric market based on power transfer distribution factor (PTDF). Liu and Gross [27] investigated the effectiveness of the distribution factors such as PTDF and Injection Shift Factor for large power system network. The impacts of errors in distribution factors in the area of congestion modelling have been assessed. Nimura and Niu [28] presented indices to select the load curtailment option at different locations for congestion relief. Generation re-dispatch has been optimized with minimal cost and bilateral transaction approach. Lee et al. [29] presented index-based method namely Congestion Cost Index (CCI) and Average Total Congestion Cost Index (ATCCI) to promote fair competition for short-term transmission planning. CCI has been used to minimize the distance between the market equilibrium point and operating point by re-dispatching the generation based on historical data bidding. ATCCI determines the optimal location and capacity of transmission line considering its line constraints. Chung et al. [30] proposed small-signal stability index for contingency screening and determining the small-signal security status of a system. Power transfer capability under small-signal security constraint has been increased by rescheduling the generation using sensitivity-based dispatched method.

14.5.1.2 Optimal Placement of FACTS Device-Based Approach

Tiwari and Sood [9] proposed an optimization technique for optimal location and parameter setting of the TCSC with a view to maximize social welfare, minimize TCSC investment cost, recover the TCSC investment cost and enhance the trading capability in deregulated power system. The success of this approach is tested by comparing with the Sequential Quadratic Programming (SQP) and Fuzzy-GA approach. Bavafa et al. [31] presented congestion relieving method by optimal placement of SSSC in the system using Genetic Algorithm (GA) and Ant Colony Search (ACS) algorithm, and comparison of the result shows that the ACS is better than GA. Kumar and Venkaiah [32] applied GA and Particle Swarm Optimization (PSO)-based OPF with SSSC to reduce the congestion management. It has been proved that PSO gives better result than GA. In Ref. [33], optimal rescheduling of generation and optimal placement and sizing of UPFC have been determined by Coordinated Aggregated-based Particle Swarm Optimisation (CAPSO) algorithm. The effectiveness of the approach has been verified by comparing the result of the CAPSO algorithm with the SQP method. It has also been shown that optimal placement of UPFC influences the generation level and load of different buses and significant decrease in system total cost. Chong et al. [34] investigated the optimal rating of UPFC located at congested lines using interior point optimization method. Economic analysis of algorithm indicates that congestion cost reduces with UPFC during heavy loading condition. Yu and Lusan [35] proposed a model to optimize the placement of FACTS devices by considering the multiple time periods with losses. It has also been shown that the proposed approach is not trustworthy as it may give optimal solution for one time period but may not be for other time periods. In this work, demands have been modelled as function of price to reflect the operating principle of deregulated power system. Hashemzadeh and Hosseini [36] presented line outage sensitivity factor-based methodology for locating the series FACTS devices. The PSO has been applied to minimize the TCC and total generation cost.

In Ref. [37], the optimal location of single- and multi-type FACTS devices has been determined using PSO to enhance the system loadability with minimum cost of installation. It has also been analysed that the system loadability cannot be further improved after placing certain number of FACTS devices. UPFC gives the maximum system loadability with higher cost, and TCSC gives the better performance with minimum installation cost. SVC offers the lowest cost of installation. Wibow et al. [38] presented a hybrid PSO-based method for multi-FACTS device allocation for congestion relief and voltage stability. Different factors such as control coordination among FACTS devices, generation rescheduling and load shedding have been considered under contingency state in this analysis. Detailed cost and benefit analysis has been carried out under normal and contingency state. It has also been shown that the installed capacity of FACTS devices is determined by the capacity required under contingency condition. In Ref. [39], Fuzzy-based GA algorithm for optimal location of TCSC and SSSC has been presented. In this work, optimal rescheduling of generators, maximization of the social welfare and relieve congestion in deregulated market environment considering the valve point loading condition have been analysed. The proposed Fuzzy-GA approach is compared with SQP approach and found that it improves the convergence, reduces the number of iterations and achieves better solution with and without TCSC/SSSC. Nagalakshmi and Kamaraj [10] analysed that the loadability of the power system can be enhanced by placing different FACTS devices like SVC, TCSC and TCPST optimally in the network. Different optimization methods such as composite differential evolution (CoDE), Differential Evolution (DE) and PSO have been applied. It has been found that the DE with TCSC enhanced the maximum loadability with less computation time and faster convergence. It has also been shown that the computational effort further reduced with CoDE.

Esmaili et al. [11] proposed a multi-objective optimization technique for locating the series FACTS device using LMP method to relieve the congestion with enough margins of voltage and transient stability. Modified augmented ϱ-constrained method has been used for generating pareto solution, and most preferred pareto solution has been selected using fuzzy decision maker method. Khatavkar et al. [40] presented a method of FACTS devices placement using strength pareto evolution algorithm and GA for multi-objective and single-objective optimization problems, respectively.

Congestion has been created by uniform loading of the system. Two FACTS devices, SVC and TCSC, have been considered to improve voltage stability index, branch loading index and loss reduction. Reddy et al. [12] proposed single- and multi-objective genetic algorithms for optimal location of FACTS (TCSC and SVC) devices considering voltage stability, branch loading and loss minimization. The developed algorithms have been tested by creating the congestion in the system by either uniform loading or line outage or bilateral or multilateral transaction. Gitizadeh and Kalantar [41] investigated combined optimization process of Simulated Annealing (SA) and SQP to locate the TCSC and SVC optimally. In the first stage, SQP-based optimization method with FACTS devices has been applied to reduce the congestion with the evolution of static security margin. In the second stage, SA-based optimization method has been applied to find the optimal solution. The proposed approach is used to relieve the congestion and also enhance the distance of voltage collapse point. In Ref. [42], Reddy et al. observed the reduction in congestion by placing FACTS devices such as UPFC and TCSC using GA-based optimization. Two out of three FACTS devices are considered, and it is observed that reduction in congestion with one TCSC and two UPFC is more compared with the one UPFC and two TCSC. Hooshmand et al. [43] implemented the hybrid optimization which is a combination of bacterial foraging (BF) with Nelder-Mead (NM) method to solve the OPF problems by considering the penalty cost of emission and non-smooth fuel cost function. The proposed algorithm has been applied to minimize the cost of emission, cost of generation and cost of TCSC. Optimal placement of TCSC has been determined using congestion contribution rent method, and hence LMP or nodal price at each bus is required to find out. Mishra and Gundavrapu [44] proposed disparity line utilization factor (DLUF)-based placement of IPFC to improve the congestion management. IPFC has a capability to compensate multi-transmission line, and it has been placed with highest value of DLUF. Fire fly algorithm has been applied for the optimal tuning of IPFC parameter for multi-objective function consisting of total voltage deviation, active power loss, size of the IPFC and security margin. The results of the proposed techniques have been compared with GA techniques and have been found that they outperform than GA.

14.5.1.3   Price-Based Approach

Srivastava and Verma [45] demonstrated location-based pricing concept. Real and reactive power pricing scheme is determined using linear programming formulation with maximizing social welfare. The effect of placement of FACTS devices on pricing scheme has also been discussed. Shreshtha and Feng [17] analysed the effect of TCSC on spot price of power market. As the placement of TCSC in the network redistributes the power flow, the effect of TCSC on various aspects such as total social benefit, consumption, total generation and transmission congestion and benefit to individual supplier and customer has been analysed. It has been also found that the effect of TCSC compensation on various aspects of electricity market is not linear. Joorabian et al. [46] addressed the issue regarding the LMP-based method for locating TCSC and its parameter in the deregulated power market. The priority list of the lines for placement of TCSC has been prepared based on LMP value of each bus, and the generator cost and congested cost have been reduced. In Ref. [47], the effect of TCSC and its compensation level on congestion and spot price under different loading condition such as varying pool and bilateral loading condition has been analysed. The placement of TCSC in the network reduces losses and LMP at the congested node and increases social welfare. It has been suggested that the operation of TCSC should be controlled by ISO because the compensation level of TCSC has created new bottleneck and affects some participants adversely. Acharya and Mithulanantham [19] proposed a method for optimal location of series FACTS devices based on LMP difference and congestion rent contribution method in deregulated power system. Results of the proposed methods have been compared with the sensitivity-based approach and show that LMP difference method is less promising than congestion rent contribution method. It has also been shown that the sensitivity-based approach often fails to capture the optimal location correctly. Verma and Gupta [48] analysed the impact of UPFC placement on reactive and real power pricing in deregulated power system. The placement of UPFC has been determined based on sensitivity factor. It has also been concluded that the placement of UPFC reduces the real and reactive power loss, generation cost, congestion and spot price significantly. Tiwari and Sood [49] presented an efficient approach for location, number and sizing of multiple UPFC in a pool and bilateral power market with a view to maximize the social welfare under the deregulated environment. It has also been investigated that the system's real power losses decrease with installation of UPFC. Nabavi et al. [50] presented GA-based optimal generation schedule in deregulated power market. LMP is determined at each bus using nodal pricing method. Huang and Yan [51] addressed the pricing scheme for utilization of FACTS devices and penalty for users to operate at their limits. Hamoud and Bradly [52] described the method to compute TCC and LMP at any selected bus in the network by hydro one PROCOSE program. Sensitivity analysis to evaluate the change in LMP with change in system operating condition and system parameter has been carried out. LMP can be used to split the network in different zones, and each zone has its own LMP which has been calculated at hourly as well as interval bases. Conejo et al. [53] analysed sensitivity of LMP with various parameters such as reactive and active power demand, voltage and generator cost parameter. This sensitivity analysis provides the information regarding the behaviour and functioning of the power system. It also assists the seller and buyer for strategic bidding and ISO for continuous assessment of the system. Sood et al. [54] proposed a generalized optimal dispatch model to determine the size of the non-firm transaction and pool. LMP has been calculated using marginal cost theory. It has been proved that all load centres and GENCOs get social benefits by LMC pricing methods.

Momoh et al. [55] calculated the LMP for reactive and real power pricing with the framework of full AC optimal power flow. The decomposition of LMP in three components such as energy, loss and congestion has been computed. Interior point method with branch and bound has been used to solve OPF problem with discrete variable. Leevongwat and Rastgoufard [56] developed a methodology for forecasting locational marginal price in deregulated electricity market. The price of electricity has been determined using unit commitment, economic load dispatch and OPF which includes LMP calculation. Murali et al. [18] proposed bat algorithm for evaluating the spot price using DCOPF in pool-based electricity market. Different loss cases have been considered in transmission pricing scheme during the congestion to recover the congestion cost, loss cost and system cost. Wang et al. [57] developed a risk-based locational pricing and congestion management approach to enhance the system security level. Risk-based LMP is an additional price signal reflecting the impact of change in nodal injection on the system risk level. It gives more secure operating conditions and reduces the volatility through time. Chellam and Kalyani [58] proposed a scheme to evaluate the transmission congestion pricing using tracing principle in deregulated power system. The various costs are calculated in pool, bilateral and multilateral transaction market and proved that congestion in pool market is lesser than bilateral and multilateral market. The congestion cost is obtained from the generator's fixed cost and cost due to loss. Kang et al. [59] proposed zonal marginal pricing approach for congestion management by partitioning the network sequentially. Congestion contribution nodes to the congested lines have been identified using PTDF.

In general, this approach discussed the placement of FACTS devices using sensitivity-based approach and optimization techniques-based approach with respect to single- or multi-objective functions. Its impact on LMP has been analysed in detail.

14.5.2 GENERATION RESCHEDULING-BASED APPROACH

Due to thermal limit of the transmission line, a customer cannot get the power at cheaper rate even though generators having power at cheaper rates are ready to sell. Congestion in the network also restricts the convergence of all the available transaction. Rescheduling of generator is one of the approaches for relieving the congestion in the network. The selection of the generator for rescheduling is determined by using factors such as Generator Sensitivity Factors (GSF), PTDF and Relative Electrical Distance (RED) concept. Rescheduling of generator increases the energy rate, and therefore it is needed to reduce the rescheduling cost of generators. Different bio and nature inspired search-based optimization techniques have been proposed in the literature to minimize the rescheduling cost of generation.

The available literature shows authors using generator rescheduling by sensitivity factors-based approach for relieving congestion. The different search-based optimization techniques such as PSO, modified PSO and GA have been applied to reduce the rescheduling cost of generator [60–68]. Talukdar et al. [69] described the generation rescheduling and load shedding-based approach for congestion management. The generator and load buses have been ranked by considering the sensitivities of the overloaded line to bus injection, cost of generation and amount of load shedding. Based on the ranking, system operator can plan the generator rescheduling by considering the securities of the system with minimum load shedding and rescheduling cost. Venkaiah and Kumar [70] presented the generation rescheduling-based congestion management approach using fuzzy adaptive BF method. The participating generators for rescheduling have been selected based on their generator sensitivity to the congested line. The effectiveness and robustness of the algorithm have been tested by comparing the obtained result with the PSO and Simple Bacterial Foraging. It has been found faster with less computational burden. Dutta and Singh [13] suggested PSO techniques to reschedule generation for congestion management optimally. Generators for rescheduling have been selected based on generator's sensitivity analysis. It has been proved that if the choice of PSO parameters is not appropriate, then the result may diverge from the desired value. Hazra and Sinha [14] presented a method to relieve the congestion using fast decoupled load flow with PSO. This method gives a set of pareto solution instead of global solution. In order to consider the major contingency like generator outage and line outage, voltage- and frequency-dependent load flow model has been employed. Kumar and Shekhar [15] used three bid block structures that ensured voltage limit and static security, for generator rescheduling-based congestion management. The congested cost of each hour of the day, real and reactive power loss and loadability factor with and without FACTS devices have been calculated, and it has been found that the loadability limit with UPFC is higher than other FACTS devices. Using the RED concept and generator sensitivity shift factor, Yesuratnam and Pushpa [71] presented the rescheduling of the generator. The generator shift sensitivity factors have been applied to find the desired proportions of generators for overloading and the curtailment of generators for relieving congestion. Reddy [16] presented voltage-dependent load modelling, an approach of congestion management using load shedding and rescheduling of generation. Objective function consists of minimization cost of generation and load shedding, maximization of social welfare including minimization of load shedding, demand response offers, maximization of load served and minimization of load served error. Multi-objective strength pareto evolutionary algorithm has been employed to solve the proposed congestion management problem. Verma and Mukherjee [72] investigated the effectiveness and robustness of Ant Lion Optimizer (ALO) algorithm for congestion management. The main advantages of ALO are that it takes lesser time for function evaluation, not stuck into local minima and gives promising convergence characteristic. The result obtained with this algorithm has been compared with other latest algorithms such as SA, PSO, FPA, RSM, CBA, BA and FF, and it has been found that it outperformed over other algorithms.

Gope et al. [73] used firefly algorithm for rescheduling generators with the integration of pumped storage hydro unit. Bus sensitivity factors determine the optimal location of pumped storage hydro unit thereby improving voltage profile and reducing losses and total cost. The result obtained with this proposed algorithm has been compared with other optimization techniques such as ABC and PSO and found that it performed better. Deb et al. [74] integrated wind farm in the system and analysed its generator rescheduling. The wind farm location is selected based on PTDF. The generator rescheduling is done by generator sensitivity factor (GSF). Artificial Bee Colony (ABC) algorithm has been employed to mitigate congestion. Huang and Yan [75] investigated the impact of TCSC and SVC as a re-dispatch method with the objective to minimize the total amount of transaction curtailed. Meena and Selvi [76] proposed a method of combination rescheduling of generation and installation of TCSC for relieving the congestion in deregulated electric market. The modified PSO-based OPF has been applied for congestion management. The participation of generator for rescheduling has been limited using sensitivity factor. Muneender and Kumar [77] presented optimal rescheduling of real and reactive power generator for zonal congestion management. The selection of generator for rescheduling is based on Real Transmission Congestion Distribution Factors (PTCDFs) and Reactive Transmission Congestion Distribution Factors (QTCDFs). Fast Distance Ratio Particle

Swarm Optimization-based OPF techniques have been used for congestion management problem. The results obtained with this approach have been compared with conventional PSO, real coded GA and binary coded GA and found more economical.

The main objective of this approach is to reduce the rescheduling cost and converge all the transactions to meet the load without congestion in the line. Different heuristic and meta-heuristic optimization techniques have been applied to satisfy the different single or multi-objective functions.

14.5.3 ZONAL CONGESTION MANAGEMENT-BASED APPROACH

In zonal congestion management, power system network is divided into different zones by finding the nodes having similar locational marginal price. In some countries, zonal pricing method has been adopted, and it is determined using OPF. In this method, generators are paid the zone price of energy and loads must pay the zone price for energy. Zone price will remain same throughout the system if there is no congestion, and under this condition generators are paid the cost for their energy as the loads pay. Congestion within a zone is relieved by rescheduling of those generators in the zone which are responsible for contributing the congestion.

Kumar et al. [78] proposed a zonal congestion management in which different zones have been decided using real and reactive power flow index. The congested cost has been minimized by rescheduling of generators in the congested zone. The impact of reactive power support either from generators or capacitor on congestion cost has been analysed and shown that the congestion cost with reactive power support has been reduced. In Ref. [79], transmission congestion distribution factor-based zonal congestion management method is derived from the ac load flow Jacobian sensitivity. Rescheduling of generator and loads has been carried out in the congested zone to relieve the congestion. Sarwar and Siddiqui [80] identified the potential location of Distributed Generation (DG) in the different congested zones to relieve the congestion using the zonal congestion management method based on LMP across different links. The LMP difference between the lines provides useful information to identify the different congested zones. Wawrzyniak et al. [81] divided the energy market into different zones using LMP by taking into account variable weather conditions. OPF and two-fold clustering approach have been used for forming a zone. In Ref. [82], different zones are determined based on congestion relief index. Demand-side and supply-side management have been applied in the most sensitive zone for rescheduling the generation for congestion management. Imran and Bialek [83] investigated that zonal congestion management is not effective in European electricity market. It is almost impossible to form zones, and if the zones are actually formed, they create market inefficiency and arbitrage possibilities. Alomoush and Shahidehpour [84] combined the features of both inter- and intra-zonal schemes. The LMP has been used to define the zonal boundaries and facilitates the way to deal with congestion charges and credits. It is also observed that the Fixed Transmission Rights (FTR) that originate and end in the same zone contribute negligibly to congestion charges and credits.

This approach describes how the congestion within a zone is handled. Different methodologies such as placement of DG, distribution factor method, congestion relief index and fixed transmission rights have been applied within a zone to relieve congestion.

14.5.4 Renewable/Distributed Resources-Based Approach

With high penetration and uncertainty associated with the renewable/distributed resources in the network, the occurrence of congestion is frequent and indefinite. Because of these, the renewable resources have a significant impact on the congestion cost and congestion management. Different algorithms have been presented in the literature to assess the impact of the renewable/distributed resource on the congestion management.

In congestion management, optimal utilization of renewable energy sources (RES) was analysed by Nesamalar et al. [85]. This analysis also includes the impact of time and seasonal constraints of RES. To minimize the cost of rescheduling of generators, PSO is applied with a time-varying acceleration coefficient. The generator is selected for rescheduling using apparent power congestion index. In Ref. [86], the impact of renewable resources on congestion management has been assessed using Contribution Factor (CF)-based method derived from ac power flow analysis. The most desirable and undesirable Distributed Resources (DR) for congestion management have been identified by partitioning the CF space by Index of Total DR Amount space using fuzzy cluster approach. This approach identifies the proper clusters in the CF space for individual operating limits and the correlation among the multiple constraints. Keshtkar et al. [87] analysed how the reserve generation can be useful for congestion management. Further, this work also analysed the variation in total cost of power generation due to use of reserve generation in total generation. The cost of power generation and total generation after and before congestion alleviation is compared with and without existence of reserve generation. It has been shown that power generation cost decreased and total generation increased due to the use of reserve generation for congestion management. Vergnol et al. [88] analysed the impact of renewable resources and comparison of various rules for congestion management. The two different compensation methods used are (i) partial compensation and (ii) market mechanism methods for reducing the production of renewable resources for congestion management. In the first partial compensation method, it will reduce the loss of renewable production taking into account the priority of wind farm, while in second market mechanism method it allows renewable producers to set their own price for curtailment. Afkousi-Paqaleh et al. [89] proposed an approach for congestion management by optimal placement and sizing of DR considering the cost/benefit analysis. Different economic factors such as congestion revenue rent, upgrade investment deferred and DR costs are considered for cost or benefit analysis. Different load level and load growth have also been considered to assess the impact of DR. Bae and Uchida [90] developed a new approach to mitigate the power flow issues with integrated renewable resources in deregulated electric market. The frequency and power flow issues have been solved by adjusting the price signal based on the demand response. Zwaenepoel et al. [91] investigated the use of frequency restoration reserves in a congested area to reduce the congestion. As the frequency restoration reserves are limited, it cannot solve the large congestion management problem, and hence priority is given in the congested areas. Huang et al. [92] reviewed existing congestion management methods for distribution network with high penetration of distributed energy resources. These methods have been classified into two categories: market methods and direct control methods.

Generally the market method is used as congestion management, and direct control method can be treated as the backup for market method. The priority list has been prepared for selecting the appropriate method for congestion management. Vergnol et al. [93] proposed a real-time supervisor-based congestion management method with high penetration of wind farms. It is based on the efficient, automatic and dynamic re-dispatching mechanism using both conventional and wind generation. This method reduces the re-dispatching costs and also increases the network reliability with maximization of renewable production.

This approach discusses how the renewable/DR are utilized in relieving the congestion. The impact of renewable/DR on rescheduling cost and fuel cost has also been analysed.

14.5.5   AUCTION-BASED APPROACH

Fukutome et al. [94] developed the single price auction model for electricity market in Japan to maximize social surplus. The market clearing price (MCP) and trade volume of participant have been derived using proposed algorithm and congestion has been relieved. The congestion contributed transaction has been determined by physical flow basis. ISO relieves the congestion by removing congestion attributed transaction from the network. de Vries [95] categorized the different congestion management methods in two ways: congestion pricing methods and remedial methods. The explicit auctioning, implicit auctioning and market splitting can be categorized under the congestion pricing methods and allocate scarce transmission capacity by charging for access to the congested link. The remedial method consists of re-dispatching and counter trading. The congestion pricing method sends a signal to market while the remedial method to system operator. Yang et al. [96] proposed a method based on the nodal price deviation to implement the dynamic zoning according to different load levels. A coordinated auctioning method which is an extension of explicit auctioning eliminates the congestion by considering the relation of power systems and economic benefits of electricity market. Wachi [97] presented a method of decomposition of MCP based on single price auction model. MCP can be decomposed into a variety of parts corresponding to the associated factors such as bidding curves, transmission congestion and voltage limitation. Tang et al. [98] introduced the Pennsylvania New Jersey Maryland FTR auction market for congestion management and proposed an integrated process of congestion management for both contract and spot markets. Alomoush and Shahidehpour [99] addressed transmission congestion issues by proposing generalized model for FTR auction. These rights can buy, sell and trade through proper auction-based mechanism and also hedge congestion charges on constrained path. It gives the financial and operational certainty and maximizes the efficient use of system.

14.5.6   MISCELLANEOUS METHOD

Kirthika and Balamurugan [100] suggested that the overloading and reliability of the line is improved by using a dynamic control algorithm which changes the line impedance of the transmission line by placing series FACTS devices in the transmission line. Singh et al. [101] presented a congestion relief method by optimally placing DG using a zbus-based CF method in the deregulated power system. This method gives the sensitivities of a line flow to bus injection, and it is independent of selection of slack bus. The line flow will be redistributed due to placement of DG in such a way that congested line gets relieved. The proposed approach has been compared with the LMP and consumer payment approach, and a significant improvement in social welfare and a decrease in congestion have been noted. Yesuratnam and Thukaram [102] proposed a voltage stability-based RED concept for congestion management. It identifies the generators which are responsible for overloading, and the desired change in generation for congestion relief has been determined using this concept. It estimates the relative location of load bus with respect to generator bus. The desired change in generation will improve stability margin and minimize the transmission losses and transmission tariff. Granelli et al. [103] presented an optimal network reconfiguration-based approach for congestion management. The optimal reconfiguration of network for congestion network is a large-scale mixed integer programming problem, and it has been solved by branch and bound algorithm and GA. System operator can relieve congestion in the line by switching operation to reduce the curtailment of load and generation. Alomoush [104] presented different performance indices to measure the severity of congestion and system usage under different dispatch scenarios. It identifies the optimal dispatch among the different dispatch scenarios. In Ref. [105], an optimal coordination between demand response and FACTS device controllers for re-dispatch has been proposed using mixed integer optimization technique. Wu [106] analysed the variation in LMP and MCP by implementing price-based demand response. It is formulated by considering demand response participation level and energy pay back rates. Morais et al. [107] proposed a methodology to manage the resource using demand response program. The impact of demand response on LMP has been evaluated. During high demand condition, the LMP and operating cost have been reduced with demand response program. Dehnavi and Abdi [108] determined the optimal buses and time for implementation of demand response program to reduce the cost and improve the load curve characteristic, reliability and security of the network. The optimal buses have been identified using PTDF, Available Transfer Capacity and dynamic direct current optimal power flow (DCOPF). Singh et al. [109] presented the comparison between centralized and decentralized approach for congestion management. The OPF using interior point method has been applied to maximize the social welfare function considering transmission loss. Among the two approaches, the decentralized multi-utility power markets provide better independent dispatch. Various optimization algorithms, key challenges and issues related to congestion management have been elaborately discussed in Ref. [110]. In Ref. [111], the coordinated exchange of DSO-TSO is implemented in managing the congestion in both transmission and distribution systems. The congestion in distribution system is alleviated by coordinated energy management of consumers. The economic interest and private information of prosumers have also been taken care by ISO [112]. The line congestion in distribution network can also be alleviated by local energy sharing among the network [113]. The maximum PEV penetration and its V2G and G2V operation to alleviate the congestion in the distribution network is analysed in Ref. [114]. In Ref. [115], the congestion is eliminated by applying energy storage. The robust optimization is carried out to reduce the system peak when the system is having uncertain renewal resources, load and state of charge of energy storage system. In Ref. [116], the decentralized transmission switching operation is carried out to manage the multi-area power system congestion. A decentralized congestion is managed in smart distribution network by collaborating electrical vehicle [117]. The congestion and voltage profile issues have been solved by novel flexibility exchange strategy [118]. In Ref. [119], the congestion in distribution network having PEV and heat pumps is managed by distributed optimization-based dynamic tariff method.

Limited literature is available on energy management with RES in a congested deregulated power system. In Ref. [120], the authors investigate the congestion conditions in an existing hydropower plant upon utilization of wind power. It discusses a generalized model of congestion management after consideration of RES into the system. The aim is to determine non-firm and firm multilateral and bilateral transactions and generations for a given demand. The social benefit is maximized, and the locational marginal price is determined using marginal cost theory [121]. A combined operation of a thermal and a hydro generator leads to the preparation of a re-dispatching schedule resulting in alleviation of the system congestion. The reactive and active power balance constraints are considered to ensure the voltage limits are not violated [122]. The congestion on the network is relieved by rescheduling of the generator by considering different constraints such as line loading and bus voltage. The contingency such as line outage and load variation is also accounted for simulation and relieving the congestion [123]. The forced outage rate and demand response of the advance metering infrastructure are considered for carrying out reliability constrained congestion management [33]. The stochastic approach using chance constrained optimization for congestion management has been carried out in Ref. [124]. In this approach, both the uncertainty of wind and demand response has been considered in managing the congestion. In [125], multi-objective-based transmission congestion is analysed using demand response program. The demand response techniques shift the non-essential demand from peak hours to valley or off peak hours and manage the congestion in the transmission network. In Ref. [126], the real-time congestion management in distribution network is alleviated by using two-tier demand response with flexible demand swap and transactive control (Table 14.1).

14.6 CONCLUSION

In deregulated power system, the congestion management is a crucial and challenging task as the market is cleared without considering the constraints of network. In this chapter, a critical and exhaustive review of different approaches for congestion management such as optimal placement of FACTS devices, rescheduling of generators, zonal congestion and the impact of congestion on price at congested and non-congested buses have been presented at length. Emerging trends like impact of renewable resources, various auction methodologies and strategy for congestion

TABLE 14.1

Summary of Different Approach

Name of the Approach	Description	References	Optimization Techniques Applied	List of the Objective Functions	Comments
Sensitivity based	Different sensitivity indexes have been used for relieving congestion management	[22–30]	Coordinated Aggregated PSO, Fast Distance Ratio PSO, Genetic Algorithm, Particle Swarm Optimization, Ant Colony Search, Fuzzy-GA, Differential Evolutionary, CODE, Simple Bacterial Foraging, Strength Pareto Evolutionary Algorithm, Ant Lion Optimizer, Simulated Annealing, Bacterial Foraging, Nelder-Mead, Fire Fly, Artificial Bee Colony.	Maximization of Social Welfare, Enhancement of Voltage Stability, Enhancement of transient stability, Minimization of Fuel Cost, Improvement of Voltage Profile, Minimization of Load Shedding, Minimization of Emission, Maximization of Load Served Error, Minimization of loss reduction	Although it is an efficient approach, the non-linearity associated with the system cannot be captured.
Optimal placement of FACTS device based	Optimal location, rating and setting of control parameters of different single or multiple type FACTS devices such as TCSC, SSSC, and UPFC has been found out for relieving the congestion	[9–12,31–34,36–44]			It provides an alternative solution where the installation of new transmission line is not possible. *(Continued)*

TABLE 14.1 (*Continued*)
Summary of Different Approach

Name of the Approach	Description	References	Optimization Techniques Applied	List of the Objective Functions	Comments
Price based	Determination of LMP and different approach to reduce the LMP has been demonstrated	[18,19,45–58]			Degree of congestion across the network can be found out by determining the value of LMP at different nodes.
Generation rescheduling based	Rescheduling of generators using factors such as PTDF, GSF and RED for congestion management has been discussed. Rescheduling cost of generators has been reduced using different optimization algorithms.	[13–16,60–77]			Although it is a common approach and implemented frequently, it increases the energy price due to rescheduling of generator. *(Continued)*

TABLE 14.1 (*Continued*)
Summary of Different Approach

Name of the Approach	Description	References	Optimization Techniques Applied	List of the Objective Functions	Comments
Zonal congestion management based	Forming of zone by dividing a network into different zones and carrying out congestion management procedure within a zone. LMP of each zone has also been determined.	[78–84]			It can be implemented if it is possible to divide a network in different zones.
Renewable/DR based	Impact of renewable resources on congestion management and congestion cost has been discussed.	[85–93]			This approach can be applied when different types of distributed/renewable resources are available for generation along with conventional generation. (*Continued*)

TABLE 14.1 (*Continued*)
Summary of Different Approach

Name of the Approach	Description	References	Optimization Techniques Applied	List of the Objective Functions	Comments
Auction based	Market-based technique, such as explicit and implicit auctioning, single price auctioning, market splitting, physical flow-based method, etc., has been discussed for congestion management.	[94–99,127]			In this approach, congestion management is carried out at the time of market clearing process by including different auction methods.
Miscellaneous approach: (a) DG placement, (b) demand response, (c) network reconfiguration, (d) relative electrical distance concept	Different miscellaneous approaches such as DG placement, demand response, network reconfiguration, etc. have been presented.	[100–109]			These approaches are considered when modification in generation, network configuration and customer demands is possible.

management have been included. Congestion in the network makes power system inefficient, unreliable, insecure and costly. The detailed information regarding the relieving of congestion can be useful to aspirants who would like to work in the field of deregulated power system.

REFERENCES

1. K. Bhattacharya, M. H. J. Bollen, and J. E. Daalder, *Operation of Restructured Power Systems, ser. Power Electronics and Power Systems*. New York: Kluwer Academic Publishers, 2001.
2. S. A. Khaparde and A. R. Abhyankar, *Restructured Power Systems*, Volume 1. Oxford: Alpha Science International Ltd, May 2015.
3. V. Nanduri and T. K. Das, "A survey of critical research areas in the energy segment of restructured electric power markets," *International Journal of Electrical Power and Energy Systems*, vol. 31, no. 5, pp. 181–191, 2009.
4. A. Pillay, S. P. Karthikeyan, and D. Kothari, "Congestion management in power systems a review," *International Journal of Electrical Power and Energy Systems*, vol. 70, pp. 83–90, 2015.
5. A. Kumar, S. Srivastava, and S. Singh, "Congestion management in competitive power market: A bibliographical survey," *Electric Power Systems Research*, vol. 76, no. 1–3, pp. 153–164, 2005.
6. M. B. Nappu, A. Arief, and R. C. Bansal, "Transmission management for congested power system: A review of concepts, technical challenges and development of a new methodology," *Renewable and Sustainable Energy Reviews*, vol. 38, pp. 572–580, 2014.
7. S. Frank, I. Steponavice, and S. Rebennack, "Optimal power flow: A bibliographic survey-I," *Energy Systems*, vol. 3, no. 3, pp. 221–258, 2012.
8. S. Frank, I. Steponavice, and S. Rebennack, "Optimal power flow: A bibliographic survey-II," *Energy Systems*, vol. 3, no. 3, pp. 259–289, 2012.
9. P. K. Tiwari and Y. R. Sood, "An efficient approach for optimal allocation and parameters determination of TCSC with investment cost recovery under competitive power market," *IEEE Transactions on Power Systems*, vol. 28, no. 3, pp. 2475–2484, Aug 2013.
10. S. Nagalakshmi and N. Kamaraj, "Comparison of computational intelligence algorithms for loadability enhancement of restructured power system with FACTS devices," *Swarm and Evolutionary Computation*, vol. 5, pp. 17–27, 2012.
11. M. Esmaili, H. A. Shayanfar, and R. Moslemi, "Locating series FACTS devices for multi-objective congestion management improving voltage and transient stability," *European Journal of Operational Research*, vol. 236, no. 2, pp. 763–773, 2014.
12. S. S. Reddy, M. S. Kumari, and M. Sydulu, "Congestion management in deregulated power system by optimal choice and allocation of FACTS controllers using multi-objective genetic algorithm," in *IEEE PES T D*, April 2010, pp. 1–7.
13. S. Dutta and S. P. Singh, "Optimal rescheduling of generators for congestion management based on particle swarm optimization," *IEEE Transactions on Power Systems*, vol. 23, no. 4, pp. 1560–1569, November 2008.
14. J. Hazra and A. Sinha, "Congestion management using multiobjective particle swarm optimization," *IEEE Transactions on Power Systems*, vol. 22, no. 4, pp. 1726–1734, November 2007.
15. A. Kumar and C. Sekhar, "Congestion management with FACTS devices in deregulated electricity markets ensuring loadability limit," *International Journal of Electrical Power and Energy Systems*, vol. 46, pp. 258–273, 2013.
16. S. Reddy, "Multi objective based congestion management using generation rescheduling and load shedding," *IEEE Transactions on Power Systems*, vol. 32, no. 2, pp. 852–863, 2016.

17. G. Shrestha and W. Feng, "Effects of series compensation on spot price power markets," *International Journal of Electrical Power and Energy Systems*, vol. 27, no. 5–6, pp. 428–436, 2005.

18. M. Murali, M. S. Kumari, and M. Sydulu, "Optimal spot pricing in electricity market with inelastic load using constrained bat algorithm," *International Journal of Electrical Power and Energy Systems*, vol. 62, pp. 897–911, 2014.

19. N. Acharya and N. Mithulananthan, "Locating series FACTS devices for congestion management in deregulated electricity markets," *Electric Power Systems Research*, vol. 77, no. 3–4, pp. 352–360, 2007.

20. B. Singh, K. S. Verma, P. Mishra, R. Maheswari, U. Srivastava, and A. Baranwal, "Introduction to FACTS controllers: A technological literature survey," *International Journal of Automation and Power Engineering*, vol. 1, no. 9, pp. 193–234, December 2012.

21. Nemat-Talebi, M. Ehsan, and S. M. T. Bathaee, "An efficient power injection modeling and sequential power flow algorithm for FACTS devices," in *IEEE SoutheastCon Proceeding*, March 2004, pp. 97–104.

22. S. Singh and A. David, "Optimal location of FACTS devices for congestion management," *Electric Power Systems Research*, vol. 58, no. 2, pp. 71–79, 2001.

23. M. Mandala and C. P. Gupta, "Optimal placement of TCSC for transmission congestion management using hybrid optimization approach," in *International Conference on IT Convergence and Security (ICITCS)*, December 2013, pp. 1–5.

24. K. Verma, S. Singh, and H. Gupta, "Location of unified power flow controller for congestion management," *Electric Power Systems Research*, vol. 58, no. 2, pp. 89–96, 2001.

25. K.-H. Lee, "Optimal siting of TCSC for reducing congestion cost by using shadow prices," *International Journal of Electrical Power and Energy Systems*, vol. 24, no. 8, pp. 647–653, 2002.

26. F. L. Alvarado, "Solving power flow problems with a matlab implementation of the power system applications data dictionary," in *Proceedings of the 32nd Annual International Conference on Systems Sciences, (HICSS-32), Hawaii*, vol. Track3, January 1999, pp. 1–7.

27. M. Liu and G. Gross, "Effectiveness of the distribution factor approximations used in congestion modelling," in *Proceedings of the 14th Power Systems Computation Conference, Sevilla*, June 2002, pp. 24–29.

28. T. Niimura and Y. Niu, "Transmission congestion relief by economic load management," *IEEE Power Engineering Society Summer Meeting*, vol. 3, pp. 1645–1649, July 2002.

29. K. Y. Lee, S. Manuspiya, M. Choi, and M. Shin, "Network congestion assessment for short-term transmission planning under deregulated environment," in *IEEE Power Engineering Society Winter Meeting*, vol. 3, pp. 1266–1271, 2001.

30. C. Y. Chung, L. Wang, F. Howell, and P. Kundur, "Generation rescheduling methods to improve power transfer capability constrained by small-signal stability," *IEEE Transactions on Power Systems*, vol. 19, no. 1, pp. 524–530, February 2004.

31. M. Bavafa, N. Navidi, S. Hesami, and B. A. Parsa, "A new approach for security constrained congestion management using SSSC with ant colony search algorithm," in *Asia-Pacific Power and Energy Engineering Conference*, March 2010, pp. 1–5.

32. D. M. V. Kumar and C. Venkaiah, "Swarm intelligence based security constrained congestion management using SSSC," in *Asia-Pacific Power and Energy Engineering Conference*, March 2009, pp. 1–6.

33. S. Hajforoosh, S. M. H. Nabavi, and M. A. S. Masoum, "Coordinated aggregated-based particle swarm optimisation algorithm for congestion management in restructured power market by placement and sizing of unified power flow controller," *IET Science, Measurement Technology*, vol. 6, no. 4, pp. 267–278, July 2012.

34. B. Chong, X. P. Zhang, L. Yao, K. R. Godfrey, and M. Bazargan, "Congestion management of electricity markets using FACTS controllers," in *IEEE Power Engineering Society General Meeting*, June 2007, pp. 1–6.

35. Z. Yu and D. Lusan, "Optimal placement of FACTS devices in deregulated systems considering line losses," *International Journal of Electrical Power and Energy Systems*, vol. 26, no. 10, pp. 813–819, 2004.
36. H. Hashemzadeh and S. H. Hosseini, "Locating series FACTS devices using line outage sensitivity factors and particle swarm optimization for congestion management," in *IEEE Power Energy Society General Meeting*, July 2009, pp. 1–6.
37. M. Saravanan, S. M. R. Slochanal, P. Venkatesh, and J. P. S. Abraham, "Application of particle swarm optimization technique for optimal location of FACTS devices considering cost of installation and system loadability," *Electric Power Systems Research*, vol. 77, no. 3–4, pp. 276–283, 2007.
38. R. S. Wibowo, N. Yorino, M. Eghbal, Y. Zoka, and Y. Sasaki, "FACTS devices allocation with control coordination considering congestion relief and voltage stability," *IEEE Transactions on Power Systems*, vol. 26, no. 4, pp. 2302–2310, Nov 2011.
39. S. Nabavi, A. Kazemi, and M. Masoum, "Social welfare maximization with fuzzy based genetic algorithm by TCSC and SSSC in double-sided auction market," *Scientia Iranica*, vol. 19, no. 3, pp. 745–758, 2012.
40. V. Khatavkar, M. Namjoshi, and A. Dharme, "Congestion management in deregulated electricity market using FACTS multi-objective optimization," in *Indian Control Conference (ICC)*, January 2016, pp. 467–473.
41. M. Gitizadeh and M. Kalantar, "A new approach for congestion management via optimal location of FACTS devices in deregulated power systems," in *Third International Conference on Electric Utility Deregulation and Restructuring and Power Technologies, (DRPT)*, April 2008, pp. 1592–1597.
42. K. R. S. Reddy, N. P. Padhy, and R. N. Patel, "Congestion management in deregulated power system using FACTS devices," in *IEEE Power India Conference*, 2006, pp. 1–8.
43. R.-A. Hooshmand, M. J. Morshed, and M. Parastegari, "Congestion management by determining optimal location of series FACTS devices using hybrid bacterial foraging and nelder–mead algorithm," *Applied Soft Computing*, vol. 28, pp. 57–68, 2015.
44. A. Mishra and V. N. K. Gundavarapu, "Line utilisation factor-based optimal allocation of IPFC and sizing using firefly algorithm for congestion management," *IET Generation, Transmission Distribution*, vol. 10, no. 1, pp. 115–122, 2016.
45. S. C. Srivastava and R. K. Verma, "Impact of FACTS devices on transmission pricing in a de-regulated electricity market," in *International Conference on Electric Utility Deregulation and Restructuring and Power Technologies Proceedings, (DRPT)*, 2000, pp. 642–648.
46. M. Joorabian, M. Saniei, and H. Sepahvand, "Locating and parameters setting of TCSC for congestion management in deregulated electricity market," in *6th IEEE Conference on Industrial Electronics and Applications*, June 2011, pp. 2185–2190.
47. N. Acharya and N. Mithulananthan, "Influence of TCSC on congestion and spot price in electricity market with bilateral contract," *Electric Power Systems Research*, vol. 77, no. 8, pp. 1010–1018, 2007.
48. K. S. Verma and H. O. Gupta, "Impact on real and reactive power pricing in open power market using unified power flow controller," *IEEE Transactions on Power Systems*, vol. 21, no. 1, pp. 365–371, Feb 2006.
49. P. K. Tiwari and Y. R. Sood, "Efficient and optimal approach for location and parameter setting of multiple unified power flow controllers for a deregulated power sector," *IET Generation, Transmission Distribution*, vol. 6, no. 10, pp. 958–967, October 2012.
50. S. M. H. Nabavi, S. Jadid, M. A. S. Masoum, and A. Kazemi, "Congestion management in nodal pricing with genetic algorithm," in *International Conference on Power Electronics, Drives and Energy Systems, PEDES*, December 2006, pp. 1–5.

51. G. M. Huang and P. Yan, "Establishing pricing schemes for FACTS devices in congestion management," in *IEEE Power Engineering Society General Meeting*, vol. 2, July 2003, pp. 1025–1030.

52. G. Hamoud and I. Bradley, "Assessment of transmission congestion cost and locational marginal pricing in a competitive electricity market," *IEEE Transactions on Power Systems*, vol. 19, no. 2, pp. 769–775, May 2004.

53. A. J. Conejo, E. Castillo, R. Minguez, and F. Milano, "Locational marginal price sensitivities," *IEEE Transactions on Power Systems*, vol. 20, no. 4, pp. 2026–2033, November 2005.

54. Y. R. Sood, N. Padhy, and H. Gupta, "Deregulated model and locational marginal pricing," *Electric Power Systems Research*, vol. 77, no. 5–6, pp. 574–582, 2007.

55. J. A. Momoh, Y. Xia, and G. D. Boswell, "Locational marginal pricing for real and reactive power," in *IEEE Power and Energy Society General Meeting - Conversion and Delivery of Electrical Energy in the 21st Century*, July 2008, pp. 1–6.

56. I. Leevongwat and P. Rastgoufard, "Forecasting locational marginal pricing in deregulated power markets," in *IEEE/PES Power Systems Conference and Exposition PSCE*, March 2009, pp. 1–9.

57. Q. Wang, G. Zhang, J. D. McCalley, T. Zheng, and E. Litvinov, "Risk-based locational marginal pricing and congestion management," *IEEE Transactions on Power Systems*, vol. 29, no. 5, pp. 2518–2528, September 2014.

58. S. Chellam and S. Kalyani, "Power flow tracing based transmission congestion pricing in deregulated power markets," *International Journal of Electrical Power and Energy Systems*, vol. 83, pp. 570–584, 2016.

59. C. Kang, Q. Chen, W. Lin, Y. Hong, Q. Xia, Z. Chen, Y. Wu, and J. Xin, "Zonal marginal pricing approach based on sequential network partition and congestion contribution identification," *International Journal of Electrical Power and Energy Systems*, vol. 51, pp. 321–328, 2013.

60. P. Boonyaritdachochai, C. Boonchuay, and W. Ongsakul, "Optimal congestion management in an electricity market using particle swarm optimization with time-varying acceleration coefficients," *Computers and Mathematics with Applications*, vol. 60, no. 4, pp. 1068–1077, 2010.

61. M. Sarwar and A. S. Siddiqui, "An efficient particle swarm optimizer for congestion management in deregulated electricity market," *Journal of Electrical Systems and Information Technology*, vol. 2, no. 3, pp. 269–282, 2015.

62. A. S. Siddiqui, M. Sarwar, and S. Ahsan, "Congestion management using improved inertia weight particle swarm optimization," in *6th IEEE Power India International Conference (PIICON)*, December 2014, pp. 1–5.

63. J. Paul, T. Joseph, and S. Sreedharan, "PSO based generator rescheduling for relieving transmission overload," in *International Multi Conference on Automation, Computing, Communication, Control and Compressed Sensing*, March 2013, pp. 409–414.

64. S. Deb and A. K. Goswami, "Mitigation of congestion by generator rescheduling using particle swarm optimization," in *1st International Conference on Power and Energy in NERIST (ICPEN)*, December 2012, pp. 1–6.

65. S. K. Joshi and K. S. Pandya, "Active and reactive power rescheduling for congestion management using particle swarm optimization," in *21st Australasian Universities Power Engineering Conference (AUPEC)*, September 2011, pp. 1–6.

66. S. Sivakumar and D. Devaraj, "Congestion management in deregulated power system by rescheduling of generators using genetic algorithm," in *International Conference on Power Signals Control and Computations (EPSCICON)*, January 2014, pp. 1–5.

67. M. S. Kumar and C. P. Gupta, "Congestion management in a pool model with bilateral contract by generation rescheduling based on PSO," in *International Conference on Advances in Power Conversion and Energy Technologies (APCET)*, August 2012, pp. 1–6.

68. A. K. Khemani and N. K. Patel, "Generation rescheduling of most sensitive zone for congestion management," in *Nirma University International Conference on Engineering*, December 2011, pp. 1–5.
69. B. Talukdar, A. Sinha, S. Mukhopadhyay, and A. Bose, "A computationally simple method for cost-efficient generation rescheduling and load shedding for congestion management," *International Journal of Electrical Power and Energy Systems*, vol. 27, no. 5–6, pp. 379–388, 2005.
70. C. Venkaiah and D. V. Kumar, "Fuzzy adaptive bacterial foraging congestion management using sensitivity based optimal active power rescheduling of generators," *Applied Soft Computing*, vol. 11, no. 8, pp. 4921–4930, 2011.
71. G. Yesuratnam and M. Pushpa, "Congestion management for security oriented power system operation using generation rescheduling," in *IEEE 11th International Conference on Probabilistic Methods Applied to Power Systems (PMAPS)*, June 2010, pp. 287–292.
72. S. Verma and V. Mukherjee, "Optimal real power rescheduling of generators for congestion management using a novel ant lion optimiser," *IET Generation, Transmission Distribution*, vol. 10, no. 10, pp. 2548–2561, 2016.
73. S. Gope, A. K. Goswami, P. K. Tiwari, and S. Deb, "Rescheduling of real power for congestion management with integration of pumped storage hydro unit using firefly algorithm," *International Journal of Electrical Power and Energy Systems*, vol. 83, pp. 434–442, 2016.
74. S. Deb, S. Gope, and A. K. Goswami, "Generator rescheduling for congestion management with incorporation of wind farm using artificial bee colony algorithm," in *Annual IEEE India Conference (INDICON)*, December 2013, pp. 1–6.
75. G. M. Huang and P. Yan, "TCSC and SVC as re-dispatch tools for congestion management and TTC improvement," *IEEE Power Engineering Society Winter Meeting*, vol. 1, pp. 660–665, 2002.
76. T. Meena and K. Selvi, "Cluster based congestion management in deregulated electricity market using PSO," in *Annual IEEE India Conference - Indicon*, December 2005, pp. 627–630.
77. E. Muneender and D. M. V. Kumar, "Optimal rescheduling of real and reactive powers of generators for zonal congestion management based on FDR PSO," in *Transmission Distribution Conference Exposition: Asia and Pacific*, October 2009, pp. 1–6.
78. A. Kumar, S. C. Srivastava, and S. N. Singh, "A zonal congestion management approach using real and reactive power rescheduling," *IEEE Transactions on Power Systems*, vol. 19, no. 1, pp. 554–562, February 2004.
79. A. Kumar, S. Srivastava, and S. Singh, "A zonal congestion management approach using ac transmission congestion distribution factors," *Electric Power Systems Research*, vol. 72, no. 1, pp. 85–93, 2004.
80. M. Sarwar and A. S. Siddiqui, "Zonal congestion management based on locational marginal price in deregulated electricity market," in *Annual IEEE India Conference (INDICON)*, December 2015, pp. 1–4.
81. K. Wawrzyniak, G. Oryn´czak, M. Klos, A. Goska, and M. Jakubek, "Division of the energy market into zones in variable weather conditions using locational marginal prices," in *39th Annual Conference of the IEEE on Industrial Electronics Society, IECON*, November 2013, pp. 2027–2032.
82. M. H. Moradi, S. Dehghan, and H. Faridi, "Improving zonal congestion relief management using economical technical factors of the demand side," in *IEEE 2nd International Power and Energy Conference, PECon*, December 2008, pp. 1027–1032.
83. M. Imran and J. W. Bialek, "Effectiveness of zonal congestion management in the european electricity market," in *IEEE 2nd International Power and Energy Conference, PECon*, December 2008, pp. 7–12.

84. M. I. Alomoush and S. M. Shahidehpour, "Fixed transmission rights for zonal congestion management," *IEE Proceedings - Generation, Transmission and Distribution*, vol. 146, no. 5, pp. 471–476, September 1999.

85. J. J. D. Nesamalar, P. Venkatesh, and S. C. Raja, "Optimal utilization of renewable energy sources for congestion management," *IFAC-PapersOnLine*, vol. 48, no. 30, pp. 264–269, 2015.

86. J. Liu, M. M. A. Salama, and R. R. Mansour, "Identify the impact of distributed resources on congestion management," *IEEE Transactions on Power Delivery*, vol. 20, no. 3, pp. 1998–2005, July 2005.

87. H. Keshtkar, A. Darvishi, and S. H. Hosseinain, "Analysis on reserve effects on congestion management results in an integrated energy and reserve power market," in *19th Iranian Conference on Electrical Engineering*, May 2011, pp. 1–6.

88. A. Vergnol, V. Rious, J. Sprooten, and B. Robyns, "Integration of renewable energy in the european power grid: Market mechanism for congestion management," in *7th International Conference on the European Energy Market*, June 2010, pp. 1–6.

89. M. Afkousi-Paqaleh, M. Rashidinejad, and K. Lee, "Optimal placement and sizing of distributed resources for congestion management considering cost/benefit analysis," in *IEEE PES General Meeting*, July 2010, pp. 1–7.

90. H. Bae and K. Uchida, "Supply and demand balance control of power system with renewable energy integration by introducing congestion management," in *IEEE Eindhoven PowerTech*, June 2015, pp. 1–6.

91. B. Zwaenepoel, T. Vandoorn, G. V. Eetvelde, and L. Vandevelde, "Virtual power plant to deliver congestion management and frequency restoration reserve," in *IEEE 5th International Symposium on Power Electronics for Distributed Generation Systems (PEDG)*, June 2014, pp. 1–4.

92. S. Hunag, Q. Wu, and A. H. Nielsen, "Review of congestion management methods for distribution networks with high penetration of distributed energy resources," in *IEEE PES Innovative Smart Grid Technologies, Europe*, October 2014, pp. 1–6.

93. A. Vergnol, J. Sprooten, B. Robyns, V. Rious, and J. Deuse, "Optimal network congestion management using wind farms," in *Integration of Wide-Scale Renewable Resources Into the Power Delivery System, CIGRE/IEEE PES Joint Symposium*, July 2009, pp. 1–9.

94. S. Fukutome, T. Wachi, L. Chen, Y. Makino, and G. Koshimizu, "Bidding market based on single price model with network constraints," in *IEEE PES Power Systems Conference and Exposition*, vol. 3, pp. 1245–1250, October 2004.

95. L. J. de Vries, "Capacity allocation in a restructured electricity market: Technical and economic evaluation of congestion management methods on interconnectors," *IEEE Power Tech Proceedings, Porto*, vol. 1, pp. 1–6, 2001.

96. W. Yang, Q. Wan, and Y. Tang, "Congestion management based on dynamic zoning and coordinated auctioning method," in *Third International Conference on Electric Utility Deregulation and Restructuring and Power Technologies, (DRPT)*, April 2008, pp. 527–532.

97. T. Wachi, S. Fukutome, Y. Makino, and L. Chen, "Decomposition of market clearing price based on single price auction model," in *IEEE Power and Energy Society General Meeting - Conversion and Delivery of Electrical Energy in the 21st Century*, July 2008, pp. 1–8.

98. Y. Tang, H. Xu, and Q. Wan, "Research on the application of financial transmission right in congestion management," in *Third International Conference on Electric Utility Deregulation and Restructuring and Power Technologies, DRPT 2008*, April 2008, pp. 364–369.

99. M. Alomoush and S. Shahidehpour, "Generalized model for fixed transmission rights auction," *Electric Power Systems Research*, vol. 54, no. 3, pp. 207–220, 2000.

100. N. Kirthika and S. Balamurugan, "A new dynamic control strategy for power transmission congestion management using series compensation," *International Journal of Electrical Power and Energy Systems*, vol. 77, pp. 271–279, 2016.

101. K. Singh, V. K. Yadav, N. P. Padhy, and J. Sharma, "Congestion management considering optimal placement of distributed generator in deregulated power system networks," *Electric Power Components and Systems*, vol. 42, no. 1, pp. 13–22, 2014.

102. G. Yesuratnam and D. Thukaram, "Congestion management in open access based on relative electrical distances using voltage stability criteria," *Electric Power Systems Research*, vol. 77, no. 12, pp. 1608–1618, 2007.

103. G. Granelli, M. Montagna, F. Zanellini, P. Bresesti, R. Vailati, and M. Innorta, "Optimal network reconfiguration for congestion management by deterministic and genetic algorithms," *Electric Power Systems Research*, vol. 76, no. 6–7, pp. 549–556, 2006.

104. M. I. Alomoush, "Performance indices to measure and compare system utilization and congestion severity of different dispatch scenarios," *Electric Power Systems Research*, vol. 74, no. 2, pp. 223–230, 2005.

105. A. Yousefi, T. Nguyen, H. Zareipour, and O. Malik, "Congestion management using demand response and FACTS devices," *International Journal of Electrical Power and Energy Systems*, vol. 37, no. 1, pp. 78–85, 2012.

106. L. Wu, "Impact of price-based demand response on market clearing and locational marginal prices," *IET Generation, Transmission Distribution*, vol. 7, no. 10, pp. 1087–1095, October 2013.

107. H. Morais, P. Faria, and Z. Vale, "Demand response design and use based on network locational marginal prices," *International Journal of Electrical Power and Energy Systems*, vol. 61, pp. 180–191, 2014.

108. E. Dehnavi and H. Abdi, "Determining optimal buses for implementing demand response as an effective congestion management method," *IEEE Transactions on Power Systems*, vol. 32, no. 2, pp. 1537–1544, 2016.

109. B. Singh, R. Mahanty, and S. Singh, "Centralized and decentralized optimal decision support for congestion management," *International Journal of Electrical Power and Energy Systems*, vol. 64, pp. 250–259, 2015.

110. "Congestion management approaches in restructured power system: Key issues and challenges," *The Electricity Journal*, vol. 33, no. 3, p. 106715, 2020.

111. "Dso-tso cooperation issues and solutions for distribution grid congestion management," *Energy Policy*, vol. 120, pp. 610–621, 2018.

112. J. Hu, J. Wu, X. Ai, and N. Liu, "Coordinated energy management of prosumers in a distribution system considering network congestion," *IEEE Transactions on Smart Grid*, vol. 12, pp. 468–478, 2020.

113. X. Ai, J. Wu, J. Hu, Z. Yang, and G. Yang, "Distributed congestion management of distribution networks to integrate prosumers energy operation," *IET Generation, Transmission Distribution*, vol. 14, no. 15, pp. 2988–2996, 2020.

114. S. Deb, A. Goswami, P. Harsh, J. Sahoo, R. Chetri, R. Roy, and A. Shekhawat, "Charging coordination of plug-in electric vehicle for congestion management in distribution system integrated with renewable energy sources," *IEEE Transactions on Industry Applications*, vol. 56, pp. 5452–5462, 2020.

115. X. Yan, C. Gu, X. Zhang, and F. Li, "Robust optimization-based energy storage operation for system congestion management," *IEEE Systems Journal*, vol. 14, no. 2, pp. 2694–2702, 2020.

116. M. Khanabadi, Y. Fu, and C. Liu, "Decentralized transmission line switching for congestion management of interconnected power systems," *IEEE Transactions on Power Systems*, vol. 33, no. 6, pp. 5902–5912, 2018.

117. A. Asrari, M. Ansari, J. Khazaei, and P. Fajri, "A market framework for decentralized congestion management in smart distribution grids considering collaboration among electric vehicle aggregators," *IEEE Transactions on Smart Grid*, vol. 11, no. 2, pp. 1147–1158, 2020.

118. H. Liao and J. V. Milanovi´c, "Flexibility exchange strategy to facilitate congestion and voltage profile management in power networks," *IEEE Transactions on Smart Grid*, vol. 10, no. 5, pp. 4786–4794, 2019.

119. S. Huang, Q. Wu, H. Zhao, and C. Li, "Distributed optimization-based dynamic tariff for congestion management in distribution networks," *IEEE Transactions on Smart Grid*, vol. 10, no. 1, pp. 184–192, 2019.

120. F. R. Førsund, B. Singh, T. Jensen, and C. Larsen, "Phasing in wind-power in norway: Network congestion and crowding-out of hydropower," *Energy Policy*, vol. 36, no. 9, pp. 3514–3520, 2008.

121. Y. R. Sood and R. Singh, "Optimal model of congestion management in deregulated environment of power sector with promotion of renewable energy sources," *Renewable Energy*, vol. 35, no. 8, pp. 1828–1836, 2010.

122. K. Singh, N. Padhy, and J. Sharma, "Congestion management considering hydro–thermal combined operation in a pool based electricity market," *International Journal of Electrical Power and Energy Systems*, vol. 33, no. 8, pp. 1513–1519, 2011.

123. S. Verma and V. Mukherjee, "Firefly algorithm for congestion management in deregulated environment," *Engineering Science and Technology, an International Journal*, vol. 19, no. 3, pp. 1254–1265, 2016.

124. "Chance-constrained stochastic congestion management of power systems considering uncertainty of wind power and demand side response," *International Journal of Electrical Power and Energy Systems*, vol. 107, pp. 703–714, 2019.

125. "Multi-objective transmission congestion management considering demand response programs and generation rescheduling," *Applied Soft Computing*, vol. 70, pp. 169–181, 2018.

126. "Two-tier demand response with flexible demand swap and transactive control for real-time congestion management in distribution networks," *International Journal of Electrical Power and Energy Systems*, vol. 114, p. 105399, 2020.

127. S. Tao and G. Gross, "Congestion management allocation in multiple transaction networks," *IEEE Power Engineering Society Winter Meeting*, vol. 1, pp. 168–176, 2002.

15 Restructuring of Power System Network to Mitigate Renewable Energy Evacuation Constraints
A Comprehensive Study

Kiran Yadav, Om Prakash Mahela,
Sushma Lohia, and Baseem Khan

CONTENTS

DOI: 10.1201/9781003278030-15

15.1 INTRODUCTION

Transmission system network in Rajasthan is constructed, operated and maintained by the state transmission utility (STU). STU maintains transmission network associated with voltage levels of 33, 132, 220, 400 and 765 kV. Generation of power is achieved using the thermal power plants, nuclear power plants, hydel power plants and renewable energy (RE) sources. RE potential (solar and wind) is mainly concentrated in western parts of Rajasthan. Generated energy is required to be transmitted for long distances to reach the load centre(s) of Rajasthan which are mainly in the central and eastern parts of Rajasthan and also to be transmitted to northern and western grids of India through Inter State Transmission System. Government of India has ambitious plan for integration of 175 GW of RE into grid by the end of year 2022. The predominant of this RE capacity shall be through solar plants for which Rajasthan already has a strategic advantage. Government of Rajasthan has also stated to strengthen the transmission infrastructure to harness full potential of RE in Rajasthan. Existing installed capacity of RE power is 7,384 MW. Further, the under execution RE projects owned by the state utilities have the capacity of 10,000 MW. The RE projects under execution of the central transmission utility (CTU) have the capacity of 20,330 MW in the central sector projects category [1,2].

The intermittency of RE, poor plant load factor for renewable generators, etc. pose various challenges in operating the transmission system. The integration level of existing RE (only considering state sector renewable generation) against existing handled peak load by transmission system is around 32%. At this integration level of RE, the transmission system is experiencing high voltages on various extra high-voltage buses at substations, overloading of few transmission corridors and constrained capability to transfer power towards load centre(s). Buses of 400 and 220 kV levels are already experiencing the overvoltages in the range of 420–450 kV and 230–250 kV, respectively [3]. The above-mentioned challenges create problems in the evacuation of RE power. However, this can be obviated by optimal restructuring of the existing transmission system by addition of GSS and transmission lines of different voltage levels. A wide range of models have been reported for restructuring of power system network in India for meeting load requirements as well as requirements of policies of both the state and central governments [4–18]. This also helps to meet out the increased load demand and evacuation of increased generation, either conventional or RE. The philosophy and approaches used for the purpose of planning as well as operation which are well established during the past decades and are changing. The need has also been arising that India should recognize and meet these forthcoming challenges. To make the market as competitive, there are various methods adopted for the purpose of restructuring the industry of power [19]. Brief overview of related work and contribution of the chapter are described in the following subsections.

15.1.1 RELATED WORK

Restructuring of power system network has been carried out in different countries of the world due to various reasons. Romanian government has taken a step towards restructuring of network of state utilities and assets of state-owned power generation system, which had been considered in 2010 [20]. In Ref. [21], the authors analysed the welfare-related complicacies associated with implications of reforms of power sector during a condition where there is both the public sector undertakings and utilities of private sector. This study is related to the restructuring of state power sector in West Bengal, India. In Ref. [22], the authors examined the process adopted for the implementation of different governmental schemes based on the network perspective in Haoqiao which is a poor village situated in northern China. In Ref. [23], the authors described a report on restructuring of electricity market in Iran. Different changes have been undertaken in world electricity industry. Initial steps for restructuring of electricity industry in Iran had been considered in the year 1990. This had resulted in launching the wholesale electricity market of Iran in November 2003. In Ref. [24], a scheme for restructuring power system and self-adaptive differential evolutionary approach are proposed for improving and control of power flow with the help of Unified Power Flow Controller with the constraints in the practical security. A new formula for parameters associated with the tuning of differential evolutionary schemes has been designed in a way that these have become adaptive for whole iteration range. In Ref. [25], the authors proposed a technique for optimal allocation of the multi-type flexible AC transmission system equipment which can be deployed in the restructured network of the power system in the presence of wind generation. The main objective of the proposed technique is to maximize the present quantity of profit in long term. Different studies related to the reactive power compensation, protections and power quality have been reported in literature [26–41].

15.1.2 Contribution of the Chapter

Restructuring of the transmission network in real transmission network needs to be invested to mitigate RE evacuation constraints. Main contributions of the chapter are detailed below:

- This work is aimed to suggest the restructuring of the Rajasthan transmission system to develop Bulk Power Handling capability to transfer power from western Rajasthan to Load Centre(s).
- Development of adequate transmission capacity by restructuring the network to evacuate the additional expected renewable generation of around 5,000 MW by year 2023.
- To explore the possibility of utilization of the RE power at different load pockets in the western parts of Rajasthan.
- To control power flow in few transmission corridors for obtaining optimum resource utilization.

15.2 TEST UTILITY GRID

The proposed study is performed on the transmission network of the STU network of Rajasthan, India. Network of STU is maintained by the Rajasthan Rajya Vidhyut Prasaran Nigam Ltd. (RVPN) and generation of electrical power is managed by the

Rajasthan Rajya Vidhyut Utpadan Nigam Ltd. There are three distribution companies in the state which include Jaipur Vidhyut Vitaran Nigam Ltd., Ajmer Vidhyut Vitaran Nigam Ltd. and Jodhpur Vidhyut Vitaran Nigam Ltd. Jurisdiction of the distribution companies of Rajasthan is shown in Figure 15.1. However, the STU has the jurisdiction over the entire state. The power map of the transmission network in western parts of Rajasthan is shown in Figure 15.2. The 400 kV GSSs are situated at Kakani (Jodhpur New), Jodhpur (Surpura), Barmer, Ramgarh, Bhadla, Jaisalmer-I (Akal) and Jaisalmer-II (Bhesda). There are 765 kV GSSs at the Ajmer, Bhadla-I, Bhadla-II, Fatehgarh-I and Fatehgarh-II. In the Bhadla region there is solar power integration system on the 220 kV GSS and 132 kV GSS. In the Barmer and Jaisalmer regions, there are wind power generation plants, and installation of solar power plants is under progress.

Existing practical network of transmission system of Rajasthan STU including the system of voltage levels 132, 220, 400 and 765 kV is considered for the study. The conventional generators such as thermal power plants, nuclear power plants and hydro power plants are integrated into this system. RE is continuously being integrated into the existing network specifically in western Rajasthan. The existing network consists of transmission lines and GSSs rated at voltage levels such as 765, 400, 220, 132 and 33 kV. This system contains the power system network developed by the CTU and STU. Following are the important 765 kV GSS owned by the CTU and STU operating in the territory of Rajasthan [3,41].

- 765 kV GSS Anta owned by RVPN
- 765 kV GSS Phagi owned by RVPN
- 765 kV GSS Ajmer owned by CTU

FIGURE 15.1 Geographical regions of Rajasthan distribution companies.

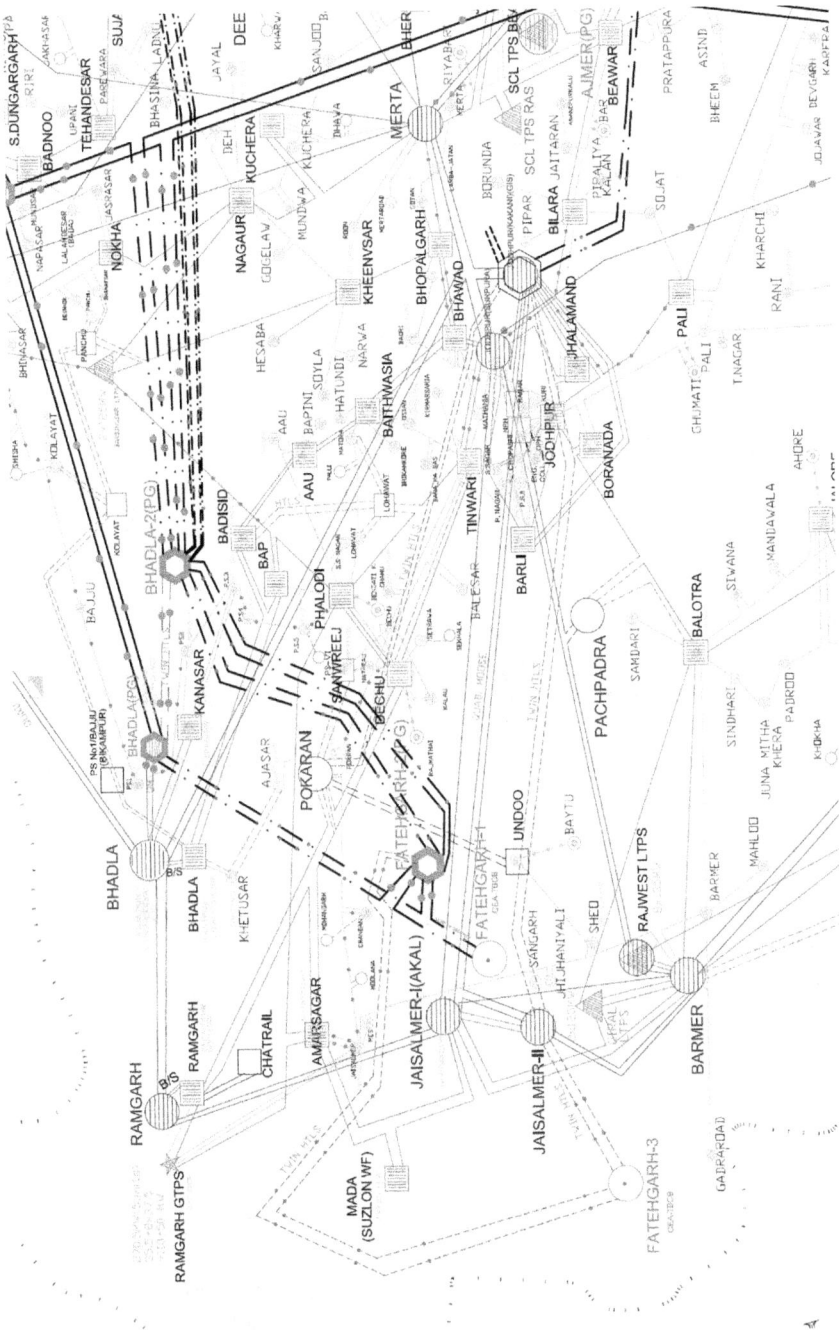

FIGURE 15.2 Power map of western parts of Rajasthan.

- 765 kV GSS Chhitorgarh owned by CTU
- 765 kV GSS Bikaner owned by CTU and under construction
- 765 kV GSS Khetri owned by CTU and under construction
- 765 kV GSS Bhadla-I and II owned by CTU and under construction
- 765 kV GSS Fatehgarh-I under construction by tariff-based competitive bidding
- 765 kV GSS Fatehgarh-II under construction and owned by CTU
- 765 kV GSS Kucheri under construction and owned by CTU
- 765 kV GSS Sikar under construction and owned by CTU

The existing GSSs in Rajasthan and the length of the total circuits of transmission lines as on 31.03.2020 are provided in Table 15.1. These include the 765, 400, 220 and 132 kV GSSs. The 33 and 11 kV GSSs are owned by the distribution companies, and these were constructed according to the load demand. Extra high-voltage substations are constructed by RVPN, PGCIL and under scheme of public private partnership. Generation schedule of Rajasthan is included in Table 15.2 [3,41].

15.2.1 Base Transmission System

Existing practical network of transmission system of Rajasthan STU including the system of voltage levels 132, 220, 400 and 765 kV discussed in the above section is considered as base transmission system for the proposed study. All the conventional generators such as thermal power plants, nuclear power plants and hydro power plants that are integrated into this system are considered in the base transmission system. RE integrated into the existing network and the existing network of the power system network developed by the CTU and STU in Rajasthan are considered for this study.

15.3 PROPOSED TRANSMISSION NETWORK RESTRUCTURING METHODOLOGY

The study is considered for restructuring of the transmission system to obviate the RE evacuation constraints from western parts of Rajasthan.

TABLE 15.1
Grid Substation Statistics Lengths of Transmission Lines

S. No.	Voltage Level (kV)	Number of GSS	Circuit Length of Transmission Lines (km)
1	765	6	425.498
2	400	27	7,604.444
3	220	124	15,443.394
4	132	459	18,245.566
Total		616	41,718.902

TABLE 15.2
Generation Schedule of Rajasthan

S. No.	Power Plants	Installed Unit Wise Capacity (Number of Units × MW Capacity)	Total Installed Capacity (MW)
1	Suratgarh power plant	6 × 250 + 2 × 660	2,820
2	Chhabra power plant	4 × 250 + 2 × 660	2,320
3	Kawai power plant	2 × 660	1,320
4	KTPS power plant	2 × 110 + 3 × 210 + 2 × 195	1,240
5	Kalisindh power plant	2 × 600	1,200
6	Giral LTPS	2 × 125	250
7	Mahi HEP	2 × 25 + 2 × 45	140
8	RPS HEP	4 × 43	172
9	JS HEP	3 × 33	99
10	Ramgarh GTPS	1 × 35.5 + 2 × 37.5 + 1 × 50 + 1 × 110	270.50
11	Dholpur GTPS	3 × 110	330
12	Barsingsar LTPS	2 × 125	250
13	RAPP A	1 × 100 + 1 × 200	300
14	Rajwest LTPS	8 × 135	1,080
15	VSLP LTPS	135	135
16	Ramgarh GTPS (Stage IV)	1 × 50 + 1 × 110	160
17	Anta GTPS	1 × 153.2 + 3 × 88.71	419.33
18	RAPP B	2 × 220	440
19	RAPP C	2 × 220	440
20	RAPP D	2 × 700	1,400
21	Shri cement	2 × 150	300
	Total		**27,727.795**

15.3.1 TEST SYSTEM

The study is performed on the transmission system of Rajasthan, and the following issues have been considered:

- Envisaged 4,885 MW new solar power projects have been considered for grid connectivity with intra-state transmission system.
- Envisaged 1,426 MW new Wind Power Projects have also been considered for grid connectivity in intra-state transmission system. In case, the DISCOMS of Rajasthan/STU plan to purchase wind generation from Inter State Transmission System Wind Power Projects, then the transmission system identified for evacuation of 1,426 MW WPP would not be considered for implementation.
- In the Load Flow Studies, the net dispatch from Solar and Wind generators has been assumed as 75% of total installed solar/wind capacity (15,052 MW), i.e. the studies have been conducted for 11,289 MW solar/wind generation.
- 600 MW Distributed Solar generators would be connected at Discom's 33/11 kV S/S; hence, new transmission system is not required. These Distributed Solar

generators have been discounted for by considering net dispatch as 6,618 MW, which is 75% of installed capacity of 8,823 MW for FY 2022–2023.
- The additional transmission system needs to be planned for approx. 5,000 MW (75% of 6,311 MW) solar/wind capacity.
- Total generation of the 17,584 MW is considered for the study.
- Load of 15,169 MW is considered for the study.

15.3.2 LOAD FLOW STUDY

The load flow study is carried out considering the various alternate schemes. All the schemes are tested with respect to loadings on the transmission lines and transformers. The schemes where the loadings are observed to be normal are considered as the most feasible schemes. Further, all the identified schemes are tested simultaneously to check the stability of the network.

15.3.3 SHORT CIRCUIT STUDY

The short circuit study is carried out using the MiPower software to check the suitability of the proposed studies during the faulty conditions. The protection equipment installed in the transmission system of the STU are rated for short circuit current of 50 kA. Hence, the three-phase short circuit fault current on all the buses should not exceed 50 kA. This is achieved by performing the short circuit studies.

15.4 RESULTS OF LOAD FLOW STUDY AND DISCUSSION

The results related to load flow study to design a suitable restructuring of the transmission system in the western parts of Rajasthan to mitigate RE evacuation constraints are detailed in this section.

15.4.1 LOAD FLOW STUDY PROCEDURE

The load flow study has been performed using the fast decoupled load flow method. The study is carried out for Base Case of the Rajasthan transmission system and various schemes designed for the RE power evacuation. Load flow study for base case and all feasible schemes has been carried out considering the high wind and solar power generation (75%) to test the strength of the proposed transmission system. Load flow studies have been carried out for the total system load of 15,169 MW corresponding to FY 2022–2023. The projected load is decided by the Central Electricity Regulatory Commission for a particular state. For Rajasthan state, the projected load corresponding to the year 2022–2023 is equal to 15,169 MW [42,43].

15.4.2 CREATION OF 765 kV GSS

A pooling substation is required to collect the RE power from different parts of western Rajasthan through the network of 132, 220 and 400 kV voltage levels. This power collected in bulk amount is required to be transmitted over long distance. This can be achieved by creation of a 765 kV GSS at Jodhpur and connected to the existing 765 kV GSS Phagi (Jaipur) from where power will be dispersed to the load centres in the Jaipur and Alwar regions of Rajasthan. This network will also help to transmit power from the Kawai-Kalisindh-Chhabra generation complex to the western parts of Rajasthan when there is low RE power generation. The following additional transmission systems are investigated for the technical feasibility using the load flow studies:

- 3 × 1,500 MVA, 765/400 kV transformer.
- 300 km 765 kV D/C Jodhpur-Phagi line.
- 200 km 400 kV D/C Jodhpur (765 kV GSS)-Jaisalmer-II twin HTLS line.
- 400 kV D/C Pokaran-Jodhpur (765 kV GSS) HTLS line.
- Existing 400, 220 and 132 kV Transmission system.

The power flows observed during the load flow study on the important transmission lines and transformers considering the proposed 765 kV GSS Jodhpur with 75% RE generation are tabulated in Table 15.3.

TABLE 15.3
Results of Load Flow Study with Creation of 765 kV GSS

S. No.	Transmission Lines/Transformers	Power Flows (MW)
A	**Existing Elements**	
1	4,500 MVA, 765/400 kV transformer at 765 kV GSS Phagi	2,766
2	815 MVA, 400/220 kV transformer at 400 kV GSS Kankani	323
3	765 kV S/C Phagi-Anta Ckt-I	−649
4	765 kV S/C Phagi-Anta Ckt-II	−654
5	765 kV S/C Phagi-Bhiwani Ckt-I	1,753
6	765 kV S/C Phagi-Bhiwani Ckt-II	1,721
7	765 kV S/C Phagi-Gwalior line Ckt-I	457
8	765 kV S/C Phagi-Gwalior line Ckt-II	447
9	765 kV D/C Phagi-Ajmer Line	−3,246
10	400 kV D/C Phagi-Ajmer line	−727
11	400 kV D/C Phagi-Heerapura line	**911**
12	400 kV D/C Phagi-Bassi line	1,602
13	400 kV S/C Kankani-Jodhpur (Surpura) Ckt.-I l	−147
14	400 kV S/C Kankani-Jodhpur (Surpura) Ckt.-II	−109
15	400 kV S/C Kankani-Merta line	549
16	400 kV S/C Kankani-Pachpadra line	−92
17	400 kV D/C Kankani-Akal Line	−617
B	**Proposed Elements**	
18	4,500 MVA 765/400 kV transformer	−2,631
19	765 kV D/C Jodhpur-Phagi line	2,619
20	400 kV D/C Jodhpur-Pokaran Twin HTLS line	−630
21	400 kV D/C Jodhpur -Jaisalmer-II line	−1,287
C	**Total System Losses (MW)**	**1,612.18**

After a detailed analysis of the results of the load flow provided in Table 15.3, it is observed that transmission system included in the proposed case with upgradation of 400 kV GSS Kankani (Jodhpur) to 765 kV GSS will be sufficient to evacuate the renewable power reaching the Jodhpur from various locations in western Rajasthan. It is also observed that the proposed transmission scheme will also help to meet load demand in the western parts of Rajasthan during the conditions of low RE power generation by transmitting the thermal power from the Kawai-Kalisindh-Chhabra generation complex. It is concluded that the proposed 765 kV GSS Jodhpur is technically viable under high RE scenario as pooling substation for pooling renewable power from western Rajasthan.

15.4.3 CREATION OF 400 kV GSS

There is high potential of renewable power generation in the Pokaran region of the western parts of Rajasthan. This power may be collected and transmitted to the pooling substation at Jodhpur by creation of a 400 kV GSS in the Pokaran region. Presently, solar power projects or wind farms are connected to the existing 220 and 132 kV substations in the Pokaran and Sanwreej area. This generated RE power is being evacuated through existing 132 and 220 kV transmission networks. During peak wind and solar power generation conditions the existing transmission lines in and around Pokaran and Sanwreej areas are loaded to thermal loading. Existing 220 kV lines do not have any spare capacity to evacuate envisaged Solar and Wind generation. Therefore in order to evacuate the renewable power and reduce loading on the existing transmission lines in and around Dechu, Pokaran and Sanwreej areas, a new 400 kV GSS is essential in that area as a pooling substation. Hence, the 400 kV GSS as pooling substation at Pokaran has been considered for Load Flow Study. The proposed 400 kV GSS at Pokaran along with proposed 400 kV, 220 kV and 132 kV interconnections have been studied through Load Flow Studies.

Existing transmission system without proposed 400 kV GSS at Pokaran and its associated proposed 400, 220 and 132 kV interconnections is considered as base case. Power flow on important transmission elements during the base case with high wind and solar power generation (75%) is shown in Table 15.4. The following transmission system is considered for the study to find optimal and most feasible transmission scheme for the pooling substation at Pokaran to pool RE power from the Pokaran region to the centralized pooling substation at Jodhpur:

- 2 × 500 MVA, 400/220 kV and 2 × 160 MVA, 220/132 kV Power Transformers.
- 150 km 400 kV D/C Twin HTLS line from 400 kV GSS Pokaran to 765 kV GSS Jodhpur.
- 25 km LILO of 220 kV S/C Ramgarh-Dechu line at 400 kV GSS Pokaran.
- 25 km LILO of 220 kV S/C Amarsagar-Dechu line at 400 kV GSS Pokaran.
- 30 km LILO of 132 kV S/C Chandan-Pokran (132 kV GSS) line at 400 kV GSS Pokaran.
- 125 MVA, 420 kV Shunt Bus Reactor.
- 25 MVA, 245 kV Shunt Bus Reactor.

- 2×50 MVA, 420 kV Line Reactor on 400 kV Pokaran-Kankani (765 kV GSS Jodhpur) Line.
- Existing transmission system of the Rajasthan STU and transmission system of CTU in the territory of Rajasthan.

Power flow on important transmission elements during the proposed case with high wind and solar power generation (75%) is tabulated in Table 15.4.

After detailed analysis of the results of the load flow study, it is observed that transmission system included in the proposed case includes a 400/220/132 kV pooling substation in the Pokaran region, and it is observed that transmission system included in the proposed case is sufficient to evacuate the renewable power from in and around the Pokaran and Sanwreej regions by stepping up to the 400 kV voltage level and transmitting power to proposed 765 kV GSS Jodhpur. After a detailed

TABLE 15.4
Results of Load Flow Study with Creation of 400 kV GSS

S. No.	Transmission Lines/Transformers	Base Case	Proposed Case
A	**Existing Elements**		
1	2,000 MVA, 400/220 kV transformers at 400 kV GSS Ramgarh	−1,398	−1,300
2	160 MVA, 220/132 kV transformer at 400 kV GSS Ramgarh	−58	−50
3	360 MVA, 220/132 kV transformer at 220 kV GSS Amarsagar	−153	−146
4	200 MVA, 220/132 kV transformer at 220 kV GSS Dechu	113	46
5	360 MVA, 220/132 kV transformer at 220 kV GSS Phalodi	101	59
6	220 kV S/C Ramgarh-Dechu line	76	-
7	220 kV S/C Ramgarh-Amarsagar line	−54	−30
8	220 kV S/C Amarsagar-Dechu line	119	-
9	220 kV D/C Dechu-Phalodi line	−255	−46
10	220 kV Dechu-Tinwari Line Ckt-I	264	162
11	220 kV Dechu-Tinwari Line Ckt-II	264	162
12	132 kV S/C Chandan-Pokaran (132 kV GSS) line	−21	-
13	132 kV S/C Pokaran-Ajasar Line	−1	−26
14	132 kV S/C Pokaran-Dechu Line	126	102
B	**Proposed Elements at 400 kV GSS Pokaran**		
15	1,000 MVA, 400/220 kV transformer	-	−633
16	320 MVA, 220/132 kV transformer	-	−69
17	400 kV D/C Twin HTLS Pokaran to 765 kV GSS Jodhpur line	-	633
18	220 kV D/C Pokaran-Dechu line	-	200
19	220 kV Pokaran-Amarsagar line	-	−216
20	220 kV Pokaran-Ramgarh line	-	−117
21	132 kV Pokaran (400 kV GSS)-Chandan line	-	−42
22	132 kV Pokaran (400 kV GSS)-Pokaran (132 kV GSS line)	-	−27
C	**Total System Losses (MW)**	1,656.40	1,612.17
D	**Total Saving in MW with Respect to Base Case (MW)**	-	44.23
E	**Total Saving in LU/ Annum Wrt Base Case**	-	1,673.80

analysis of the results of the load flow study discussed above, it is observed that the proposed case is technically viable under high RE scenario as pooling substation for pooling renewable power from in and around Pokaran and Sanwreej.

15.4.4 CREATION OF 220 kV GSS AT SANWREEJ

To analyse the requirement of 220 kV GSS at Sanwreej along with associated 220 and 132 kV, interconnecting lines under peak solar/wind scenario (75% of installed capacity) are studied for conditions corresponding to FY 2022–2023.

Presently, 132 kV GSS Sanwreej is fed through 132 kV S/C radial line from 132 kV GSS Dechu. The 40 MW solar is connected to the existing 132 kV GSS Sanwreej and new solar plants are envisaged due to availability of sufficient land bank in and around Sanwreej Village. Further, 60 MW wind generation is anticipated at Sanwreej. There is agriculture belt in and around Sanwreej, and hence the load is continuously growing. The 132 kV GSS Nathrau is fed from 220 kV GSS Dechu through 132 kV S/C line. Agriculture Load is also continuously increasing in Nathrau region.

Looking at the influx of solar/wind generation in and around Sanwreej and antici-pated load growth, it becomes essential to strengthen the transmission network in and around Sanwreej and Nathrau by creating a new 220 kV GSS in that area. New 220 kV GSS at Sanwreej would also provide 220 kV interconnection to 400 kV GSS Pokaran (Proposed), and hence depending on load generation balance in the area, the excess power would be evacuated from 400 kV GSS Pokaran. Since sufficient government land is available adjoining to 132 kV GSS Sanwreej for upgradation into 220 kV, the upgrada-tion of 132 kV GSS Sanwreej into 220 kV has been considered for load flow study.

Existing transmission system without proposed 220 kV GSS at Sanwreej and its associated proposed 220 kV and 132 kV interconnections are considered. Power flow results of load flow study for base case with high wind and solar power generation (75%) are tabulated in Table 15.5.

The following transmission system is considered for the study to find optimal and most feasible transmission scheme for creation of a 220 kV GSS at Sanwreej to evacuate RE power to the 400 kV GSS Pokaran for proposed case-1:

- 1 × 160 MVA, 220/132 kV Power Transformer at 220 kV GSS Sanwreej.
- 30 km 220 kV D/C line from 400 kV GSS Pokaran to 220 kV GSS Sanwreej.
- 38 km 132 kV D/C line from 132 kV GSS Sanwreej to Nathrau.

Power flows of load flow study for proposed case-1 with high wind and solar power generation (75%) on important transmission elements are provided in Table 15.5.

The following transmission system is considered for the study to find optimal and most feasible transmission scheme for creation of a 220 kV GSS at Sanwreej to evacuate RE power to the 400 kV GSS Pokaran for proposed case-2:

- 1 × 160 MVA, 220/132 kV Power Transformer at 220 kV GSS Sanwreej.
- 9 km LILO of one circuit of 220 KV D/C Dechu to Phalodi line 220 kV GSS Sanwreej.
- 38 km 132 kV D/C line from 132 kV GSS Sanwreej to Nathrau.

TABLE 15.5
Results of Load Flow Study with Creation of 220 kV GSS at Sanwreej

S. No.	Transmission Lines/Transformers	Base Case	Proposed Case-1	Proposed Case-2	Proposed Case-3
A	**Existing Transmission Elements**				
1	360 MVA, 220/132 kV transformer at 220 kV Phalodi GSS	65	59	60	67
2	200 MVA, 220/132 kV transformer at 220 kV Dechu GSS	66	46	57	57
3	260 MVA, 220/132 kV transformer at 220 kV Tinwari GSS	119	115	117	118
4	220 kV Phalodi-Bap line	53	54	50	66
5	220 kV Phalodi-Lohawat line	92	92	90	98
6	220 kV Phalodi-Amarsagar line	−161	−160	−162	−155
7	220 kV Phalodi-Dechu line	49	46	27	21
8	220 kV Phalodi-Dechu line	-	-	27	21
9	220 kV Dechu-Pokaran line	−222	−199	−224	−140
10	220 kV Dechu-Tinwari line ckt-I	162	162	161	168
11	220 kV Dechu-Tinwari ckt-II	162	162	162	168
12	132 kV Dechu-Dechu Ckt-I	84	73	79	77
13	132 kV S/C Dechu-Dechu Ckt-II line	73	64	69	68
14	132 kV Dechu-Phalodi line	−12	−1	−6	−5
15	132 kV Dechu-Pokaran line	−108	−98	−105	−96
16	132 kV Dechu-Setrawa line	14	15	14	15
17	132 kV Dechu-Nathrau line	4	−18	−6	−12
18	132 kV Dechu-Rajmathai line	11	11	11	11
19	220 kV Tinwari-Jodhpur Ckt.-I line	168	170	168	175
20	220 kV Tinwari-Jodhpur Ckt.-II line	168	170	168	175
21	220 kV Tinwari-Lohawat line	−85	−85	−84	−87
22	132 kV Tinwari-Chamu line	−32	−40	−36	−39
23	132 kV Tinwari-Soorsagar line	37	39	38	39
24	132 kV Tinwari-Balesar line	−17	−22	−20	−21
25	132 kV Tinwari-Mathania line	75	78	76	79
26	132 kV Tinwari-Bana ka base line	9	10	9	9
27	132 kV Tinwari-Kiramarsariya line	14	14	14	14
28	132 kV Tinwari-PS 8 line	7	9	7	9
29	132 kV Sanwreej-Dechu line	32	65	47	56
B	**Proposed Transmission Elements**				
30	160 MVA, 220/132 kV transformer at 220 kV Sanwreej GSS	-	55	25	40
31	132 kV S/C Sanwreej-Nathrau line	-	29	17	23
32	220 kV D/C Sanwreej-Pokaran line	-	−55	-	−153
33	220 kV S/C Sanwreej-Phalodi line	-	-	10	56
34	220 kV S/C Sanwreej-Dechu line	-	-	−35	58
B	**Total System Losses (MW)**	1,611.87	1,612.17	1,611.45	1,611.48

(Continued)

TABLE 15.5 (*Continued*)
Results of Load Flow Study with Creation of 220 kV GSS at Sanwreej

S. No.	Transmission Lines/Transformers	Base Case	Proposed Case-1	Proposed Case-2	Proposed Case-3
C	Total Saving in MW with Respect to Base Case (MW)	-	−0.3	0.42	0.39
D	Total Saving in LU/ Annum Wrt Base Case	-	−11.35	15.89	14.75

Power plots of load flow study for proposed case-2 with high wind and solar power generation (75%) on important transmission elements are provided in Table 15.5.

The following transmission system is considered for the study to find optimal and most feasible transmission scheme for creation of a 220 kV GSS at Sanwreej to evacuate RE power to the 400 kV GSS Pokaran for proposed case-3:

- 1 × 160 MVA, 220/132 kV Power Transformer at 220 kV GSS Sanwreej.
- 9 km LILO of one circuit of 220 KV D/C Dechu to Phalodi line 220 kV GSS Sanwreej.
- 30 km 220 kV D/C line from 400 kV GSS Pokaran to 220 kV GSS Sanwreej.
- 38 km 132 kV D/C line from 132 kV GSS Nathrau to Sanwreej.

Power plots of load flow study for proposed case-3 with high wind and solar power generation (75%) on important transmission elements are provided in Table 15.5.

It is observed that transmission system included in proposed case-1 is sufficient to evacuate the renewable power in the Sanwreej region and to cater the load demand of the Sanwreej and Nathrau areas. Overloading is not observed on transmission lines. Looking above, all the proposed cases are economically comparable, but technically LILO of one circuit of 220 kV D/C Dechu-Phalodi is not found feasible due to circulation of power between 220 kV GSS Dechu and 220 kV GSS Sanwreej through 220 and 132 kV lines. Hence, proposed case-1 is found to be more technically feasible.

15.5 TESTING OF SUITABILITY OF THE PROPOSED RESTRUCTURING OF TRANSMISSION SYSTEM USING SHORT CIRCUIT STUDY

The results related to short circuit study for restructuring of Rajasthan transmission system to evacuate RE power from the western parts of Rajasthan to the load centres are obtained for the base case and after considering the entire proposed network. It can be observed that the short circuit level at all buses is normal and below 50 kA for the base and after incorporating the proposed network which is the maximum permissible values of the fault level. This will be considered as the base for proposed study of restructuring of transmission system to evacuate RE power from western parts of Rajasthan.

15.6 CONCLUSIONS

This research work presents a detailed study to propose optimized restructuring of Rajasthan transmission system to mitigate the constraints of evacuation of RE power from western parts of Rajasthan. The study is carried out for the base transmission system of Rajasthan without considering the restructuring and loadings on the transmission elements. The overloaded elements are identified and different schemes are designed to relieve such overloading. Different schemes are analysed and most feasible schemes are identified. Different schemes identified are detailed in the following subsections.

15.6.1 765 kV GSS at Jodhpur

It is concluded that creation of 765 kV GSS at Jodhpur is technically viable under high RE scenario as pooling substation for pooling renewable power from western Rajasthan. Therefore, the following transmission system is recommended at proposed 765 kV GSS Jodhpur (Kanknai).

- 3 × 1,500 MVA, 765/400 kV Substation by upgrading 400 kV GSS Kankani.
- 300 km, 765 kV D/C Jodhpur-Phagi line with 2 × 330 MVAR switchable reactors at Jodhpur end and 2 × 240 MVAR switchable reactors at Phagi end of the line.
- 5 km 765 kV D/C Line with Hex Zebra Conductor in each phase in the approach section of proposed 765 kV GSS Jodhpur (Kankani).
- 200 km 400 kV D/C Jodhpur (765 kV GSS)-Jaisalmer-II twin HTLS line with 2 × 50 MVAR, 420 kV switchable reactors at Jodhpur end of the line.
- 400 kV GSS at Pokaran.

It is concluded that creation of 400 kV GSS at Pokaran is technically viable under high RE scenario as pooling substation for pooling renewable power from in and around Pokaran and Sanwreej. Therefore, the following transmission system is recommended at proposed 400 kV GSS Pokaran (pooling substation).

- 2 × 500 MVA, 400/220 kV and 2 × 160 MVA, 220/132 kV Power Transformers, 125 MVA, 420 kV Shunt Bus Reactor, 25 MVAR, 245 kV Bus Reactor at Pokaran.
- 150 km 400 kV D/C Twin HTLS line from 400 kV GSS Pokaran to 765 kV GSS Jodhpur (Kankani) with 2 × 50 MVAR, 420 kV switchable line reactors at Pokaran end.
- 25 km LILO of 220 kV S/C Ramgarh-Dechu line at 400 kV GSS Pokaran.
- 25 km LILO of 220 kV S/C Amarsagar-Dechu line at 400 kV GSS Pokaran.
- 30 km LILO of 132 kV S/C Chandan-Pokran (132 kV GSS) line at 400 kV GSS Pokaran.

15.6.2 220 kV GSS at Sanwreej

It is concluded that creation of 220 kV GSS at Sanwreej is technically viable under high RE scenario. Therefore, the following transmission system is recommended at proposed 220 kV GSS Sanwreej.

- 1 × 160 MVA, 220/132 kV Power Transformer and 25 MVAR, 245 kV Bus Reactor.
- 30 km 220 kV D/C line from 400 kV GSS Pokaran to 220 kV GSS Sanwreej.
- 38 km 132 kV D/C line from 132 kV GSS Sanwreej to Nathrau.

REFERENCES

1. https://energy.rajasthan.gov.in/content/raj/energy-department/rrecl/en/home.html#.
2. https://niti.gov.in/writereaddata/files/175-GW-Renewable-Energy.
3. https://energy.rajasthan.gov.in/rvpnl.
4. B. Khan, G. Agnihotri, S. E. Mubeen, and G. Naidu, "A TCSC Incorporated Power Flow Model for Embedded Transmission Usage and Loss Allocation," *AASRI Procedia*, vol. 7, pp. 45–50, 2014.
5. B. Khan, G. Agnihotri, G. Gupta, and P. Rathore, "A Power Flow Tracing based Method for Transmission Usage, Loss & Reliability Margin Allocation," *AASRI Procedia*, vol. 7, pp. 94–100, 2014.
6. B. Khan, G. Agnihotri, P. Rathore, A. Mishra, and G. Naidu, "A Cooperative Game Theory Approach for Usage and Reliability Margin Cost Allocation under Contingent Restructured Market," *International Review of Electrical Engineering*, vol 9, no. 4, pp. 854–862, 2014.
7. B. Khan and G. Agnihotri, "A Comprehensive Review of Embedded Transmission Pricing Methods Based on Power Flow Tracing Techniques," *Chinese Journal of Engineering*, vol. 2013, Article ID 501587, 13 pages, 2013.
8. B. Khan, G. Agnihotri, and A. S. Mishra, "An Approach for Transmission Loss and Cost Allocation by Loss Allocation Index and Co-operative Game Theory," *Journal of The Institution of Engineers (India): Series B*, vol. 97, pp. 41–46, 2016.
9. B. Khan and G. Agnihotri, "A Novel Transmission Loss Allocation Method Based on Transmission Usage," *2012 IEEE Fifth Power India Conference*, 2012, pp. 1–3.
10. P. Rathore, G. Agnihotri, B. Khan, and G. Naidu, "Transmission Usage and Cost Allocation Using Shapley Value and Tracing Method: A Comparison," *Electrical and Electronics Engineering: An International Journal (ELELIJ)*, vol. 3, pp. 11–29, 2014.
11. B. Khan and G. Agnihotri, "An Approach for Transmission Usage & Loss Allocation by Graph Theory," *WSEAS Transactions on Power Systems*, vol. 9, pp. 44–53, 2014.
12. S. Khare, B. Khan, and G. Agnihotri, "A Shapley Value Approach for Transmission Usage Cost Allocation under Contingent Restructured Market," *2015 International Conference on Futuristic Trends on Computational Analysis and Knowledge Management (ABLAZE)*, Noida, 2015, pp. 170–173
13. B. Khan, G. Agnihotri, and G. Gupta, "A Multipurpose Matrices Methodology for Transmission Usage, Loss and Reliability Margin Allocation in Restructured Environment," *Electrical & Computer Engineering: An International Journal*, vol. 2, no. 3, p. 11, September 2013.
14. T. F. Agajie, B. Khan, H. H. Alhelou, and O. P. Mahela, "Optimal Expansion Planning of Distribution System Using Grid-Based Multi-Objective Harmony Search Algorithm," *Computers & Electrical Engineering*, vol. 87, p. 106823, 2020.
15. B. Khan, H. H. Alhelou, and F. Mebrahtu. "A Holistic Analysis of Distribution System Reliability Assessment Methods with Conventional and Renewable Energy Sources," *AIMS Energy*, vol. 7, no. 4, pp. 413–429, 2019.
16. D. Anteneh and B. Khan, "Reliability Enhancement of Distribution Substation by Using Network Reconfiguration a Case Study at Debre Berhan Distribution Substation," *International Journal of Economy, Energy and Environment*, vol. 4, no. 2, pp. 33–40, 2019.

17. R. K. Pachauri et al., "Impact of Partial Shading on Various PV Array Configurations and Different Modeling Approaches: A Comprehensive Review," *IEEE Access*, vol. 8, pp. 181375–181403, 2020.

18. R. K. Pachauri, O. P. Mahela, B. Khan, A. Kumar, S. Agarwal, H. H. Alhelou, and J. Bai, "Development of Arduino Assisted Data Acquisition System for Solar Photovoltaic Array Characterization under Partial Shading Conditions," *Computers & Electrical Engineering*, vol. 92, p. 107175, 2021.

19. M. M. Tripathi, A. K. Pandey, and D. Chandra, "Power System Restructuring Models in the Indian Context," *The Electricity Journal*, vol. 29, pp. 22–27, 2016.

20. V. Popovici, "2010 Power Generation Sector Restructuring in Romania—A Critical Assessment," *Energy Policy*, Elsevier, vol. 39, pp. 1845–1856, 2011.

21. D. Saha and R. N. Bhattacharya, "Analysis of the Welfare Implications of Power-Sector Restructuring in West Bengal, India," *Utilities Policy*, Elsevier, vol. 56, pp. 62–71, 2019.

22. C. Chen, J. Gao, and J. Chen, "Behavioral Logics of Local Actors Enrolled in the Restructuring of Rural China: A Case Study of Haoqiao Village in Northern Jiangsu," *Journal of Rural Studies*, Elsevier, 2019.

23. G. Reza Yousefi, S. M. Kaviri, M. A. Latify, and I. Rahmati, "Electricity Industry Restructuring in Iran," *Energy Policy*, vol. 108, pp. 212–226, 2017.

24. P. Acharjee, "Optimal Power Flow with UPFC Using Security Constrained Self-Adaptive Differential Evolutionary Algorithm for Restructured Power System," *Electrical Power and Energy Systems*, vol. 76, pp. 69–81, 2016.

25. A. Elmitwally and A. Eladl, "Planning of Multi-Type FACTS Devices in Restructured Power Systems with Wind Generation," *Electrical Power and Energy Systems*, vol. 77, pp. 33–42, 2016.

26. O. P. Mahela and A. G. Shaik. "Recognition of Power Quality Disturbances Using S-Transform Based Ruled Decision Tree and Fuzzy C-Means Clustering Classifiers," *Applied Soft Computing*, vol. 59, pp. 243–257, October 2017. doi:10.1016/j.asoc.2017.05.061.

27. O. P. Mahela, O. P. Mahela, H. H. Alhelou, and P. Siano, "Power Quality Assessment and Event Detection in Distribution Network with Wind Energy Penetration Using Stockwell Transform and Fuzzy Clustering," *IEEE Transactions on Industrial Informatics*, Early Access, January 2020. doi:10.1109/TII.2020.2971709.

28. A. G. Shaik and O. P. Mahela, "Power Quality Assessment and Event Detection in Hybrid Power System," *Electric Power Systems Research*, vol. 161, pp. 26–44, March 2018. doi:10.1016/j.epsr.2018.03.026.

29. G. S. Chawda, A. G. Shaik, O. P. Mahela, S. Padmanaban, and J. B. Holm-Nielsen, "Comprehensive Review of Distributed FACTS Control Algorithms for Power Quality Enhancement in Utility Grid with Renewable Energy Penetration," *IEEE Access*, vol. 8, pp. 107614–107634, 2020. doi:10.1109/ACCESS.2020.3000931.

30. G. S. Chawda, A. G. Shaik, M. Shaik, P. Sanjeevikumar, J. B. Holm-Nielsen, O. P. Mahela, and K. Palanisamy, "Comprehensive Review on Detection and Classification of Power Quality Disturbances in Utility Grid with Renewable Energy Penetration," *IEEE Access*. doi:10.1109/ACCESS.2020.3014732.

31. O. P. Mahela, B. Khan, H. H. Alhelou, and S. Tanwar, "Assessment of Power Quality in the Utility Grid Integrated with Wind Energy Generation," *IET Power Electronics*, January 2020. doi:10.1049/iet-pel.2019.1351.

32. O. P. Mahela and A. G. Shaik, "Recognition of Power Quality Disturbances Using S-Transform and Fuzzy C-Means Clustering," *IEEE International Conference and Utility Exhibition on Co-generation, Small Power Plants and District Energy (ICUE 2016)*, BITEC, Bang Na, Bangkok, Thailand, September 14–16, 2016. doi:10.1109/COGEN.2016.7728955.

33. M. Meena, O. P. Mahela, M. Kumar, and N. Kumar. "Detection and Classification of Complex Power Quality Disturbances Using Stockwell Transform and Rule Based Decision Tree," *IEEE PES International Conference on Smart Electric Drives and Power System (ICSEDPS-2018)*, G H Raisoni College of Engineering, Nagpur, India, June 12–13, 2018. doi:10.1109/ICSEDPS.2018.8536028.

34. O. P. Mahela and A. G. Shaik, "Recognition of Power Quality Disturbances Using S-transform and Rule-Based Decision Tree," *2016 IEEE First International Conference on Power Electronics, Intelligent Control and Energy Systems (ICPEICES 2016)*, DTU, New Delhi, India, July 4–6, 2016. doi:10.1109/ICPEICES.2016.7853093.

35. B. Rathore, O. P. Mahela, B. Khan, H. H. Alhelou, and P. Siano, "Wavelet-Alienation-Neural Based Protection Scheme for STATCOM Compensated Transmission Line," *IEEE Transactions on Industrial Informatics*, Early Access, June 2020. doi:10.1109/TII.2020.3001063.

36. O. P. Mahela, J. Sharma, B. Kumar, B. Khan, and H. H. Alhelou, "An Algorithm for the Protection of Distribution Feeder Using Stockwell and Hilbert Transforms Supported Features," *CSEE Journal of Power and Energy Systems*, 2020. doi:10.17775/CSEEJPES.2020.00170.

37. A. Kulshrestha, O. P. Mahela, M. K. Gupta, N. Gupta, N. Patel, T. Senjyu, M. S. S. Danish, and M. Khosravy, "A Hybrid Protection Scheme Using Stockwell Transform and Wigner Distribution Function for Power System Network with Solar Energy Penetration," *Energies*, vol. 13, no. 14, p. 3519, 2020. doi:10.3390/en13143519.

38. G. S. Yogee, O. P. Mahela, K. D. Kansal, B. Khan, R. Mahla, H. H. Alhelou, and P. Siano, "An Algorithm for Recognition of Fault Conditions in the Utility Grid with Renewable Energy Penetration," *Energies*, vol. 13, no. 9, p. 2383. doi:10.3390/en13092383.

39. S. R. Ola, A. Saraswat, S. K. Goyal, S. K. Jhajharia, B. Khan, O. P. Mahela, H. H. Alhelou, and P. Siano, "A Protection Scheme for Power System with Solar Energy Penetration," *Applied Sciences*, vol. 10, no. 4, Chapter No. 1516, pp. 1–22, February 2020. doi:10.3390/app10041516.

40. S. R. Ola, A. Saraswat, S. K. Goyal, V. Sharma, B. Khan, O. P. Mahela, H. H. Alhelou, and P. Siano, "Alienation Coefficient and Wigner Distribution Function Based Protection Scheme for Hybrid Power System Network with Renewable Energy Penetration," *Energies*, vol. 13, no. 5, Chapter No. 1120, March 2020. doi:10.3390/en13051120.

41. S. R. Ola, A. Saraswat, S. K. Goyal, S. K. Jhajharia, B. Rathore, and O. P. Mahela, "Wigner Distribution Function and Alienation Coefficient Based Transmission Line Protection Scheme," *IET Generation, Transmission and Distribution*, vol. 14, no. 10, pp. 1842–1853, 22 May 2020. doi:10.1049/iet-gtd.2019.1414.

42. S. K. Lakwal, O. P. Mahela, M. Kumar, and N. Kumar, "Optimized Approach for Restructuring of Rajasthan Transmission Network to Cater Load Demand of Rajasthan Refinery and Petrochemical Complex," *IEEE International Conference on Computing, Power and Communication Technologies (GUCON 2019)*, September 27–28, 2019, Greater Noida, India.

43. D. K. Sharma, O. P. Mahela, and S. Agarwal, "Design and Implementation of System Protection Scheme for Kawai-Kalisindh-Chhabra Thermal Complex in Rajasthan, India," *IEEE International Conference on Computing, Power and Communication Technologies (GUCON 2019)*, September 27–28, 2019, Greater Noida, India.

Index

For Product Safety Concerns and Information please contact our EU
representative GPSR@taylorandfrancis.com
Taylor & Francis Verlag GmbH, Kaufingerstraße 24, 80331 München, Germany

www.ingramcontent.com/pod-product-compliance
Lightning Source LLC
Chambersburg PA
CBHW052011230326
41598CB00078B/2623

* 9 7 8 1 0 3 2 2 3 5 1 1 0 *